THE POWER OF α

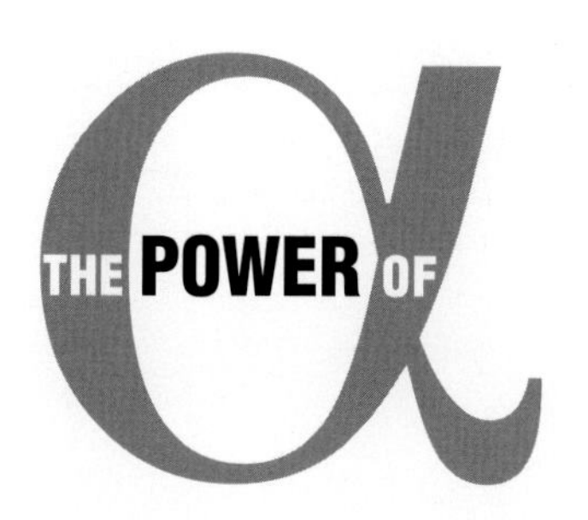

Electron Elementary Particle Generation
with α-Quantized Lifetimes and Masses

Electron Elementary Particle Generation with α-Quantized Lifetimes and Masses

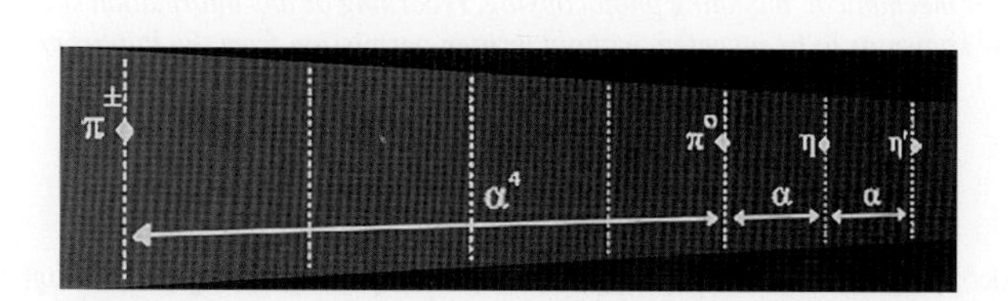

MALCOLM H. MAC GREGOR

Formerly of Lawrence Livermore National Laboratory, USA

World Scientific

NEW JERSEY • LONDON • SINGAPORE • BEIJING • SHANGHAI • HONG KONG • TAIPEI • CHENNAI

Published by

World Scientific Publishing Co. Pte. Ltd.
5 Toh Tuck Link, Singapore 596224
USA office: 27 Warren Street, Suite 401-402, Hackensack, NJ 07601
UK office: 57 Shelton Street, Covent Garden, London WC2H 9HE

British Library Cataloguing-in-Publication Data
A catalogue record for this book is available from the British Library.

THE POWER OF α
Electron Elementary Particle Generation with α-Quantized Lifetimes and Masses

ISBN-13 978-981-256-961-5
ISBN-10 981-256-961-8

Printed in Singapore by B & JO Enterprise

"The one thing everyone who cares about fundamental physics seems to agree on is that new ideas are needed. ... We are missing something big."

"... In the history of science, there have been many instances of discoveries that surprised scientists because they were not anticipated by theory. Are there observations today that we theorists have not asked for, that no theory invites — observations that could move physics in an interesting direction? Is there a chance that such observations have already been made but ignored because, if confirmed, they would be inconvenient for our theorizing?"

Lee Smolin, *The Trouble with Physics*, pp. 308 and 203–4
(Houghton Mifflin, Boston, 2006)

Preface: The Training of an Elementary Particle Phenomenologist

This book is about the *phenomenology* of the elementary particle. Phenomenology is not experiment, and it is not really theory; it is the intermediate region that ties these two areas of physics together. Academically, physicists who are interested in elementary particles carry out theses in *experimental* particle physics or in *theoretical* particle physics, but not in phenomenological particle physics. There is no formal academic program that prepares a person to become a phenomenologist.

When a physicist decides to learn a new area of physics, he or she often does a phenomenological survey to get up to speed — to learn the "big picture." One asks: "What are the experimental facts that have been established in this field, and how well has theory succeeded in tying these facts together? What are the unanswered questions? Where is the frontier? Is there some place here where I can make a contribution?" My personal venture into the field of elementary particle phenomenology started in 1969. This preface is a brief summary of the route I followed to get there.

At the time when I started into physics at the University of Michigan, right after WWII, there was no field of specialization labeled as "elementary particle physics." There were not enough particles known to justify it. Of course, that situation changed very quickly in the 1950's. In those days we graduated in physics with the label of "experimentalist" or "theorist," and that was pretty much it. I was the former. The *Physical Review* itself was a rather slender journal that encompassed all of physics. Its division into several thick sections came much later.

My training in science began with my induction into the U.S. Navy as a 17-year-old volunteer in the spring of 1944, right after completing a shortened high-school curriculum. I enrolled in training school to

Fig. 1. The Institute of Theoretical Physics, Copenhagen, Denmark, Fall 1960. The occasion was the gathering for the official photograph of the year's attendees at the Institute. Professor Niels Bohr is addressing us, and he is delivering a short lesson on optics. He points out that "if you can't see the camera, the camera can't see you." Professor Bohr's index finger is aimed in the direction of the present author.

become a radio technician, which included both radio communications and also the newly-implemented radar systems. We were taught the fundamentals of electromagnetic circuit theory, and we spent several months finding and repairing problems in radios and radars that our instructors purposely created in the equipment. The entire program was 11 months long, and included (in my case) pre-radio school in Chicago, Illinois, primary school in Monterey, California, and secondary school (as a member of the Navy Air Corps) in Corpus Christi, Texas. I ended up with the classification of Aviation Electronics Technician's Mate First Class. After completing my training, I served as an instructor and repairman with a Naval air squadron stationed in the US, and later as an instructor in radar navigation. After the war ended in August 1945, my specialty was one of the last ones to be discharged, since they needed AETM's to keep the planes flying phase shift analyses had now reached a natural stopping point for me. The next step, which involved going up into the inelastic scattering region, was a whole new venture, and it was not clear just how reliable theory would be in handling inelastic scattering. Potential models were in a transitional stage between schemes based on the analytic continuation of scattering amplitudes and quite different schemes based on the use of discrete quarks. Thus it seemed to be the time for a little reflection about the best thing to do next. The nucleon–nucleon scattering studies had been carried out in a professional manner, and, as one user commented in print, they served as the world standard for a decade. But the work, in my opinion, was missing one desirable ingredient — it contained no really new ideas or concepts. Every physicist likes to think that he or she can come up with something original — some new idea which can be added to other great ideas — ideas that contribute to the complex body of information we denote as the "field of physics."

This story of my personal adventures along the physics path I followed before getting into particle phenomenology leads up to one decisive "event" (a personal "aha") that occurred during the third week of June, 1969. For some time I had been promising myself that I would set aside one week of full-time concentrated effort during which I would study the elementary particle database, as contained in *Review of Particle Physics* (RPP) 1969, to see what if anything I could deduce about the field. Finally the time arrived. Armed with the RPP and some scratch pads and pencils, I sat in isolation at my office desk for five full days, eight hours a day. (I actually lost eight pounds in weight during this week of "doing nothing except thinking.") At the end of this time I had reached two conclusions: (a) the muon mass serves as a fundamental elementary particle mass quantum; (b) quark internal binding energies are modest (a few percent). These two conclusions, which were both outside of the conventional thinking of the time (and still are), have formed the basis of the work I have done in the area of elementary particle physics since that date. This work has for the most part stayed at the level of phenomenology, but if the phenomenology is not correct, there is little point in building an elaborate theoretical superstructure: the foundation for theoretical work must be rock solid. The present book, *The Power of α*, displays the results of my efforts in their most recent form. The research path I have followed between June 1969 and April 2006 (the date as I write these words) is documented in the Postscript to this book, and in the annotated bibliography displayed in Appendix D. The path I have taken has not been an easy one, but it is one I would repeat again if it were necessary and possible to do so. The end of the path is still obscure, but it is the journey and not the destination that remains in my mind, and in my heart. Thus far there have been only a few companions on this path, but these few have made all the difference.

Malcolm Mac Gregor
Santa Cruz, California, 2006

Contents

List of Figures

Preface: Figures

Introduction: Lifetime Figures

Introduction: Mass Figures

Chapter 1 Figure

Chapter 2 Figures

Chapter 3 Figures

Chapter 4 Figures

Chapter 5 Figures

Postscript Figures

Introduction A Pictorial Journey through the Landscape of α-Quantized Elementary Particle Lifetimes and Masses

0.1 The Experimental Journey

One of the most impressive achievements of elementary particle physicists during the past century has been the creation of a very comprehensive and detailed experimental database. The motivation for carrying out these complicated and expensive experiments springs in large part from our desire to learn how the universe is composed, which devolves into the question of the mass structure of the elementary particle. Many important questions about elementary particles have been answered, but the problem of elementary particle masses remains unsolved. In the present book we examine two facets of the elementary particle database, particle mass *values* and particle mass *stability* (particle lifetimes), in order to see if we can find the solution. Particle masses have been well studied, but particle lifetimes have not — at least in a global sense. In the main body of the book we go through this analysis in detail, but in this Introduction we simply use displays of the experimental data to exhibit the mass patterns and lifetime patterns that can be found in the data. This is a view of the mass and lifetime *forest*, and it supplies motivation for later spending time in studying the *trees*. Interestingly, it is the mass stability that leads us to new information about the mass structure, so we start this journey by analyzing the collection of well-measured elementary particle lifetimes, which range over 27 orders of magnitude. Then we move on to the elementary particle masses, which range over almost six orders of magnitude. The experimental data tell their own story, so our aim in this Introduction is to keep the words to a minimum and let the graphic displays speak for themselves. The references to this work are contained in the main body of the text.

0.2 Global Lifetime α-Quantization

A total of 157 elementary particle lifetimes τ have been well measured to date. These lifetimes are listed in Appendix A. The longer lifetimes were obtained by measuring their flight paths, and the shorter lifetimes were

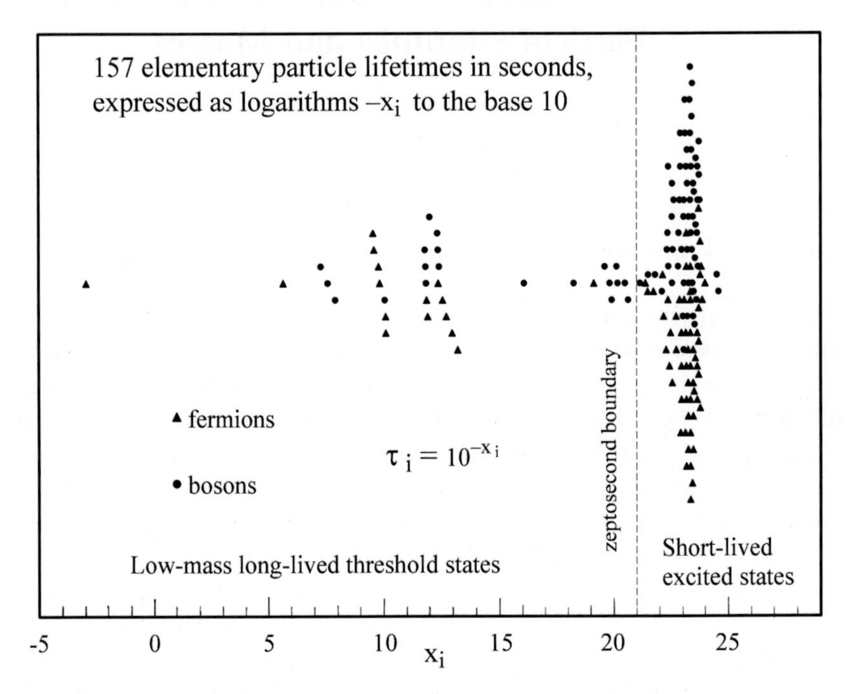

Fig. 0.1.1 The 157 well-measured elementary particle "lifetimes" or "mean lives" τ measured in units of seconds, and expressed here in powers of 10 via the equation $\tau_i = 10^{-x_i}$. The logarithms x_i of the lifetimes τ_i in this equation are plotted along the horizontal x-axis (the abscissa), and for clarity are spread out arbitrarily along the vertical y-axis (the ordinate). Triangles represent half-integer-spin *fermions*, and circles represent integer-spin *bosons*. There are 36 long-lived threshold-state particles that have lifetimes $\tau_i > 1$ zs (zeptosecond), and 121 short-lived excited-state particles that have lifetimes $\tau_i < 1$ zs, where 1 zs $\equiv 10^{-21}$ s. As can be seen in the figure, the threshold-state lifetimes occur in well-separated groups which are scattered over many orders of magnitude, whereas the excited-state lifetimes occur as a continuum of values over a relatively compact range. Fermions and bosons appear together in the same lifetime groups, showing that they share a common set of particle basis states. These 157 lifetimes were taken from the latest Particle Data Group (PDG) data compilation, *Review of Particle Physics* (RPP) 2006, and they are summarized in Appendix A.

deduced from the *full widths* of their mass resonance shapes. A global display of all of these lifetimes at the same time requires a logarithmic representation, which we can obtain by writing $\tau_i = 10^{-x_i}$ s, where x_i is the lifetime logarithm of the ith particle. These lifetimes are displayed in Fig. 0.1.1. As can be seen, the 121 particle lifetimes τ that are shorter than 10^{-21} s (1 zeptosecond (zs)) form essentially a continuum of values, whereas the 36 longer lifetimes are separated into widely-spaced groups. The particles in Fig. 0.1.1 are identified as (integer-spin) bosons or (half-integer-spin) fermions. These two types of particles combine together in the same lifetime groupings, which indicates that they share a common substructure whose fermion or boson form does not drastically affect the lifetime.

Figure 0.1.2 is the same lifetime plot as Fig. 0.2.1, but the particles are now identified by their dominant quark content (quark *flavor*) in addition to their boson and fermion labels. As can be seen, these quark flavors dictate the lifetimes of the long-lived threshold-state particles. In the case of the excited-state lifetimes, by way of contrast, the quark flavors are more-or-less randomly distributed.

Figure 0.1.3 is a study of just the 36 long-lived threshold states. The lifetimes are plotted here in the form of ratios to the reference $\pi^{\pm}$ lifetime, and they are displayed using factors of 100 instead of 10 for the logarithmic scaling. As Fig. 0.1.3 shows, this scaling factor is not quite large enough. Since the fine structure constant α (a pure number) has the reciprocal value $\alpha^{-1} \cong 137$, and since the spectacular successes of the calculations in quantum electrodynamics (QED) are expressed in powers of the coupling constant α, it seems logical to find out if these threshold-state lifetimes are in fact exhibiting an α-dependence. This possibility is investigated in Fig. 0.1.4, which is the same plot as Fig. 0.1.3, but uses 137 (or its reciprocal, if we move in the direction of increasing x_i values) instead of 100 as the lifetime scaling factor. As can be seen, the lifetime groups in this figure tend to fall on the α-spaced grid lines, and this tendency persists over the entire range of lifetime values — a range that encompasses 23 orders of magnitude. Hence the fine structure constant α appears have relevance for all of the elementary particles, and not just the leptons (electrons and muons) of quantum electrodynamics (QED), where it is known to apply.

In addition to the global spacing of lifetime groups in powers of α that is displayed in Fig. 0.1.4, there is also a hyperfine (HF) distribution of lifetimes within each of the groups that is superimposed on the overall group spacings. This HF structure is not random, and it conveys additional

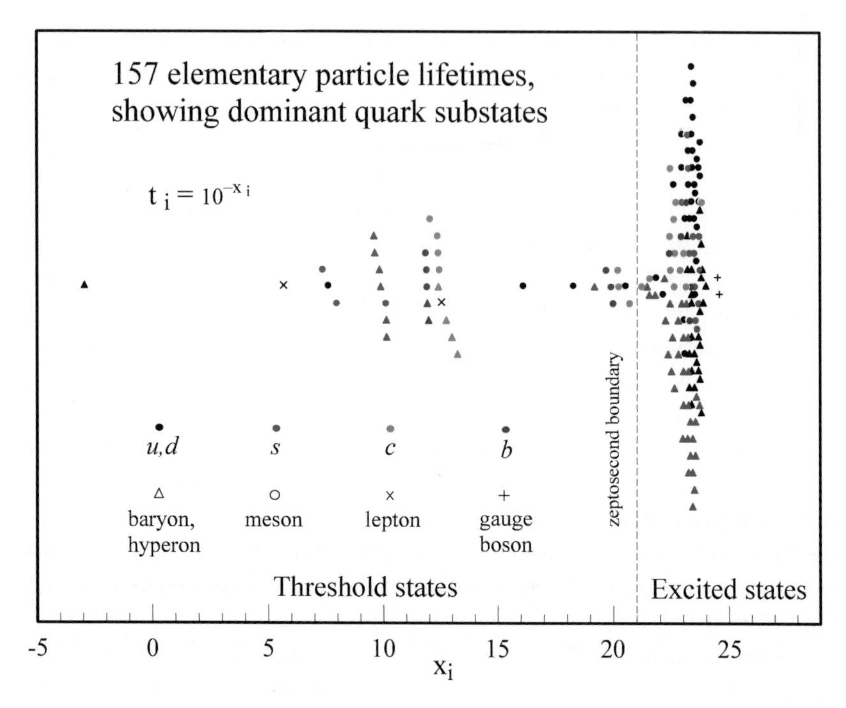

Fig. 0.1.2 This is the same 157-particle lifetime plot as in Fig. 0.1.1, but with each particle identified by its dominant quark *flavor*. The five basic quark flavors are as follows: u = up, d = down, s = strange, c = charm and b = bottom. (A sixth quark flavor, t = top, has a lifetime of roughly 10^{-25} s.) The baryons and mesons that are formed out of these quarks are denoted as *hadrons* (strongly interacting particles). The two *leptons* in the figure, which are denoted by $\times$'s, are the muon μ and tau τ; they are weakly interacting particles with no discernable quark structure. The formation of lifetime clusters in the 10^{-8} to 10^{-13} s range is due to the facts that (1) the quark decays in this lifetime range involve single *unpaired* quarks, (2) these decays are each dominated by a single flavor of quark, and (3) each quark flavor has its own characteristic decay rate. The lifetime clusters in the 10^{-16} to 10^{-21} s range involve *paired* quarks, and also a few radiative decays which do not require internal quark transformations, and they tend to echo the quark patterning of the longer-lived groups. All 36 of these long-lived $\tau_i > 1$ zs states represent *threshold excitations* where the various quark flavor combinations first appear. The 121 short-lived $\tau_i < 1$ zs states represent higher-mass *excited states* which have more-or less evenly distributed quark flavors (at least on the global time scale considered here). It is interesting to note that the tau *lepton*, which has a *mass* that is comparable to the charmed D-meson *hadron* masses, also has a *lifetime* that is comparable to the D-meson lifetimes shown near 10^{-12} s.

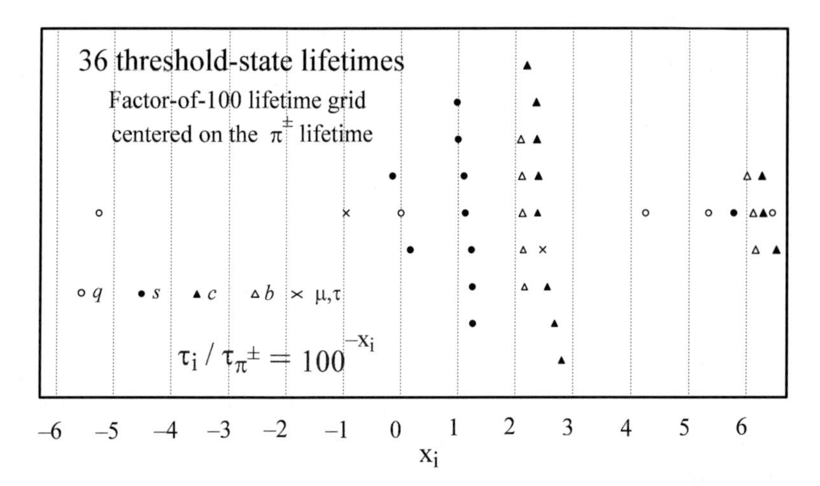

Fig. 0.1.3 The 36 long-lived threshold-state lifetimes τ_i of Figs. 0.1.1 and 0.1.2, expressed here as ratios to the $\pi^{\pm}$ lifetime (which is the lifetime of the longest-lived low-mass unstable hadron). These lifetime ratios are plotted as logarithms $-x_i$ to the base 100. As can be observed in Fig. 0.1.3, the actual lifetime group spacings are somewhat larger than the factor-of-100-spaced vertical grid lines of the lifetime plot.

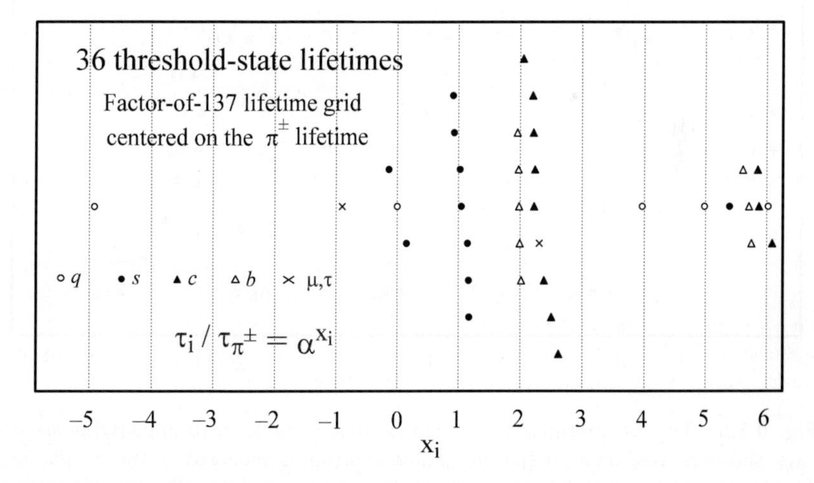

Fig. 0.1.4 This is the same plot as in Fig. 0.1.3, but with the lifetime ratios expressed here as logarithms $-x_i$ to the base 137 instead of 100. As can be seen, the factor-of-137-spaced vertical grid lines now agree closely with the observed lifetime groupings over a wide range of values. This factor-of-137 lifetime scaling represents a scaling in powers of the fine structure constant $\alpha = e^2\hbar c \cong 1/137$. The small deviations of the individual values within the lifetime groups can be attributed to a "hyperfine" (HF) structure, as we demonstrate in the following figures.

information about the mass structures of these particles. This topic is discussed in the next section.

0.3 Unpaired-Quark Lifetime Hyperfine (HF) Structure

Figure 0.1.5 is a display of the 23 lifetimes in Fig. 0.1.4 whose decays involve unpaired-quark transformations. As can be seen in Fig. 0.1.5, there are $q \equiv (u, d)$-, s-, and b-quark lifetime groups that are centered around the $x_i = 0$, 1 and 2 lifetime α-grid lines, and also a c-quark group that is displaced from the $x_i = 2$ grid line to slightly shorter lifetime values. In each of these groups, there are characteristic lifetime "steps" of about the same magnitude. As we will illustrate in the following figures, this lifetime hyperfine structure is composed mainly of a pervasive factor-of-2 lifetime substructure, which is logically related to an analogous mass substructure within these particles.

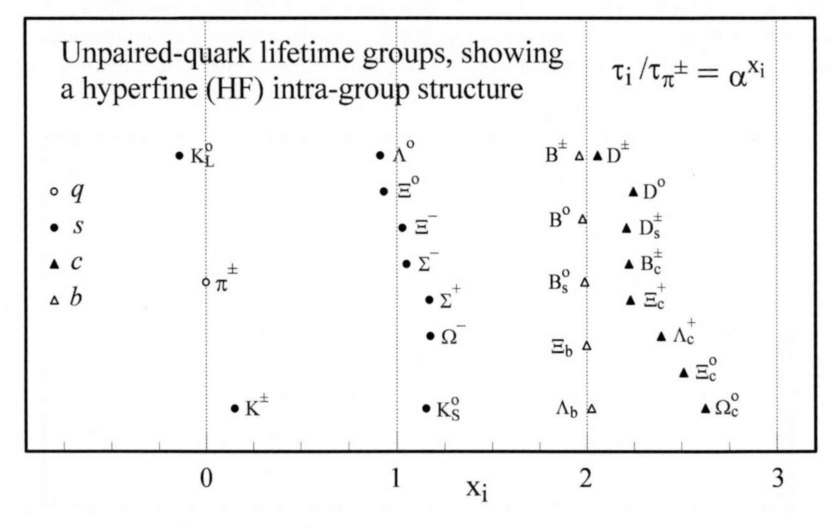

Fig. 0.1.5 The 23 threshold-state lifetimes that correspond to unpaired-quark decays, shown plotted on an α-spaced lifetime grid that is anchored on the $\pi^\pm$ lifetime. The b-quark lifetimes all lie accurately on the $x_i \cong 2$ α-grid line. By way of contrast, the other lifetime groups — the pseudoscalar (PS) mesons on the $x_i \cong 0$ grid line, the s-quark hyperons on the $x_i \cong 1$ α-grid line, and the c-quark mesons and hyperons slightly beyond the $x_i \cong 2$ α-grid line — all exhibit fairly regular hyperfine (HF) structures which are superposed on the overall group spacings in powers of α. This HF structure, which furnishes direct evidence for a common set of quantized mass subunits in these quark states, is displayed in detail in Figs. 0.1.6–0.1.10.

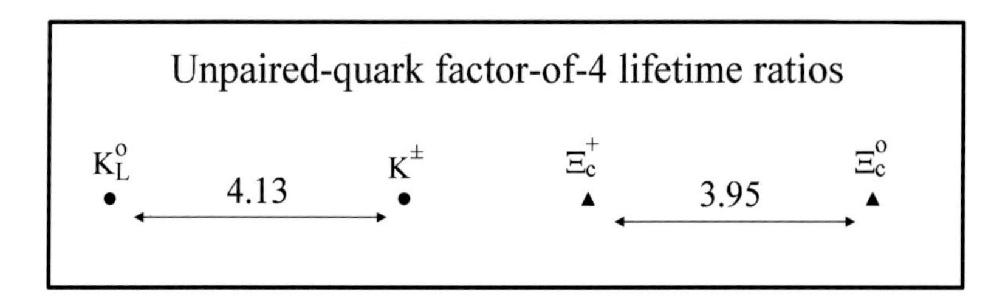

Fig. 0.1.6 Factor-of-4 lifetime ratios in the K mesons and the charmed cascade hyperons. The average value of these two lifetime ratios is 4.04, which is close enough to an integer that it suggests these results are not accidental. Furthermore, the unrelated $\pi^\pm$ and Λ_c^+ lifetimes appear as the geometric means of these two pairs, respectively, thus forming two factor-of-2 lifetime triads, as displayed in Fig. 0.1.8.

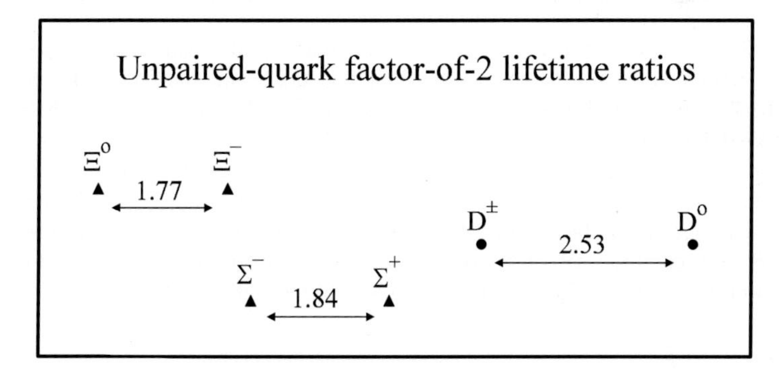

Fig. 0.1.7 Three factor-of-2 lifetime spacings between pairs of particles in the same isotopic spin multiplets. These all involve unpaired-quark decays. They are each only approximately equal to a factor of 2, but their average value is 2.05, which suggests that they represent factor-of-2 HF "decay triggers" upon which are superimposed even smaller correction factors that cause random deviations from exact values of 2.

Figure 0.1.6 displays two factor-of-4 lifetime ratios — one that involves the strange K mesons, and one that involves the strange-charmed Ξ_c baryons. These lifetime ratios between closely-related isotopic-spin pairs are close enough to the integer value 4 that this result is probably not just a coincidence. Figure 0.1.7 shows three factor-of-2 lifetime ratios — two between related Ξ and Σ hyperon isotopic-spin pairs, and one that involves an isotopic-spin pair of charmed D mesons. The ratios show significant deviations from the integer value 2, but the deviations appear to occur randomly in the directions of both smaller and larger lifetime ratios.

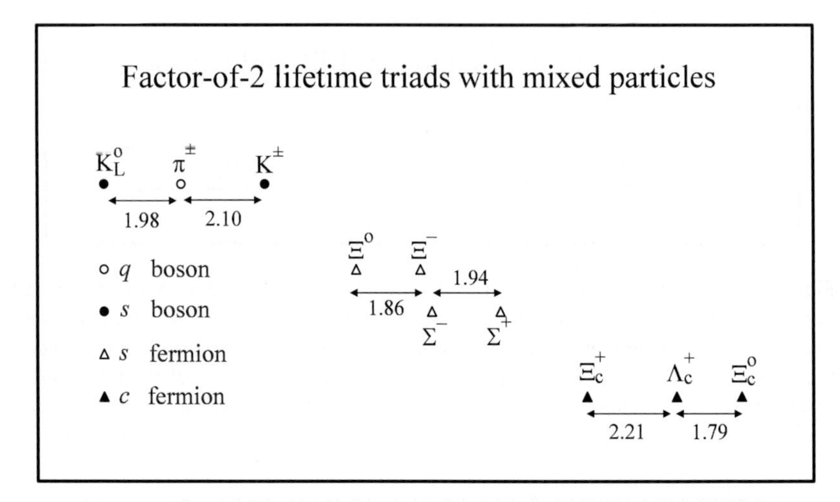

Fig. 0.1.8 Three factor-of-2 lifetime triads that include mixed sets of particles. These triads involve the factor-of-4 pairs of Fig. 0.1.6 and the factor-of-2 pairs of Fig. 0.1.7, together with the $\pi^\pm$ and Λ_c^+ intermediaries. The factors of 2 in the middle triad are defined relative to the average $\Sigma^- - \Xi^-$ lifetime, which represents the central lifetime of this triad. The six factor-of-2 relationships that are created in these mixed triads have an average lifetime ratio of 1.98, which suggests that these lifetimes have first-order spacings by factors of 2 that correspond to the number of available decay triggers in a common set of quark basis states, upon which are superposed small second-order lifetime variations that are due to other factors in the decay channels.

Figure 0.1.8 contains three factor-of-2 lifetime triads, which are of special interest in that they combine different types of particles together in the same triad. The nonstrange $\pi^\pm$ and strange K mesons bear no obvious lifetime relationship to one another, and yet their displayed factor-of-2 lifetime ratios seem too accurate to be accidental. The Ξ-Σ lifetime triad contains very similar Ξ^- and Σ^- lifetimes, which are averaged to form the central lifetime of the triad. The position of the Λ_c^+ charmed baryon roughly midway between the Ξ_c^+ and Ξ_c^0 also seems significant. What is important here is not so much the numerical accuracy of each factor-of-2 lifetime ratio, but rather the fact that the occurrence of these ratios is so pervasive in the data. Every unpaired-quark threshold-state lifetime group except that of the b mesons displays these ratios. Since each of these lifetime triads contains members of two different isotopic-spin families, it means that all of the particles in a particular triad share a common subset of basis-state masses.

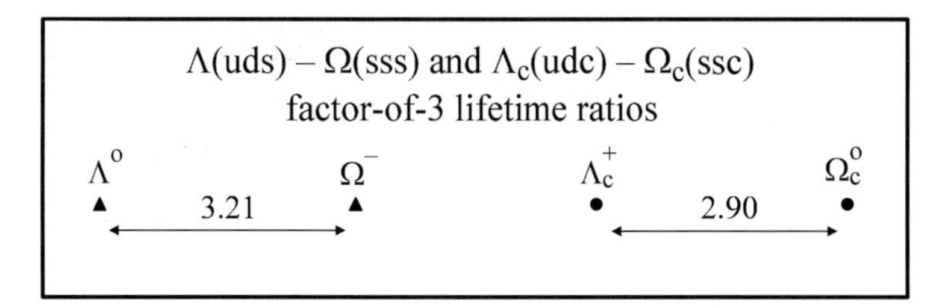

Fig. 0.1.9 Two factor-of-3 lifetime ratios that occur between related sets of Λ and Ω hyperon states. The Λ hyperons each contain one s or c quark, and the Ω hyperons each contain a total of three s and c quarks. The factor-of-3 lifetime ratios suggest that the s or c quark in the Λ triggers the decay, and that any one of the three s or c quarks in the Ω can independently trigger the decay. The replacement of an s quark by a c quark in moving from the $\Lambda^0 - \Omega^-$ pair to the $\Lambda_c^+ - \Omega_c^0$ pair shifts both the Λ^0 and Ω^- lifetimes from the $x_i \cong 1$ α-grid region to the $x_i \cong 2$ α-grid region (Fig. 0.1.5), and also increases their electric charge states by one unit of charge. This causes factor-of-2 lifetime displacements (Fig. 0.1.10), plus a characteristic factor-of-3 shift away from the α-grid for c-quarks as compared to b-quarks (Fig. 0.1.14). Hence a variety of mass and charge-state effects are occurring here, and it will be necessary to move beyond the present phenomenological approach in order to disentangle and prioritize all of these effects.

Figure 0.1.9 shows a different HF lifetime splitting. The Λ–Ω and Λ_c–Ω_c lifetime ratios are each a factor of 3, which echoes the total number of s and c quarks in each of these particles. Again, the lifetimes seem dominated by the mass substructures of these particles.

All of these factor-of-2 and factor-of-3 HF lifetime perturbations seem to be extraneous with respect to the overall lifetime scaling in powers of α. Thus it is of interest to apply phenomenological "corrections" to the lifetimes, and then see how this affects the overall scaling. Figure 0.1.10 shows the 23 unpaired-quark-state lifetimes of Fig. 0.1.5 after HF correction factors have been applied as indicated in parentheses. The "corrected" lifetimes form four distinct lifetime groups, three of which quite accurately lie on the α-spaced lifetime grid, and one (the c quarks) which is displaced toward shorter lifetime values. The question then arises as to how these corrections affect the lifetime scaling factor. An analysis of the lifetime scaling is carried out in Sec. 2.6, and it shows that the 36 uncorrected threshold-state lifetimes of Fig. 0.1.4 are most accurately fit with a lifetime scaling factor of 133.48, whereas these same lifetimes with the corrections of Fig. 0.1.10 applied yield a best-fit scaling factor of 136.09. Since $\alpha^{-1} = 137.04$ is the "theoretically-suggested" scaling factor (at least in the opinion

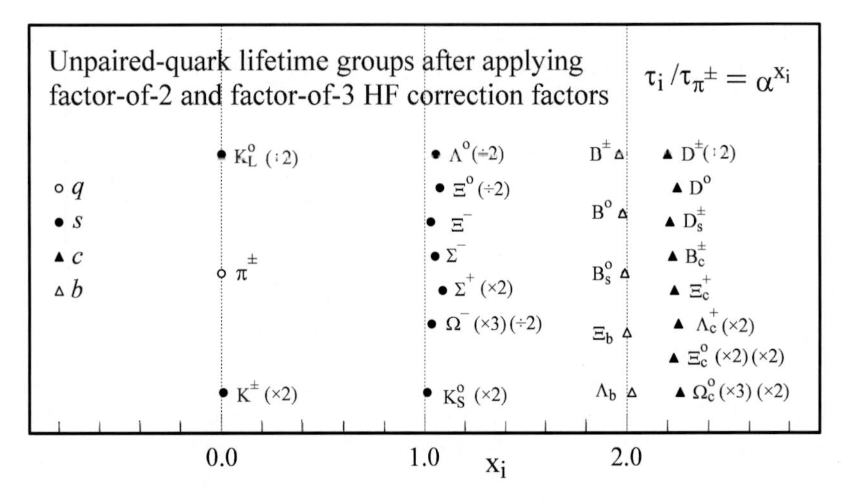

Fig. 0.1.10 The 23 unpaired-quark threshold-state lifetimes of Fig. 0.1.5 as they appear after the hyperfine "corrections" indicated in parentheses have been applied to 11 of them. The factor-of-2 and factor-of-3 HF effects are superimposed on the overall lifetime scaling in powers of α, so their phenomenological "removal" serves to enhance the clarity of the lifetime groupings on the global lifetime α-grid. When an Absolute Deviation from an Integer (ADI) spacing analysis of the lifetime logarithms x_i is carried out, in which the spacing factor S is varied so as to minimize the ADI sum, as discussed in Sec. 2.6, the ADI(S) minimum for the uncorrected lifetime logarithms of the 36 threshold states is at $S_{\text{min}} = 133.48$ (Fig. 2.6.6). After the HF corrections shown here are applied, the ADI(S) minimum shifts to $S_{\text{min}} = 136.09$ (Fig. 2.6.7). Since the "true" spacing factor for these α-spaced lifetimes is logically $S_{\text{min}} = \alpha^{-1} \cong 137.04$, the HF corrections seem justified by the results of the scaling analysis.

of the author), the HF corrections remove some extraneous factors that tend to obscure the global scaling of these lifetimes.

Figure 0.1.11 shows the entire 157-particle lifetime global array with corrections applied as shown in Fig. 0.1.10. As can be seen, the Standard Model quark structure is very clearly exhibited. Thus the scaling of lifetimes in powers of α would be sufficient by itself to reveal the existence of these quarks, although it is the way the quark states handle isotopic spins that seems to be their most striking virtue. The lifetime α-grid extends all the way from the neutron lifetime at $x_i \cong -5$ to the zeptosecond boundary at $x_i > +6$, a span of 11 powers of α.

The one unpaired-quark lifetime group in Fig. 0.1.11 that does not match the α-grid is the c-quark group, together with the τ lepton (which

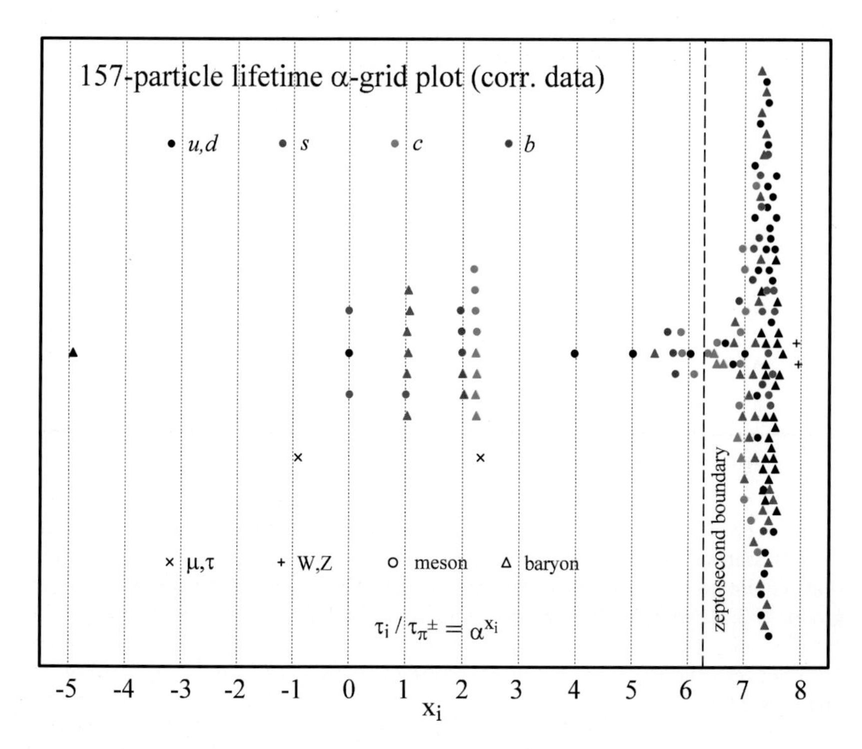

Fig. 0.1.11 The same 157-particle lifetime data set that was displayed in Figs. 0.1.1 and 0.1.2, but shown here plotted on an α-quantized lifetime grid that is anchored on the $\pi^{\pm}$ reference lifetime, and displayed with the HF corrections of Fig. 0.1.10 applied. As can be seen in Fig. 0.1.11, the threshold-state lifetimes to the left of the $\tau = 10^{-21}$ s boundary line lie quite accurately on the α-grid lines, and this accuracy is maintained over a wide range of lifetime values (23 orders of magnitude). The exceptions to this accurate α-spacing are the unpaired c-quark states, which are shifted away from the $x_i = 2$ α-grid by a factor of three in the direction of shorter lifetimes, and also the paired-quark states that lie close to the zeptosecond boundary, which are shifted in the direction of somewhat longer lifetimes. The lifetime relationships between the long-lived unpaired-quark decays and the shorter-lived paired-quark decays are displayed in more detail in Figs. 0.1.13 and 0.1.14.

has a mass that is comparable to the charmed D-meson mass). The b-quark and c-quark groups have an interesting lifetime and mass correspondence. Figure 0.1.12 displays the five (uncorrected) b-quark lifetimes together with the four c-quark lifetimes that do not require HF corrections. The ratio of their group-averaged central lifetimes is 3.21. Thus the c-quark particles decay about three times faster on average than the b-quark particles. The b-quark particles, on the other hand, are three times as massive as the

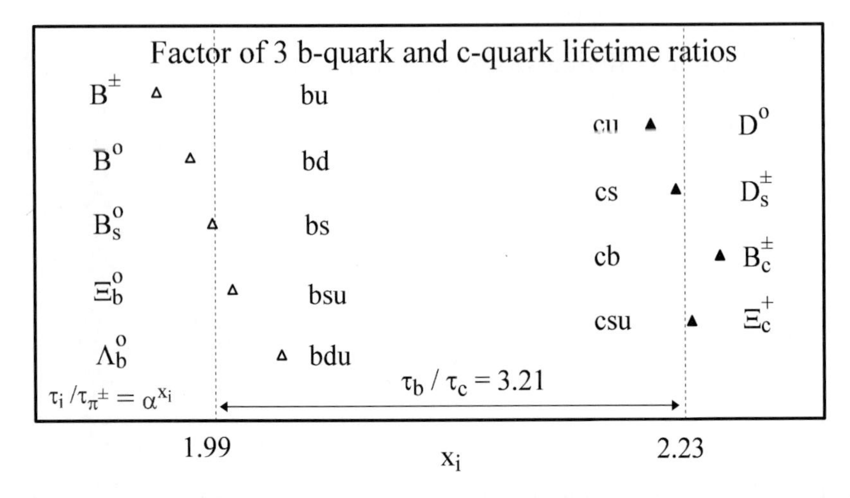

Fig. 0.1.12 The factor-of-3 flavor structure which exists between the five unpaired b-quark states (which require no HF corrections) and the four unpaired c-quark states that require no HF corrections. The group-averaged lifetime ratio between these two groups is 3.21, with the higher-mass b-quark states having the longer lifetimes. The corresponding $(\Upsilon_{1S} \equiv b\bar{b})/(J/\Psi_{1S} \equiv c\bar{c})$ mass ratio is 3.05 (Fig. 2.9.2). The lifetime ratio between the $B_s^0 \equiv bs$ and $D_s^{\pm} \equiv cs$ mesons, where a b quark is replaced by a c quark, is 2.99.

c-quark particles. We ordinarily expect the heavier unstable particles to decay faster than the lighter ones, but the opposite situation applies here. An interesting point is that the deviation of the c-quark states from the lifetime α-grid is not random, but is characteristically a factor of 3, so that we have in effect another type of HF lifetime fine structure. Perhaps the most interesting point historically is that the b-quark lifetimes, which were measured almost two decades after this lifetime α-grid was initially set forth in the literature (Sec. 2.11), and which require no HF corrections at all, accurately fall on the grid.

0.4 The α^4 "Lifetime Desert" between Unpaired and Paired Quark Decays

If Fig. 0.1.11 is examined in detail, it can be seen that the lifetime quark patterns on the α-grid lines $x_i \cong 0$, 1 and 2 are approximately repeated on the α-grid lines $x_i \cong 4$, 5 and 6. This result is brought out in more detail in Fig. 0.1.13, which shows just the 36 threshold-state particles with lifetimes

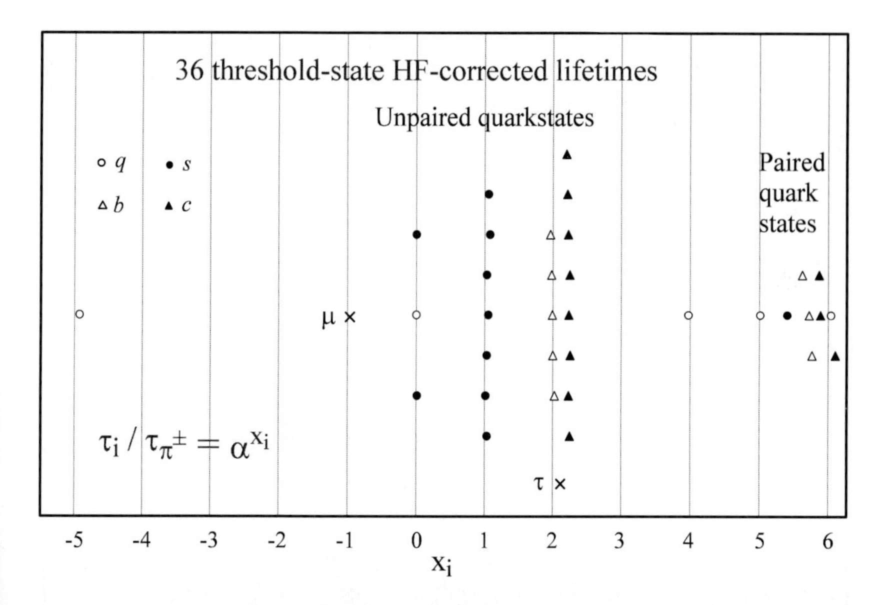

Fig. 0.1.13 The 36 threshold-state particles from Fig. 0.1.11 that have lifetimes longer than 1 zs. As can be seen, the unpaired *b*-quark and unpaired *c*-quark lifetime groups near $x_i \cong 2$ are echoed by similar paired *b*-quark and *c*-quark lifetime groups near $x_i \cong 6$. This is a lifetime separation ratio of about α^4 in magnitude, and is a characteristic result for these long-lived low-mass states. The breakdown of these groupings into specific particle families is displayed in Fig. 0.1.14.

$\tau > 1$ zs. In this figure the lifetimes in the $x_i \cong 0$, 1 and 2 region are labeled as "unpaired-quark states" and the lifetimes in the $x_i \cong 4$, 5 and 6 region are labeled as "paired-quark states" (plus a few radiative decays that do not require quark transformations in the decay process). Hence there appears to be a characteristic α^4 spacing between the decay rates when a single quark is transformed and the decay rates when a matching quark–antiquark pair is annihilated or when the decay is radiative. This fact is brought out more clearly in Fig. 0.1.14, where the various quark families are spread out along the ordinate, with the lifetime logarithms x_i displayed along the abscissa. For clarity, only the closely-related unpaired and paired quark states and isospin multiplets are included. As can be seen, there is a "lifetime desert region" between the unpaired and paired quark decays where no lifetimes seem to occur. This result is interesting in its own right, and it is also of interest in the sense that it confirms the reality of the α-dependence that exists in these hadronic lifetime decays.

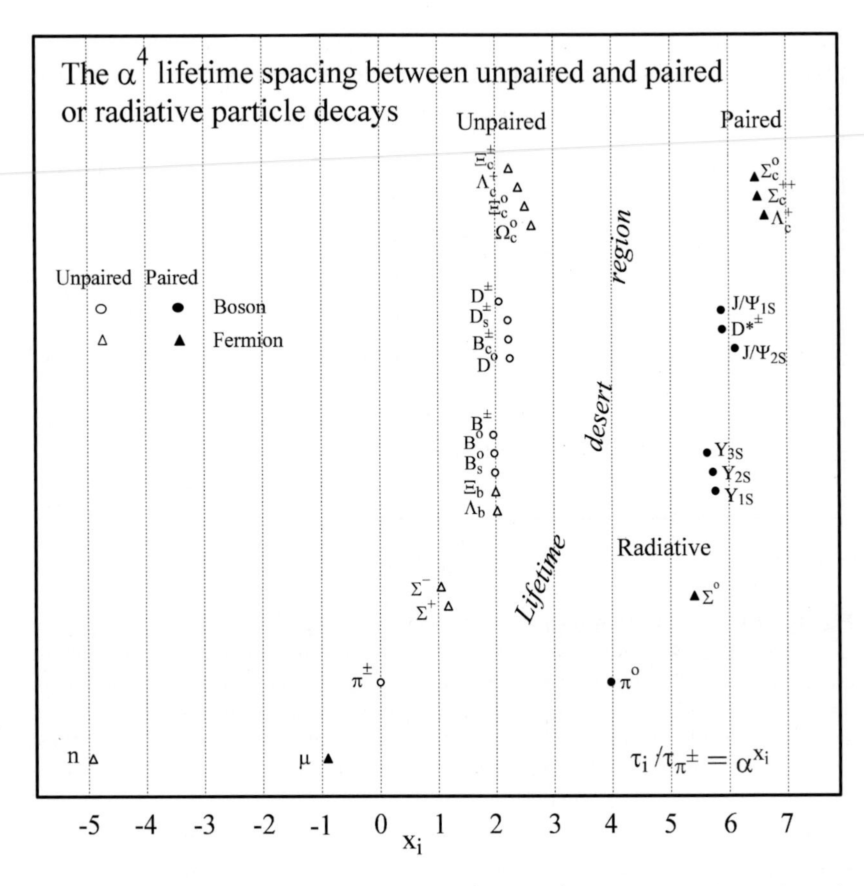

Fig. 0.1.14 The global dichotomy of slow (unpaired) single-quark decay channels and much faster (paired) quark–antiquark or radiative decay channels, as exhibited in these *uncorrected* lifetimes. A characteristic lifetime ratio of α^4 is maintained between the unpaired and paired decay modes for each type of particle family, even as the overall family lifetimes shift along the lifetime α-grid with increasing mass. The α^4 lifetime gaps that exist between these two sets of decay modes for each family combine together to form an unpopulated *lifetime desert region*. This characteristic α^4 division of related lifetime groups applies only to the low-mass, long-lived, threshold-state particles.

Our journey through the landscape of α-quantized elementary particle lifetimes is now concluded. The main goal here has been to establish the fact that the fine structure constant α plays a significant role in hadronic decays, which is a fact that has not yet been incorporated in contemporary theories of these decays. Since a particle's lifetime is a measure of the stability of its mass, and since the mass stability logically relates to the internal structure of the mass, we expect to find an α-dependence in the particle masses that in some manner accounts for the lifetime regularities. Thus we now move on to the second and final leg of our journey — a trip through the landscape of elementary particle masses.

0.5 Two α^{-1} Mass Leaps: The $m_b = 70$ MeV and $m_f = 105$ MeV Basis States

The search for an α-quantization in elementary particle *lifetimes* is phenomenologically straightforward. Since $\alpha \cong 1/137$ is a pure number, we search for lifetime ratios that contain the numerical value $1/137$ as a factor. If enough of these are found, we can provisionally conclude that the fine structure constant α is indeed playing a role in these particle decays. In order to find these ratios, we must concentrate on the long-lived threshold-state particles, since these are the only ones whose lifetimes are spread out enough to contain such a large scaling factor. Empirically, these are the 36 particles that have lifetimes $\tau > 10^{-21}$ s. When this search was carried out, as described in Secs. 0.1–0.4, it revealed a clear-cut lifetime α-scaling.

Theoretically, *lifetimes* are reciprocally coupled to *mass widths* by Heisenberg's uncertainty principle, and hence logically also to the *masses* themselves. Hence, in order to establish the validity of the lifetime α-scaling hypothesis, it becomes important to ascertain if the *masses* of these same particles exhibit a reciprocal dependence on $\alpha^{-1} \cong 137$. This mass search is also phenomenologically straightforward: we look for mass ratios that contain the numerical value 137 as a factor. Furthermore, we know which particles to examine — the 36 particles with measured lifetimes longer than 10^{-21} s. These are the particles that have α-spaced lifetimes, as displayed in Fig. 0.1.4. The masses of these 36 threshold-state particles are plotted in the energy level diagram of Fig. 0.2.1. As can be seen, they range in value from the 105 MeV mass of the muon to the 10,355 MeV mass of the Υ_{3S}. Unfortunately, this is a total mass range of less than 100, so there can be no factors of 137 among the mass ratios of any of these particles. Our search for evidence of mass α-quantization in these 36 particles is unsuccessful.

Before giving up on this mass α-quantization quest, we must ask ourselves if we are overlooking something. There are two particles we have not included here — the proton and the electron. The lifetimes of these particles are longer that 10^{-21} s. In fact they are infinite. These particles are stable, which is why they did not appear in the lifetime systematics.

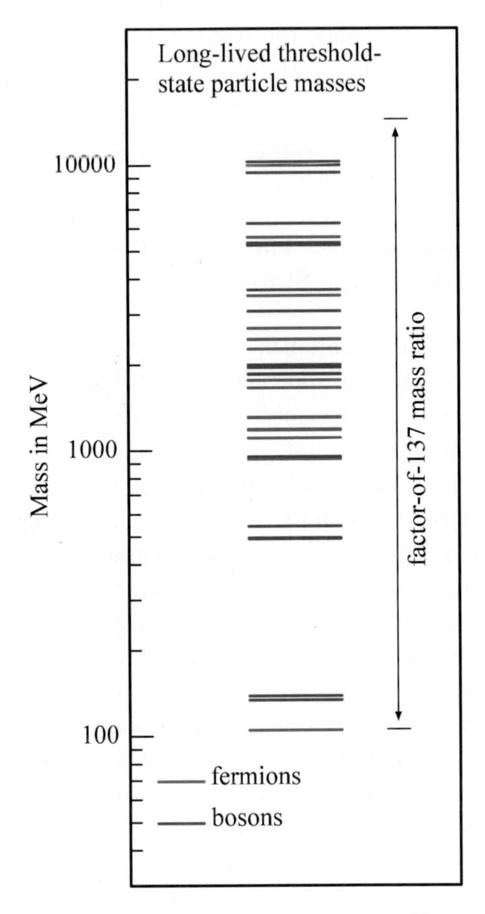

Fig. 0.2.1 A mass plot of the 36 long-lived ($\tau > 10^{-21}$ s) fermion and boson threshold-state particles that were displayed in Figs. 0.1.3 and 0.1.4. These particles are the ones that exhibit an α-dependence in their lifetimes, and therefore might exhibit a corresponding $\alpha^{-1} \cong 137$ dependence in their masses. Thus we want to examine them for any evidence of mass-ratio factors that are equal to 137. The lowest-mass state in Fig. 0.2.1 is the muon at 105 MeV, and the highest-mass state is the Υ_{3S} meson at 10,355 MeV. The ratio of these two masses is 99, so it is apparent that there are no mass ratios of 137 to be found among these particles. Hence these 36 threshold-state particles display no direct evidence of mass α-quantization.

The 938 MeV proton mass falls inside the range of masses that are displayed in Fig. 0.2.1, so its addition to the mass data set will not provide any factor-of-137 mass ratios. However, the 1/2 MeV electron mass lies well outside of that mass range, and thus opens up new possibilities for

α-quantized mass ratios. In fact, these electron possibilities are the only ones that are available among the threshold-state particles. Fortunately, they lead to success — in a very spectacular manner, as we now describe.

Figure 0.2.2 displays an energy level diagram that contains the 36-particle data set of Fig. 0.2.1 plus the proton and electron masses. The first mass level above the $m_e = 0.511$ MeV electron level is occupied by the $m_\mu = 105$ MeV muon. The m_μ/m_e mass ratio is $(3/2) \times 137$, which contains the sought-after α-quantization factor of 137. The electron and muon are both spin-1/2 leptons, and the muon decay is into the electron. Hence this is an observed α-quantization between closely related particles.

We now move on to the second mass level, which is occupied by the π^0 and $\pi^\pm$ pion isotopic-spin doublet, which has an average π mass of about 137 MeV. The π/m_e mass ratio is approximately 2×137, which again contains 137 as a factor. Moreover, we know that the pion has both particle and antiparticle substates (since the pion is its own antiparticle), so the pion mass should really be compared to the 1.022 MeV mass of an electron plus an anti-electron (positron). This mass ratio is $\pi/(e^- + e^+) \cong 137$, which is mass α-quantization in its most clear-cut form. Since the pion has spin $J = 0$, the electron–positron pair of spin-1/2 particles must be in a total spin $J = 0$ configuration for this particle–antiparticle excitation process to conserve angular momentum. The pion is a strongly interacting *hadron*, whereas electrons and positrons are weakly interacting *leptons*. Thus we do not ordinarily think of these two types of particles as being coupled together — especially in their production process. However, this is exactly what the experimental data seem to be telling us. The fine structure constant α couples electrons and muons to photons in quantum electrodynamics (QED), and it is evidently playing a similar role in quantum chromodynamics (QCD) when it couples electron pairs to the pion.

The addition of the electron to the 36 particle mass data set has yielded two different α-quantized masses (α-masses), and they appear in two different forms — *fermionic* (with half-integral spin) and *bosonic* (with, as we will see, integral spin). These two α-masses are: (1) the $m_f = (3/2)m_e/\alpha = 105$ MeV mass quantum that is created in the "α-leap" from the electron to the muon, and which is produced in conjunction with a similar α-leap in the antiparticle channel, as displayed in Fig. 0.2.2; (2) the $m_b = m_e/\alpha = 70$ MeV mass quantum that is created as part of a hadronically bound particle–antiparticle pair in the α-leap from an electron–positron pair to the pion. This is not only the beginning

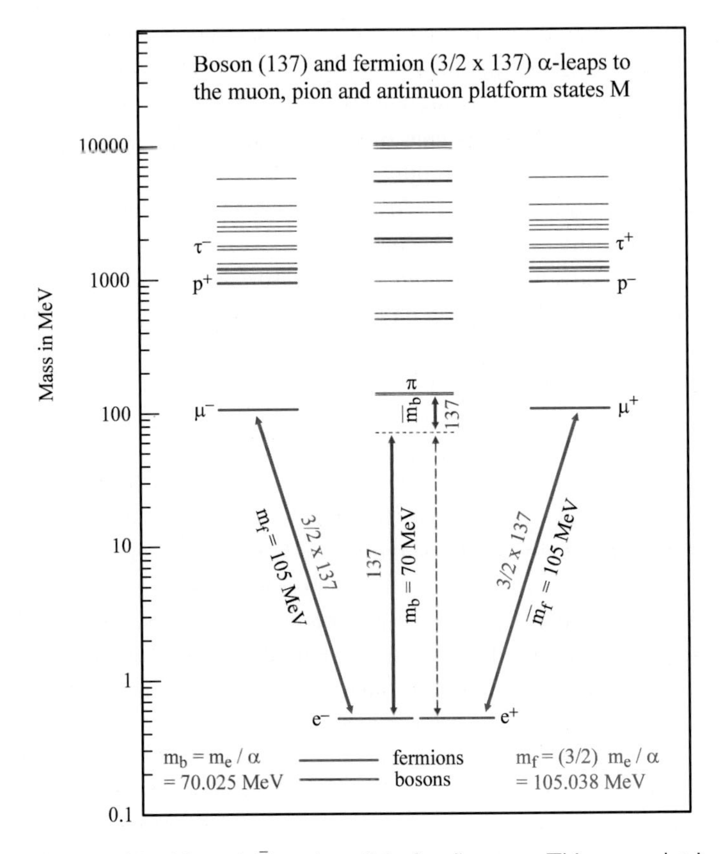

Fig. 0.2.2 The M_μ, M_π and $\bar{M}_\mu$ α-leap "platform" states. This mass plot is similar to that of Fig. 0.2.1, but also includes the two other long-lived massive particles — the stable proton and electron, which did not enter into the α-scaled lifetime systematics. Here the fermions are shown separated from the bosons, and both the particle and antiparticle fermion mass channels are displayed. Figure 0.2.2 contains two prominent 137 mass ratio factors: (1) the μ^-/e^- mass ratio of $3/2 \times 137$; (2) the $\pi/(e^- + e^+)$ mass ratio of approximately 137. The muon and pion represent the lowest-mass *fermion* and *boson* excited states, respectively, of the elementary particle mass spectrum. The particle-antiparticle-symmetric pion is reproduced as $\pi \cong m_b\bar{m}_b$, where $m_b = m_e/\alpha \cong 70$ MeV is the α-quantized *boson* mass quantum. The muon is reproduced as $\mu \cong m_f$, where $m_f = 3/2m_b \cong 105$ MeV is the α-quantized *fermion* mass quantum (Fig. 0.2.3 shows a more accurate muon calculation). The 105 MeV M_μ and $\bar{M}_\mu$ muon states and the ~137 MeV M_π pion state serve as "platforms" for higher-mass excitations. The muon-to-electron mass ratio systematics is straightforward (Fig. 0.2.3), whereas the pion-to-electron mass ratio is obscured by the 4.6 MeV mass splitting of the $\pi^\pm$ and π^0 mesons, and also by the small (~3%) binding energy of the particle and antiparticle substates in the pion. The factor of 3/2 that appears in the m_f to m_b mass ratio is not an *ad hoc* result, but is a relativistic spin effect, as described in Fig. 0.2.14.

of our search for α-quantized masses, but it is also the end, at least with respect to the low-mass long-lived threshold-state particles that we consider here. As it turns out, no further α-quantized mass quanta are needed below 12 GeV. Above 12 GeV we encounter a second α-leap that generates another α-quantized mass quantum, as described in Sec. 0.10.

It is useful at this point to formalize the concept of a particle "platform" state:

> *A particle "platform" M is a stable or metastable combination of particle substates upon which higher-mass "excitation towers" are erected. In hadrons it represents the lowest-mass state in which a particular combination of quarks occurs.*

There are two *α-leap* platforms, the 140 MeV *pion* platform M_π and 210 MeV *muon-pair* platform $M_{\mu\mu}$, which are created directly from an *electron-pair* ground state by the action of the $\alpha = e^2/\hbar c$ coupling constant. These are the platforms displayed in Fig. 0.2.2, where the muon-pair platform $M_{\mu\mu}$ is shown separated into its particle and antiparticle components M_μ and $\bar{M}_\mu$. There are also ten *meson* platforms, which are formed as quark–antiquark pair combinations of the $q \equiv (u, d)$, s, c, b set of Standard Model quarks and antiquarks. And there are a set of *baryon* platforms, which are formed as q, s, c, b quark triads, not all of which have yet been created experimentally. All of these platforms except the pion platform are composed of 105 MeV m_f mass quanta, and thus can be denoted collectively as "muon states." The Standard Model q, s, c, b quarks are denoted here as "muon quarks."

The (π, η, η', K) pseudoscalar meson nonet is a separate hadron family that requires its own set of "pion quarks," which are composed of 70 MeV m_b mass quanta. As we will demonstrate, the α-mass m_b is a spin 0 mass quantum that has no counterpart within the paradigm of the Standard Model. Hence it is a true boson, as its subscript implies. It does not seem to be observed experimentally as an individual entity. By way of contrast, the α-mass m_f is a conventional spin-1/2 fermion mass quantum that is manifested experimentally in the form of the spin-1/2 muon. In addition to reproducing the PS meson nonet, the pion quarks also combine with the q, s, c, b muon quarks to form "hybrid" hadron combinations that commonly appear as excited states above the platform states M.

We will demonstrate in our journey through the elementary particle mass spectrum below 12 GeV that the masses of the platform states

0.6 The Spin-1/2 Standard Model $q \equiv (u, d), s, c, b$ "Muon" Constituent Quarks

The concept of elementary particle mass "platforms" was introduced in Sec. 0.5, wherein a metastable platform is generated, and then an "excitation tower" is erected on this platform. The basic *α-leap platforms* M_μ, M_π and $\bar{M}_\mu$ correspond respectively to the μ^-, $(\pi^\pm, \pi^0)$, and μ^+ excitations displayed in Fig. 0.2.2. These are the lowest-mass unstable elementary particle states, and they are also the longest-lived (except for the neutron). The *meson platforms* are quark–antiquark *pairs* of q, s, c, b quarks, which form a 4×4 matrix. The *baryon platforms* are quark *triads* of q, s, c, b quarks. The Standard Model q, s, c, b quarks, as we will demonstrate, have constituent-quark masses that are multiples of the "muon" mass $m_f = 105$ MeV. The "muon mass" calculations of quark and particle masses are portrayed graphically in Figs. 0.2.4–0.2.10. The M_π pion platform excitations, which are the low-mass pseudoscalar mesons, cannot be reproduced with q, s, c b constituent quarks. They occur as multiples of the "pion" mass $m_b = 70$ MeV, and are discussed in Sec. 0.7. These "pion mass" calculations are portrayed in Figs. 0.2.11–0.2.13.

One of the most interesting platforms is the *muon* platform M_μ, which is created in conjunction with a matching *antimuon* platform $\bar{M}_\mu$, and the particles they generate are produced in matching particle–antiparticle pairs. Since these particles usually separate after being generated, the particle platforms M_μ and $\bar{M}_\mu$ can be studied individually, as shown in Fig. 0.2.2. Some quark states are also generated, which do not separate into individual states, but they can also be considered separately for discussion purposes. The excitation tower that is erected on the muon platform M_μ contains the "excitation-doubling" scheme displayed in Fig. 0.2.4, which reproduces the occupied levels in the M_μ excitation tower. In this excitation doubling process, the platform M_μ is created via an $m_f = 105$ MeV α-leap, and then 1, 2, 4 and 8 $Q_f \equiv m_f m_f = 210$ MeV excitation units are successively added to the platform mass, which results in the production of the $q = 3m_f$ and $s = 5m_f$ quarks, the $p = 9m_f = 3q$ proton, and the $\tau = 17m_f$ lepton. This excitation scenario gives the correct muon, proton and tau masses to accuracies of better than 1%, as displayed in Fig. 0.2.4. It also assigns constituent-quark masses of 315 MeV and 525 MeV to the q and s quarks. These quark masses cannot be measured directly, but their accuracies can be ascertained from the meson and baryon masses they produce, which are displayed in Figs. 0.2.7–0.2.10.

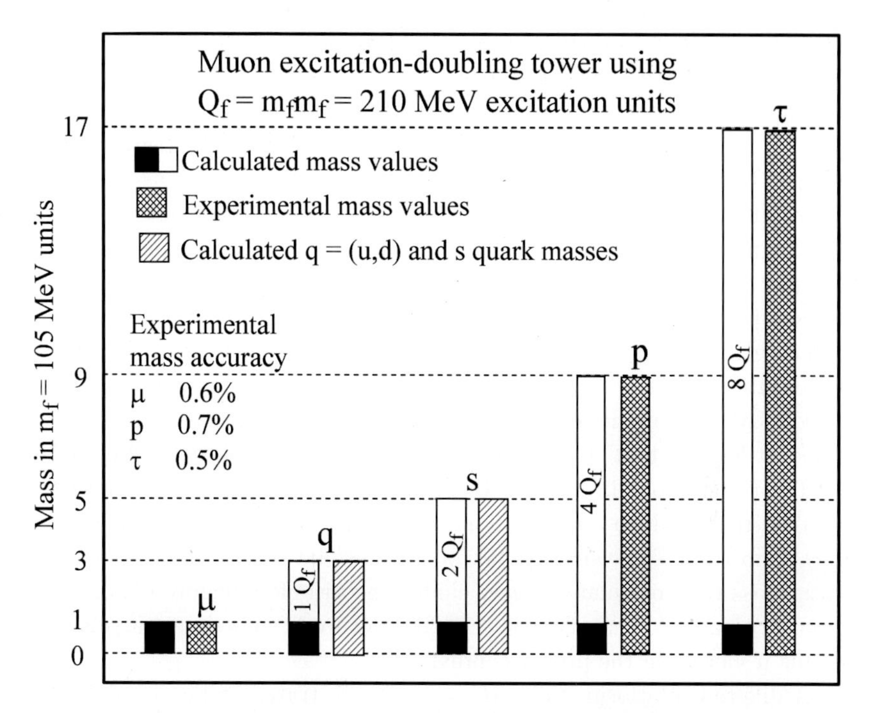

Fig. 0.2.4 A muon "excitation-doubling" production tower. The creation of M_μ, M_π and $\bar{M}_\mu$ muon and pion platform states by α-leaps from electron ground states was displayed in Fig. 0.2.2. The creation of a muon excitation column on the M_μ platform state is depicted in Fig. 0.2.4. (A matching excitation column is created simultaneously on the $\bar{M}_\mu$ platform state, and the two columns together form the $M_\mu + \bar{M}_\mu \equiv M_{\mu\mu}$ platform-state excitation tower.) The fermion mass quantum m_f is the fundamental mass unit for these excitations. (For simplicity in the present mass evaluations, we set $m_f \equiv 105$ MeV, and we ignore electron-mass contributions.) In the muon excitation column, the muon itself is first created as $\mu \equiv m_f$, and then excitation units $Q_f \equiv m_f m_f = 210$ MeV are added to the muon mass in an excitation-doubling sequence to create the following occupied levels: (1 Q_f) $q \equiv (u, d)$ quarks; (2 Q_f) the s quark; (4 Q_f) the proton; (8 Q_f) the τ lepton. As illustrated in Fig. 0.2.4, this procedure reproduces the muon, proton and lepton masses to accuracies of better than 1% (relative to the electron mass). It also generates the 315 and 525 MeV q and s constituent-quark masses (Fig. 0.2.6), which give good fits to an array of threshold-state particles, as illustrated in Figs. 0.2.7–0.2.10. The mass-tripling excitation in going from the integrally charged muon to the u and d quarks is accompanied by a fragmentation of the electric charge of the muon into 1/3 and 2/3 fragments on the quarks, which occurs only in conjunction with a matching process in the antimuon channel. The tripling of the mass and the division of the charge into thirds may be associated processes.

The first excitation level in the M_μ platform tower shown in Fig. 0.2.4 contains the $q \equiv (u, d)$ nucleon quarks, and hence represents one of the most important results in particle physics. The M_μ platform level itself is occupied by the muon, which is a structureless particle with an integral electric charge. In the first M_μ platform excitation step, the muon mass is tripled to produce the u and d quarks, and the charge is simultaneously split into 1/3 fragments (which must be done in conjunction with the corresponding $\bar{M}_\mu$ platform excitation). It seems logical to assume that the process of mass-tripling may be associated with the process of charge fragmentation into thirds. The key point we would like to bring out here is that our whole material universe is composed of u and d quarks, as particle physicists like to point out. Thus the possibility of our existence can be attributed to the innocent-looking first step of the excitation process displayed in Fig. 0.2.4.

The calculations in Fig. 0.2.4 of the μ, p and τ masses relative to the electron mass, which consist of simply adding up the excitation quanta, may seem to be somewhat trivial, but it must be kept in mind that these three mass-ratio calculations have eluded theorists for roughly 1/2 century, 1 century and 1/4 century, respectively. Nature is sometimes surprisingly simple if viewed in the proper context.

A different platform state, $M_\phi = s\bar{s}$, is illustrated in Fig. 0.2.5, where only the *particle* excitation column is shown. It consists of an $s\bar{s}$ strange quark–antiquark pair, and is manifested experimentally as the spin-1 ϕ vector meson. The s and $\bar{s}$ quarks themselves are produced in the particle and antiparticle columns of the $M_{\mu\mu}$ excitation tower (the M_μ column is displayed in Fig. 0.2.4), and they bind together hadronically to form the $s\bar{s}$ platform. The first excitation level above the M_μ platform appears at three times the $s\bar{s}$ mass, and is accomplished via the addition of an ss excitation quantum in each column of the tower, as displayed for the particle column in Fig. 0.2.5. This results in the creation of a $c\bar{c}$ quark pair, which manifests as the J/ψ vector meson. This mass tripling is then repeated with a cc excitation of each column of the $c\bar{c}$ level to produce a $b\bar{b}$ pair at three times the mass of the $c\bar{c}$ pair. This level manifests as the Υ vector meson. This type of mass tripling is analogous to the first step of the muon excitation tower shown in Fig. 0.2.4, wherein the initial mass μ of the tower in turn creates a $\mu\mu$ excitation quantum that combines with it to form the q quark at three times the mass of the μ.

During the two mass-triplings of the $s\bar{s}$ platform there is a charge exchange (CX) at each stage, so that the quark charge alternates from $s(-1/3e)$ to $c(+2/3e)$ to $b(-1/3e)$ in the particle column, with a corre-

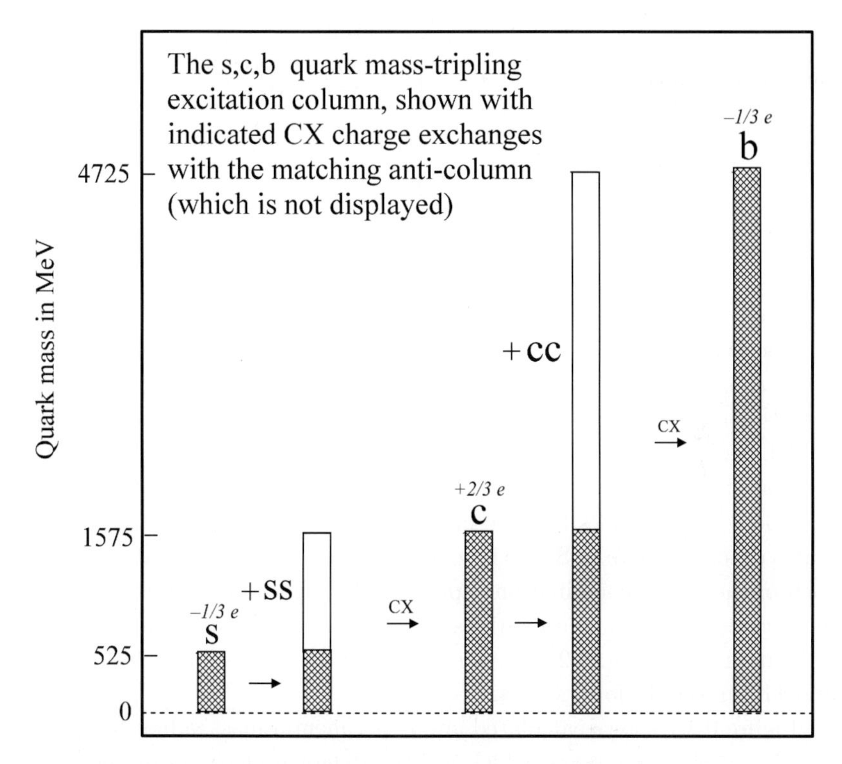

Fig. 0.2.5 The s quark "mass tripling" vector boson production tower. The "strange" 525 MeV s quark is created in the muon production column of Fig. 0.2.4, and is hadronically bound to a matching $\bar{s}$ quark that is created in the antimuon production column, thus producing the $\phi = s\bar{s}$ vector meson. The $s\bar{s}$ quark pair in turn functions as a "platform state" which is successively tripled to produce first the $J/\psi = c\bar{c}$ vector meson and then the $\Upsilon = b\bar{b}$ vector meson. Figure 0.2.5 displays the particle-column part of this mass-tripling excitation process, wherein the 525 MeV s quark leads to the 1575 MeV c quark and then to the 4725 MeV b quark. A charge exchange (CX) occurs between the particle and antiparticle columns at each tripling step, as indicated in Fig. 0.2.5, which changes the fractional charge on the quark from $-\frac{1}{3}e$ for the s quark to $+\frac{2}{3}e$ for the c quark and back to $-\frac{1}{3}e$ for the b quark. These s, c and b constituent-quark masses give good fits to the meson threshold states (Figs. 0.2.7 and 0.2.8 and Table 0.6.1), and also to the baryon and hyperon threshold states (Figs. 0.2.9 and 0.2.10 and Table 0.6.2).

sponding alternation in the antiparticle column (see Fig. 3.18.1). An analogous CX charge exchange occurs in the M_μ and $\bar{M}_\mu$ excitation columns of Fig. 0.2.4, which causes the p^+ proton to end up in the same excitation column as the μ^- muon and the τ^- lepton (see Fig. 3.18.2). As a consequence, the positively charged proton is trapped in the negatively charged M_μ excitation column and cannot decay back down to the e^- ground state. Thus the proton is stable.

The Standard Model $q \equiv (u, d)$, s, c and b quark mass values that are obtained from the sequential excitation processes of Figs. 0.2.4 and 0.2.5 are displayed in Fig. 0.2.6. These are all multiples of the $\mu \cong m_f = 105$ MeV muonic α-mass, so we can characterize these SM quarks as "muon quarks." They have "constituent-quark" masses, wherein the mass of the particle is obtained in first order by simply adding up the masses of its constituent quarks. We can test the accuracy of these quark mass values by using them in the construction of the various meson and particle states that are formed out of them. This procedure represents a tremendous success for the quark model, because it means that we can use four basic masses (which are in fact all related to one another) to fairly accurately reproduce a whole variety of particles. The particles that we include in this procedure are the *platform* states described here in Sec. 0.6, which are the lowest-mass states in which each of the quark combinations appear. The higher-mass states represent more complex quark combinations. As we will now demonstrate, the q, s, c, b quark set of Fig. 0.2.6 gives quite accurate mass values for 10 meson and 11 baryon platform states.

Figure 0.2.7 shows calculated and experimental mass values for the $q\bar{q}$, $s\bar{s}$, $c\bar{c}$ and $b\bar{b}$ quark pairs, which are the diagonal elements of the meson platform mass matrix. The observed particles that correspond to the $q\bar{q}$ pair are the ρ and ω vector mesons. However, since the calculated $q\bar{q}$ mass is about 140 MeV too low, these particles may actually be hybrid $q\bar{q}\pi$ excitations whose forms are driven by the symmetries of the production process. The other three diagonal mass values are given with good accuracy as sums of the constituent quarks, and they correspond to the mass-tripling excitation mechanism displayed in Fig. 0.2.6. Interestingly, the calculated mass accuracy improves at the higher mass values, and this result carries on to the highest mass values, as shown in Sec. 0.10.

Figure 0.2.8 displays calculated and experimental mass values for the off-diagonal elements of the meson platform mass matrix. The observed particles that correspond to the $q\bar{s}$ and $\bar{q}s$ pairs are the K^* mesons, whose calculated mass values shown here are roughly 70 MeV too low. This suggests that these particles, like the ρ and ω, may be hybrid excitations. The D_S strange-charged mesons have somewhat lower than expected experimental values. The other four platform masses are given to better than 5% accuracy. The average mass accuracy for the ten platform states displayed

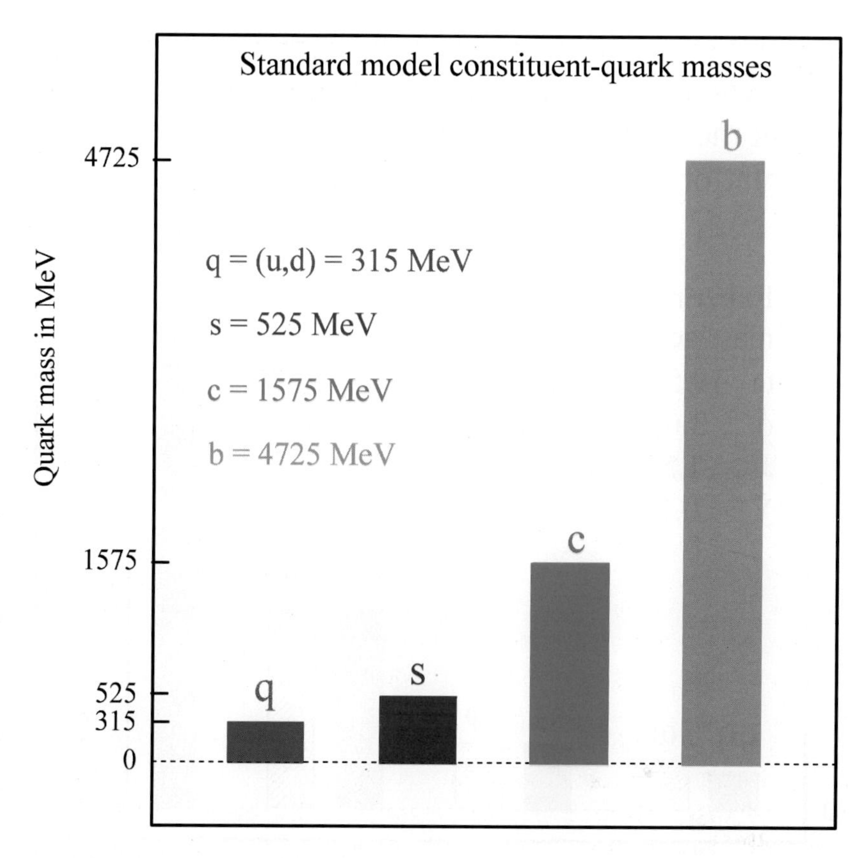

Fig. 0.2.6 The Standard Model $q \equiv (u, d)$, s, c and b quarks, shown here with their "constituent-quark" masses, wherein the observed mass of a particle is the sum of the quark masses. For simplicity in making illustrative fits to the data, we neglect small (∼3%) hadronic binding energies, and we assign the different members of an isotopic-spin multiplet the average mass of the multiplet (Appendix B). An exception is the proton, which is assigned its observed mass. The quark masses displayed here are, for simplicity, exact multiples of 105 MeV (thus neglecting any contributions from the ground-state electron mass). These q, s, c, b "muon-quark" masses give reasonable fits to all of the long-lived "platform state" particles except the pseudoscalar mesons, which follow a different excitation pattern and require a different set of "pion-quark" masses (Figs. 0.2.11–0.2.13). It is of interest to compare the "constituent-quark" masses $q \equiv (u, d) = 315$ MeV and $s = 525$ MeV that are displayed in Fig. 0.2.6 with the "effective quark" masses $u = 338$ MeV, $d = 322$ MeV and $s = 510$ MeV that are deduced from the measured baryon and hyperon magnetic moments [11]. These two completely independent determinations yield very similar q and s quark masses. By way of contrast, the corresponding u, d and s "current-quark" masses are [10]: $u = 1.5$–3.0 MeV, $d = 3$–7 MeV and $s = 95 \pm 25$ MeV, and the c and b current-quark masses are $c = 1250 \pm 90$ MeV and $b = 4200 \pm 70$ MeV.

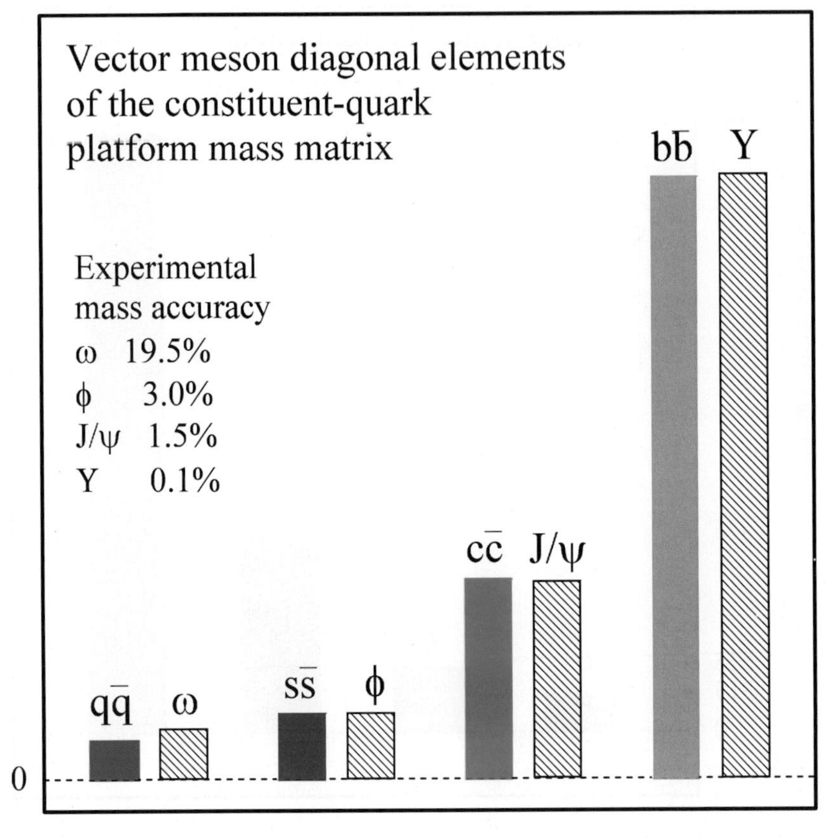

Fig. 0.2.7 The diagonal elements of the meson q, s, c, b constituent-quark platform matrix. When the q, s, c, b quarks are paired with $\bar{q}$, $\bar{s}$, $\bar{c}$, $\bar{b}$ antiquarks to produce mesons, they form a quark–antiquark platform mass matrix, where the designation "platform" denotes the lowest-mass state at which a particular quark–antiquark combination occurs (see Table 0.6.1). Figure 0.2.7 displays the diagonal elements of this quark mass matrix, which are all spin-1 vector mesons. The quark masses used here are from Fig. 0.2.6, and the gray columns in the figure represent the experimental particle masses (Appendix B). As can be seen, the calculated and experimental masses are in good agreement except for the ω meson, which is more accurately reproduced as a "hybrid" $q\bar{q}\pi$ state that combines both *muon-quark* and *pion-quark* masses. The mass agreement between calculation and experiment improves with increasing mass, which can be attributed to the vanishing of the hadronic binding energy at high masses (asymptotic freedom).

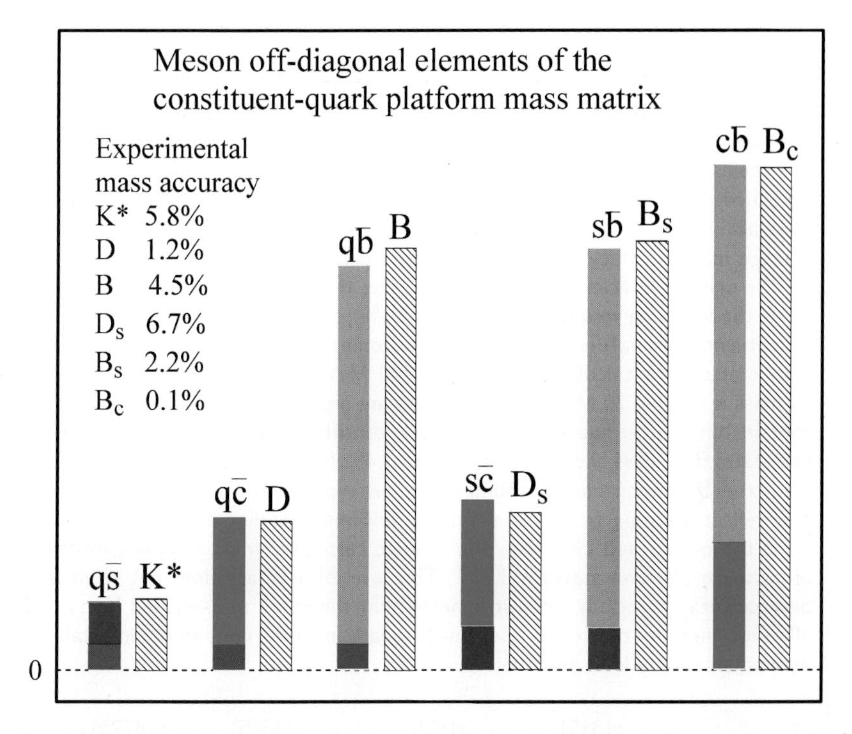

Fig. 0.2.8 The off-diagonal elements of the quark–antiquark platform mass matrix. The calculated and experimental mass values shown here are in reasonable agreement. The low-mass K^* meson is probably a hybrid excitation (Fig. 0.2.12), as is the low-mass ω meson in Fig. 0.2.7. It is interesting to note that the calculated mass accuracy improves with increasing mass in both Fig. 0.2.7 and 0.2.8. This suggests that the perturbations of the constituent-quark particle mass values are roughly independent of energy, and thus diminish in relative importance as the constituent-quark masses increase. The precision fit to the B_c meson is of special interest, both because the B_c is the platform state for the two heaviest first-order SM quarks, and hence furnishes a good test of their constituent-quark nature, and also because a very accurate mass value has recently been obtained for this meson (Appendix B).

in Figs. 0.2.7 and 0.2.8 is 4.4%, with no adjustable parameters utilized in the fits to the data. If we remove the (ρ, ω), K^* and D_S platforms from the calculations, the average mass accuracy improves to 1.8% for the seven remaining platforms. This level of accuracy seems sufficient to validate the use of constituent quarks in reproducing these meson platform states.

The results displayed in Figs. 0.2.7 and 0.2.8 are summarized in Table 0.6.1, which shows the particles that occupy the meson platform

Table 0.6.1 The meson $q \equiv (u, d)$, s, c, b constituent-quark platform mass matrix. The quark masses used here — $(q, s, c, b) = (3, 5, 15, 45) \times 105$ MeV — are displayed graphically in Fig. 0.2.6. The meson states shown here are the lowest-mass states at which these particular quark combinations appear, and their quark mass components are displayed in Figs. 0.2.7 and 0.2.8. The simplified constituent-quark matrix mass values used here give the calculated mass accuracies in % that are displayed under each quark platform state. The average mass accuracy for these ten meson platform states is 4.4%. Averaging the mass accuracies over the four particles that contain each type of quark yields the following results: q (7.6%); s (4.4%); c (2.4%); b (1.8%). Thus the constituent-mass nature of these quarks becomes more apparent in the higher-mass states. The calculated $q\bar{q}$ mass is 140 MeV too low, and the calculated $q\bar{s}$ mass is roughly 70 MeV too low. These are probably hybrid excitations that contain both muon-quark and pion-quark contributions. If we remove these two states, the mass accuracy improves to 2.4% for the remaining eight states. Also, the D_S meson has an anomalously low experimental mass value, which may reflect a strong $(c \equiv sss)\cdot(\bar{s})$ quark–antiquark binding energy. If we exclude the $q\bar{q}$, $q\bar{s}$ and $c\bar{s}$ states, the average calculated mass accuracy for the other seven platform states is 1.8%. This level of accuracy does not seem to be accidental, especially when it is noted that the quark masses used here are all exact multiples of the 105 MeV muon-quark m_f mass unit, so no adjustable parameters of any kind are involved.

	$q(315)$	$s(525)$	$c(1575)$	$b(4725)$
$\bar{q}(315)$	$\rho^{\pm,0}$, $\omega^{0,\dagger}$ 19%	$K^{*-}, \bar{K}^{*0\dagger}$ 5.8%	D^+, D^0 1.2%	B^-, $\bar{B}^0$ 4.5%
$\bar{s}(525)$	K^{*+}, K^{*0}	ϕ^0 3.0%	D_S^+ 6.7%	$\bar{B}_S^0$ 2.2%
$\bar{c}(1575)$	D^-, $\bar{D}^0$	D_S^-	J/ψ^0 1.5%	B_C^- 0.2%
$\bar{b}(4725)$	B^+, B^0	B_S^0	B_C^+	Υ^0 0.1%

† Hybrid excitation.

quark matrix and gives the accuracies of their mass calculations. Apart from numerical results, the impressive features about Table 0.6.1 are the completeness and accuracy of the experimental data, and the impressive way in which the q, s, c, b set of quarks accurately account for the isotopic spins and mass values of these meson platform states.

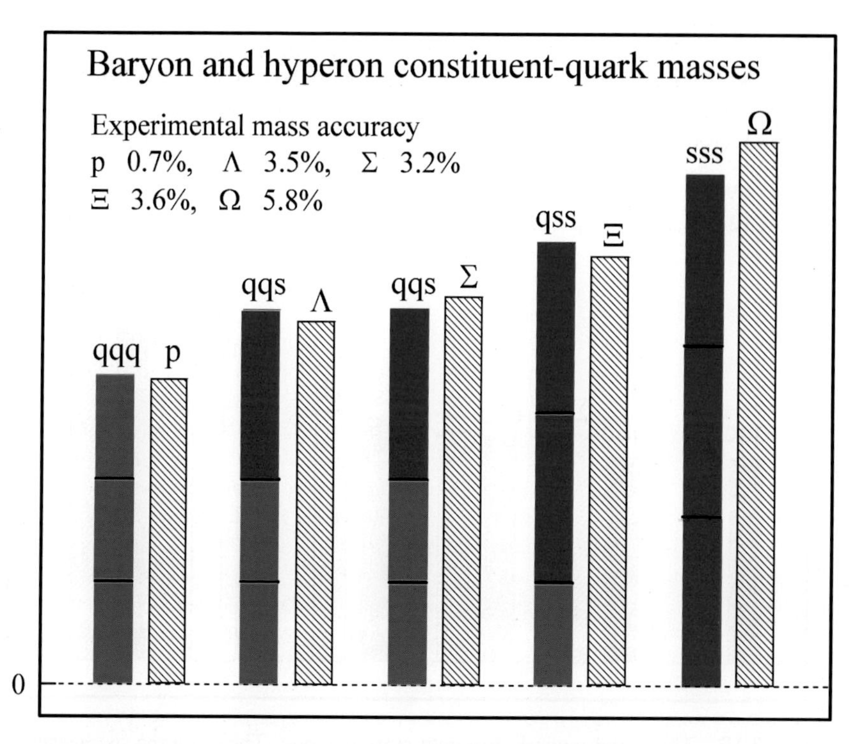

Fig. 0.2.9 The $q \equiv (u, d)$ and s quark threshold-state baryons and hyperons. The constituent-quark masses of Fig. 0.2.6 are accurate enough to reproduce all of these particles to mass accuracies of a few percent, as listed in the figure, using no adjustable parameters of any kind. In detail, the calculated Λ and Ξ masses are each about 3.5% larger than the experimental values, which suggests the presence of small hadronic binding energies for these excitations. The experimental Σ masses are roughly 70 MeV larger than the Λ mass, even though the Λ and Σ hyperons have the same basic qqs constituent-quark configuration. This suggests that the Σ excitations may be hybrid states (Sec. 3.25). The experimental Ω mass is about 140 MeV larger than the value that is obtained by extrapolating the Σ and Ξ mass values, which also suggests a hybrid excitation (Sec. 3.25), possibly caused by quantum mechanical restrictions on the fermion ground state. The accurate qqq fit to the proton mass is a strong argument for the validity of the 315 MeV constituent-quark mass value that has been assigned to the q quarks. This assignment also makes it clear that the 140 MeV pions, which are composed of quark–antiquark pairs, do *not* use u and d constituent quarks in their mass structure, even though these u and d quark assignments lead to the correct pion isotopic spin states. The pions require a different type of quark (Figs. 0.2.11–0.2.13).

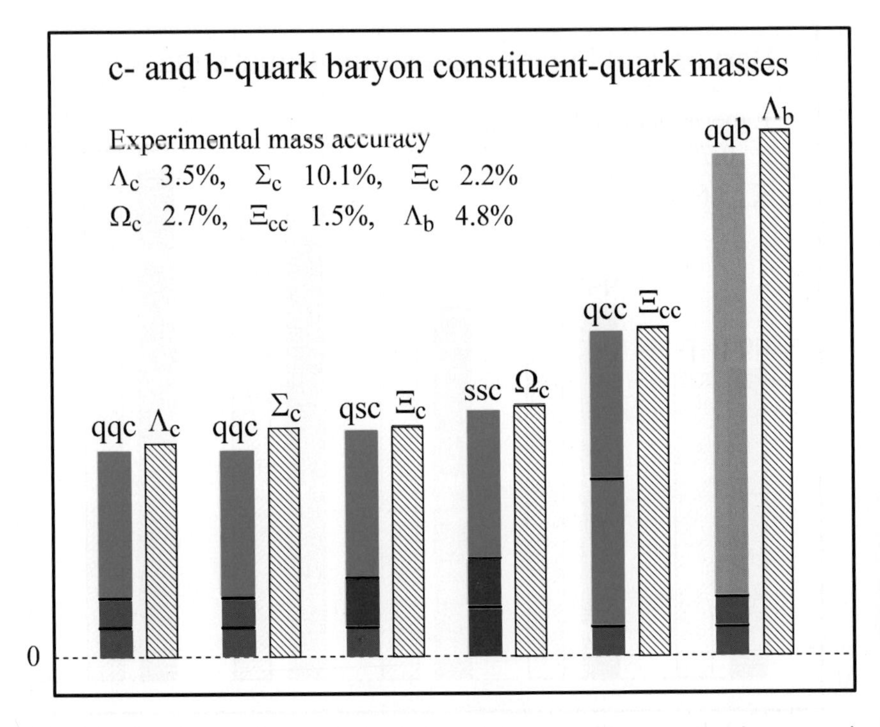

Fig. 0.2.10 The c and b quark threshold-state baryons. These are the six measured baryon platform states that contain c or b quarks. The agreement of the experimental mass values with the constituent-quark values is generally within a few percent, with the Σ_c charmed baryons showing the largest deviation. In all six cases the constituent-quark mass values are below the experimental values, which indicates that these excitations are more complex than the simple constituent-quark addition of masses suggests. But the overall mass mapping is accurate enough to verify that the same q, s, c, b set of quarks which reproduced the meson states of Figs. 0.2.7 and 0.2.8 are generally valid here. The fact that the q, s, c, b quark basis set can accurately reproduce all of the isotopic spins of the mesons and baryons that are displayed in Figs. 0.2.7–0.2.10 is one of most remarkable successes of the Standard Model. And the successful reproduction of their mass values to within a few percent with just the four related masses displayed in Fig. 0.2.6 is also a major accomplishment. The identification of these quark states has brought order into a rather chaotic collection of elementary particle states. The α-quantization of the threshold-state elementary particle *lifetimes* reveals this quark structure very clearly (Fig. 0.1.11). The main conclusion we draw from the α-quantization of the threshold-state particle *masses* is that these are constituent quarks, and not current quarks.

Table 0.6.2 The baryon q, s, c, b constituent-quark platform masses. The baryon states listed here are the lowest-mass states at which these quark combinations occur. These baryon platform states are displayed in Figs. 0.2.9 and 0.2.10. The same $q = 315$ MeV, $s = 525$ MeV, $c = 1575$ MeV, $b = 4725$ MeV constituent-quark masses are used here as in Table 0.6.1, and they give the calculated mass accuracies that are displayed with each platform state. The average mass accuracy for these 11 baryon platform states is 3.8%. The average mass accuracy for the 21 platform states listed in Tables 0.6.1 and 0.6.2 is 4.1%. This accuracy is attained by using just the four quark masses shown here, which are all generated from a common 105 MeV mass quantum, and which have no arbitrary degrees of freedom. If we remove the $q\bar{q}$ platform state from Table 0.6.1 and the $\Sigma_c = qqc$ platform state from Table 0.6.2, both of which exhibit anomalous mass values, the average mass accuracy for the remaining nineteen platform states improves to 3.0%. Thus the constituent-mass nature of these q, s, c, b quarks seems to be experimentally verified.

$p = qqq$ 0.7%		
$\Lambda = qqs$ 3.5%	$\Lambda_c = qqc$ 3.5%	$\Lambda_b = qqb$ 4.8%
$\Sigma = qqs$ 3.2%	$\Sigma_c = qqc$ 10.1%	
$\Xi = qss$ 3.6%	$\Xi_c = qsc$ 2.2%	$\Xi_{cc} = qcc$ 1.5%
$\Omega = sss$ 5.8%	$\Omega_c = ssc$ 2.7%	

Figures 0.2.9 and 0.2.10 illustrate how well this same q, s, c, b set of quarks succeeds in reproduces the masses of the baryon and hyperon platform states. The five observed baryon and hyperon platforms, which contain only q and s quarks, are shown in Fig. 0.2.9, and the six observed charm and bottom platforms, which contain c or b quarks in addition to q and s quarks, are shown in Fig. 0.2.10. (The Ξ_b platform, which contains a mixture of states, has a well-enough-determined lifetime to enter into the lifetime systematics, but does not presently have a well-determined mass value). The results displayed in Figs. 0.2.9 and 0.2.10 are summarized in Table 0.6.2. The average mass accuracy for the five platforms in Fig. 0.2.9 is 3.4%, and the average accuracy for the six platforms in Fig. 0.2.10 is 4.1%. The average for all eleven baryon platforms is 3.8%. If we remove the Σ and Σ_c baryons, which have the same basic q, s, c quark configurations as the Λ and Λ_c baryons, but have larger masses that represent more complex excitations, the mass accuracy improves to 3.1%.

Altogether, Figs. 0.2.7–0.2.10 and Tables 0.6.1 and 0.6.2 display q, s, c, b mass calculations for 21 particles and isotopic spin groups of particles. The overall average mass accuracy of the calculations is 4.1%, with no corrections, adjustments or deletions of any kind applied. This was accomplished by using mass units of 315, 525, 1575 and 4725 MeV. These mass units were not chosen or adjusted so as to give best fits to the data. Instead, they were calculated on the basis of their observed α-dependence with respect to the electron ground state. There are two relevant conclusions we would like to draw from these results, the second of which follows from the first: (1) the Standard Model q, s, c, b quarks are constituent-mass quarks; (2) the Standard Model q, s, c, b quarks cannot be used to obtain mass values for the low-mass π, η, η', K pseudoscalar meson nonet. The second conclusion serves to introduce our next topic — the pseudoscalar nonet.

0.7 The Spin-0 Generic "Pion" Constituent Quarks

The elementary particle quarks we discussed in Sec. 0.6 are multiples of the spin-1/2 fermion mass quantum $m_f = 105$ MeV, and hence can be regarded as "muon quarks." The quarks we discuss in Sec. 0.7 are multiples of the spin-0 boson mass quantum $m_b = 70$ MeV, and can be regarded as "pion quarks." These pion quarks combine together to reproduce the mass values of the PS mesons, which have always posed problems for the Standard Model. They also occur in combination with muon quarks in a variety of short-lived "hybrid" excitations of the fundamental platform states. The pion quarks we define here are "generic" quarks, which means that we assign them constituent-quark mass values but do not specify their other properties, such as the quark charge states.

Before we address the topic of pion quarks, it is useful to study the PS meson mass values, in order to see what they reveal about their mass quantization. The simplest PS mesons are the nonstrange π^0, $\pi^\pm$, η and η' particles. These particles exhibit an accurate lifetime α-quantization that extends over more that 12 orders of magnitude (Fig. 2.2.5). In order to examine their mass α-quantization, we plot the average pion mass, $\pi = 1/2$ $(\pi^\pm + \pi^0)$, together with the η and η' masses, on a 137 MeV mass grid, and we also include the 1 MeV mass of an e^-, e^+ electron pair. These particle states are displayed in Fig. 0.2.11. As can be seen, the π, η and η' masses fall exactly on the 1, 4 and 7 levels of the 137 MeV mass grid, with an average mass accuracy of 0.14%! This has to be the premier example of *linear* elementary particle mass quantization, and it brings into question the Standard Model PS meson mass formula that employs the squares of the masses rather than the masses themselves. In addition to this astonishing mass linearity, the other feature of interest is the 137 MeV value of the mass quantization unit. The reason for this serendipitous value is that the

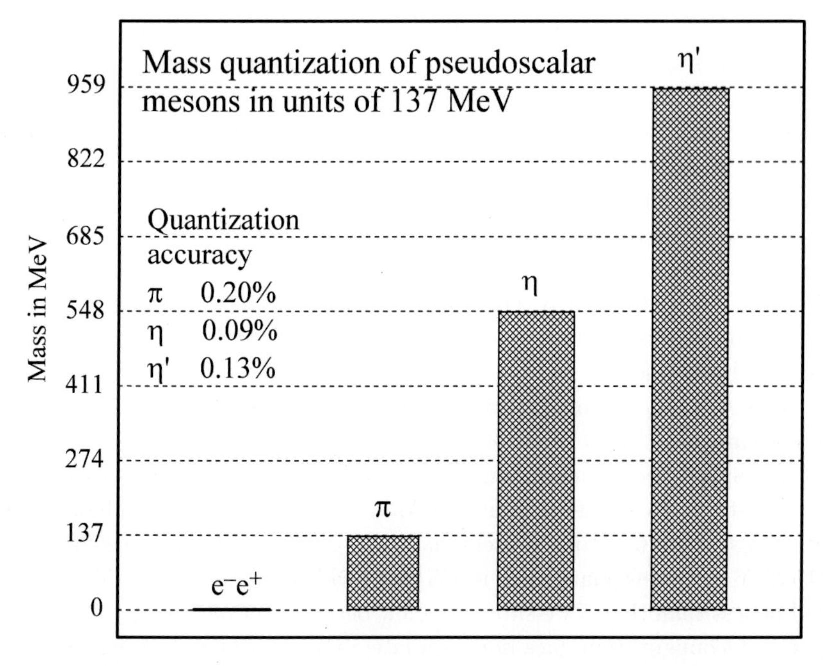

Fig. 0.2.11 A mass plot of the non-strange π, η and η' pseudoscalar (PS) mesons on a 137 MeV mass grid. The π mass shown here is the average of the $\pi^{\pm}$ and π^0 isotopic spin states. The 137 MeV quantization of these PS mesons could hardly be more accurate: they fit the mass grid at an accuracy level of a little over a tenth of a percent. Also displayed in Fig. 0.2.11 is the 1 MeV mass of an e^-e^+ electron–positron pair. This 137 MeV mass quantization is quite different from the $m_f = 105$ MeV fermion mass quantizations displayed by the "muon" constituent quarks in Sec. 0.6. It is based on the $m_b = 70$ MeV boson mass quanta that are generated in matching $\alpha^{-1} \simeq 70$ MeV mass leaps from the electron-positron "ground state" to the pion level, $\pi = m_b\bar{m}_b$ (Fig. 0.2.2). This PS meson mass plot demonstrates that the π, η and η' mesons follow a *linear* mass formula, and not the *quadratic* mass formula that is commonly invoked for them [10]. These PS meson masses are reproduced by the set of generic "pion" constituent quarks displayed in Figs. 0.2.12 and 0.2.13.

e^-e^+ ground state has a mass of 1.022 MeV, which is almost unity, and the pion — the mass-quantizing element — represents an $\alpha^{-1} \cong 137$ leap from the ground state. The calculated mass value for the $\pi = m_b\bar{m}_b$ pion is about 141 MeV if we add in the e^-e^+ ground state mass, so the mass quantization value of 137 MeV indicates a hadronic binding energy of about 2.8% between the particle and antiparticle substates in the pion.

The simplest way to reproduce the π, η and η' masses is to define the following set of generic pion quarks: $q_\pi = m_b$, $q_\eta = 4m_b$, $q_K = 7m_b$.

Figure 0.2.12 displays these quarks in the form of an excitation-doubling tower that is similar to the one shown in Fig. 0.2.4 for the muon quarks q and s. In the muon tower of Fig. 0.2.4 we have an $m_f = 105$ MeV platform state M_μ upon which $Q_f \cong m_f m_f = 210$ MeV excitation quanta are successively added to form the spin-1/2 q and s quarks and the spin-1/2 p and τ fermions. In the pion tower of Fig. 0.2.12 we have an $m_b = 70$ MeV platform state M_1 upon which $Q_b \cong m_b m_b m_b = 210$ MeV excitation quanta are successively added to form the spin 0 q_η and q_K quarks and K meson, and the spin-1 $K^*(894)$ and $K^*(1717)$ vector bosons. The excitation-doubling generation process and $Q_f = Q_b = 210$ MeV excitation mass unit are the same in both cases.

The $K^*(894)$ meson appears in the systematics of Fig. 0.2.12 in an interesting way. The excitations up to the K meson are spinless, but the K^* is a spin-1 vector meson, and hence must possess spinning quarks in its substructure. In terms of Standard Model quarks, it is the configuration $K^* = q\bar{s}$. But with our present mass assignments, the q quark mass is 315 MeV and the s quark mass is 525 MeV, which is a total of 840 MeV, and will be less than that if we subtract a few percent binding energy. Thus the $q\bar{s}$ quark configuration does not adequately reproduce the $K^*(894)$ meson. The $K^*(894)$ excitation $K^* = m_b + 4Q_b$ shown in Fig. 0.2.12 corresponds to a total excitation mass of $13m_b = 910$ MeV, which matches the $K^*(894)$ value if a 2% binding energy is imposed. This suggests that the K^* is the configuration $K^*(894) = q\bar{s} \cdot m_b \equiv q\bar{s}^*$, where $\bar{s}^* = 595$ MeV is an "excited state" of the s quark that is caused by the dynamics of the mass-doubling excitation process displayed in Fig. 0.2.12 (also see Fig. 0.2.20). The s^* quark also serves to explain the 70 MeV mass difference between the Λ and Σ hyperons (Fig. 0.2.9), so that $\Lambda = qqs$ and $\Sigma = qqs^*$, and it may account for the 140 MeV discrepancy in the Ω hyperon mass (Fig. 0.2.9) by invoking the quark representation $\Omega = ss^*s^*$.

Figure 0.2.13 displays the manner in which the π, η, η' and K mesons are reproduced by the 70, 280 and 490 MeV q_π, q_η and q_K generic quarks. The mass accuracies are in the 2% range, with most of the discrepancy in mass values due to the fact that we did not include binding energy effects. One significant feature about Fig. 0.2.13 is that the spin-0 kaon is composed of an odd number of $m_b = 70$ MeV mass units. This indicates that the α-mass m_b is a spin-0 mass quantum. Spinless boson mass quanta represent a concept that is outside of the paradigm of Standard Model physics, wherein all of the quark states are spin-1/2 fermion entities.

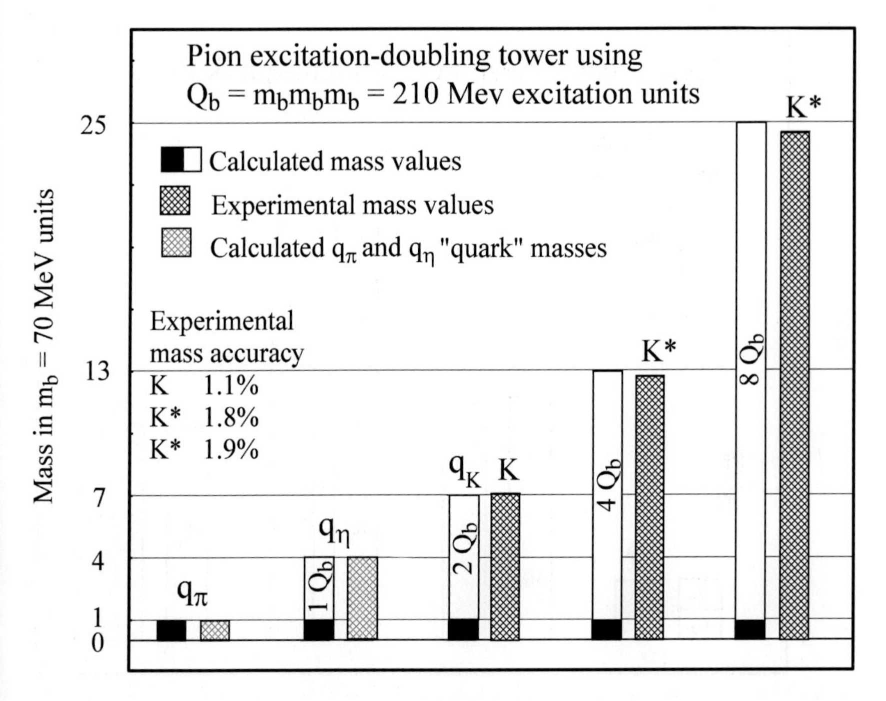

Fig. 0.2.12 The pion "excitation-doubling" production tower. This figure is the analogue of the muon "excitation-doubling" production tower displayed in Fig. 0.2.4. The basic mass units employed in these two towers are the pion α-mass $m_b = 70$ MeV and the muon α-mass $m_f = 105$ MeV, respectively. The platform masses for these towers are $M_K = 70$ MeV in Fig. 0.2.12 and $M_\mu = 105$ MeV in Fig. 0.2.4. The platform M_K corresponds to the "generic pion quark" q_π, as defined in the text. The platform M_μ corresponds to the muon. In each tower, $Q = 210$ MeV excitation quanta are added in an excitation-doubling manner to produce higher-mass states. $Q_b = m_b m_b m_b = 210$ MeV is the excitation unit in the pion tower, and $Q_f = m_f m_f = 210$ MeV is the excitation unit in the muon tower. The first 210 MeV excitation in the pion tower produces the q_η generic pion quark, and the second produces the q_K quark and spin-0 K meson. The first 210 MeV excitation in the muon tower produces the $q \equiv (u, d)$ quarks, and the second produces the s quark. The third and fourth excitation doublings in the pion tower produce the spin-1 $K^*(894)$ and $K^*(1717)$ mesons. The third and fourth excitation doublings in the muon tower produce the spin-1/2 proton and τ lepton. The anomalously large mass of the $K^*(894)$ (as compared to its expected $q\bar{s} \simeq 840$ MeV constituent-quark mass) can be attributed to the dynamics of the production mechanism, as is discussed in the text and in Sec. 3.26. The pion excitation tower, like the muon excitation tower, has both particle and antiparticle columns (only one of which is displayed in Fig. 0.2.12), which are produced simultaneously in order to preserve particle–antiparticle symmetry.

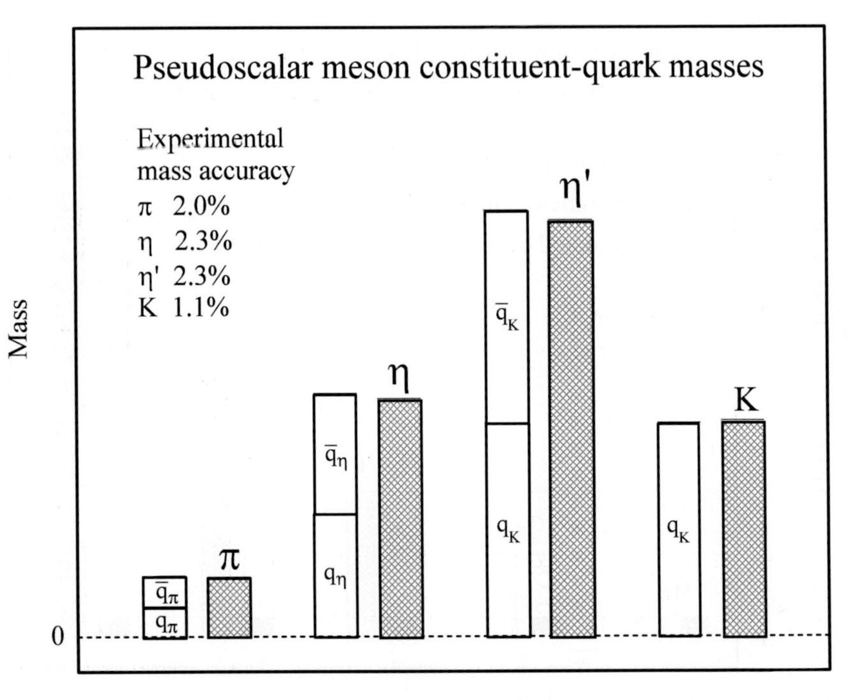

Fig. 0.2.13 The pseudoscalar meson nonet. This figure shows how the generic pion quarks of Fig. 0.2.12 combine to form the (π^+, π^-, π^0, η, η', K^+, K^0, $\bar{K}^0$, K^-) *PS* nonet of particles. By simply adding up 70 MeV mass quanta, we reproduce the masses of these states to an accuracy of about 2%. These 2% deviations, with the calculated masses for the paired quarks consistently higher than the experimental masses, reflect a hadronic binding energy of about 2% that should be applied to these states (except for the unpaired kaons), as displayed in Fig. 0.2.19. The fact that the *K* mesons contain an odd number of 70 MeV quanta is significant, because it indicates that the α-mass $m_b = 70$ MeV is a *spin-0 mass quantum*, which has no counterpart in the Standard Model formalism. Hence it is truly a boson (integer-spin) mass unit, as we have labeled it here (see Fig. 0.2.14). The *PS* mesons have posed long-standing difficulties for the Standard Model quarks, as evidenced for example by the use of a *quadratic* mass formula [10] after it became clear that the *linear* mass-interval formulas employed for baryons and hyperons did not work very well for mesons in the SU(3) representations. Also, recent refinements in the mass (Fig. 0.2.11) and lifetime (Fig. 2.2.5) values for the η' meson have made it clear that the η' appears in the *PS* mesons on an equal footing with the η, and not as an SU(3) singlet that has to be arbitrarily attached to the SU(3) octet. Experimentally, we really have a *PS nonet*, and not the *PS octet* that was originally envisioned in SU(3). The SU(3) representation works remarkably well for reproducing isotopic spins and strangeness quantum numbers, but not for reproducing masses.

In Sec. 0.6 we defined a set of q, s, c, b quark masses as multiples of the fermion α-mass $m_f = 105$ MeV, and we used these "muon quarks" to reproduce all of the long-lived platform particles except the pseudoscalar mesons. Then here in Sec. 0.7 we defined a set of q_π, q_η and q_K quark masses as multiples of the boson α-mass $m_b = 70$ MeV, and we used these generic "pion quarks" to reproduce the PS mesons. Hence we now have an α-quantized constituent-quark model that encompasses all of the long-lived threshold-state particles, which are the ones that exhibit α-quantized lifetimes. To accomplish this, we employed two basic α-masses, $m_b = 70$ MeV and $m_f = 105$ MeV. In Sec. 0.8 we demonstrate that these two masses are in fact different spin modes of the same basic mass.

0.8 The Relativistically Spinning Sphere and the $m_f/m_b = 3/2$ Mass Ratio

We have come a long way in this phenomenological journey through the global landscape of elementary particle lifetimes and masses. The lifetime studies revealed an α-scaling in the lifetimes of the long-lived threshold-state particles. Then a subsequent study of the masses of these same particles unearthed a reciprocal α^{-1} scaling in their masses. The mass α-quantization is in the form of two mass-generating α-leaps from a 1 MeV electron–positron ground state, with the fine structure constant $\alpha = e^2/\hbar c \cong 1/137$ serving as the coupling constant. The first α-leap is to the 140 MeV pion level, and the second is to the 211 MeV muon–antimuon level. These two α-leaps define two basic α-mass excitation quanta — the 70 MeV spinless boson mass quantum m_b, and the 105 MeV spin-1/2 fermion mass quantum m_f. These two α-masses serve as the "bricks" for the creation of a set of spinless "pion quarks" and a set of spin-1/2 "muon quarks," respectively. The muon quarks are the familiar $q \cong (u, d)$, s, c, b Standard Model quarks, but with constituent-quark masses rather than the SM current-quark masses. The pion quarks are spinless basis states which serve to reproduce the low-mass pseudoscalar mesons, and which also appear in "hybrid" combinations of pion and muon quarks as excitation quanta for producing short-lived excited states of underlying "platform" states.

The threshold states we have been studying here are all in the mass range below 12 GeV. There are also three known particle states with mass values well above 12 GeV — the W and Z gauge bosons and the top quark t,

The relativistically spinning sphere

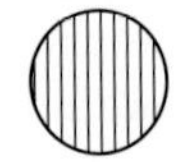

Start with a spinless J = 0 solid sphere of matter that has a rest mass M_o = 70 MeV.

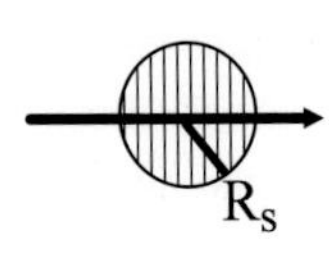

Set the mass spinning at the relativistic limit, where the equator is moving at or just below the velocity c. The calculated relativistic mass of the spinning sphere is M_S = 105 MeV, and is independent of the value for the radius R_S.

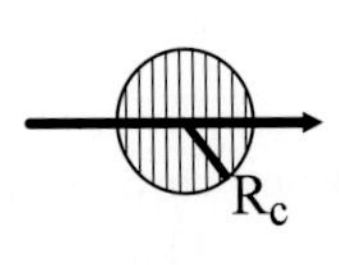

Set the radius R_S of the spinning sphere equal to the Compton radius, $R_C = \hbar/M_S c$. The calculated relativistic angular momentum of the spinning sphere is now J = 1/2 $\hbar$.

Conclusion #1. A spin 1/2 particle is half again as massive as its nonspinning counterpart.

Conclusion #2. The m_{boson} = 70 MeV and $m_{fermion}$ = 105 MeV quark mass quanta are spin 0 and spin 1/2 forms of the universal mass quantum $m_b = m_e / \alpha$ = 70 MeV.

Fig. 0.2.14 The relativistically spinning sphere model. This model is discussed in detail in Chapter 4. It establishes a mass relationship between the spin 0 boson mass quantum m_b = 70 MeV and the spin-1/2 fermion mass quantum m_f = 105 MeV, which were initially identified on the basis of the α-mass systematics displayed in Fig. 0.2.2. These two α-quantized masses are the nonspinning and relativistically spinning forms of a single fundamental 70 MeV mass unit: $m_b \equiv m_e/\alpha$.

which are discussed in Sec. 0.10. These states lead to the identification of a third mass-generating α-leap, but this time not from the electron ground state, but rather from quark states already created in the muon α-leap. Thus with respect to the electron ground state, it is actually a second-order α^{-2} generation process, which leads to a third important α-mass, the quantum $q^{\alpha} = 43{,}182$ MeV.

The first-order mass α-quantization process, with its α^{-1} pion and muon α-leaps, has led to the identification of two basic elementary particle mass quanta: $m_b = 70$ MeV ($J = 0$) and $m_f = 105$ MeV ($J = 1/2$). The question then arises as to whether there is a relationship between these two α-masses, or whether they are two completely independent mass quanta. The answer to this question leads us into the field of "mathology" — mathematical phenomenology. We can see that the spinning mass quantum m_f has a larger energy than the spinless mass quantum m_b. Can this difference in energies be attributed to the m_f spin energy? Mathologically speaking, can we find a mass model based on known mathematical principles that ties these two α-quanta together? The experimental phenomenology of particle lifetimes and masses has defined these two quanta for us. Is there a corresponding mathematical phenomenology that encompasses both of them? This is the question we explore in Chapter 4, where an affirmative answer is found in the relativistically spinning sphere (RSS). The salient mechanical properties of the RSS are summarized in Fig. 0.2.14. It turns out that the RSS has interesting electromagnetic properties as well. These properties, which are also discussed in Chapter 4, lie beyond the lifetime and mass studies that are considered in Chapters 2 and 3, but they are important elementary particle properties which help to further our understanding of what the elementary particle really *is*.

0.9 The M^X Threshold-State Particle Excitation Mechanism

The α-quantized elementary particle mass excitations that have emerged in the present studies occur as multiples of two basic mass units — 70 MeV and 105 MeV. Since these two mass units bear a 2-to-3 numerical relationship to one another, the measurement of the mass of a particle does not unambiguously determine its substructure. An observed mass of 210 MeV, for example, can be accounted-for as composed of either three 70 MeV mass quanta or two 105 MeV mass quanta, although observed spin values limit some of these choices. Thus if we are on the phenomenological level when dealing with these states, we may find more than one way of describing the same set of particle excitations. In the present section we recast some of the particle excitations of the previous sections in terms of the "supersymmetric" excitation quantum $X = 420$ MeV, which occurs in platform excitations in several different (essentially isoergic) forms. We use X quanta to construct excitation towers on the pion and muon platform states M. This is the M^X excitation process. The M^X excitation levels accurately reproduce the mass values of the basic "M^X octet" of particles — π, η, η', K, μ, p, τ, ϕ. This formalism provides a simple way of applying hadronic binding energies to the π, η, η' and ϕ excitations, as we will demonstrate.

Figure 0.2.15 gives pictorial renderings of (1) the electron and positron "ground states" in the mass α-quantization process, (2) the m_b and m_f α-masses that are generated by the coupling constant α, and (3) the M_b and M_f generic boson and fermion "platform" states that combine the electron masses and α-masses and contain their quantum numbers. Figure 0.2.16 displays the excitation sequences for the symmetric M_π, M_ϕ and $M_{\mu\mu}$ platforms, which are labeled with the names of the particles on the lowest occupied level of each platform excitation tower. Figure 0.2.17 shows the asymmetric M_K and M_μ platforms that are created by separating the symmetric M_π and $M_{\mu\mu}$ platforms into their particle and antiparticle components. The excitation towers that are erected on these platforms are displayed together in Fig. 0.2.19.

Figure 0.2.18 portrays the "supersymmetric" mass quantum $X = 420$ MeV. It can occur in a variety of symmetric and asymmetric configurations, as discussed in the caption. In particular, it can function as a "crossover" excitation in which spinless boson mass quanta are isoergically converted into spin-1/2 fermion mass quanta, and vice versa. The quantum X seems to play a special role in platform excitations, and in the production channels that lead to these excitations. The XX excitation unit occurs prominently in the M_μ and M_ϕ excitation towers of Fig. 0.2.19. The XXX excitation unit occurs in the production channel of the Ω hyperon (see Table 3.23.2), and may account for its anomalously large mass (Fig. 0.2.9).

Figure 0.2.19 displays the M^X excitation towers that contain the M^X octet of fundamental low-mass particle states. These towers are constructed entirely of X quanta on the various platform states M. The green X's represent hadronically bound excitations, which are assigned an average binding energy of 2.6%, and the black X's represent mass shell excitations that have no hadron binding energies. With this one adjustable parameter,

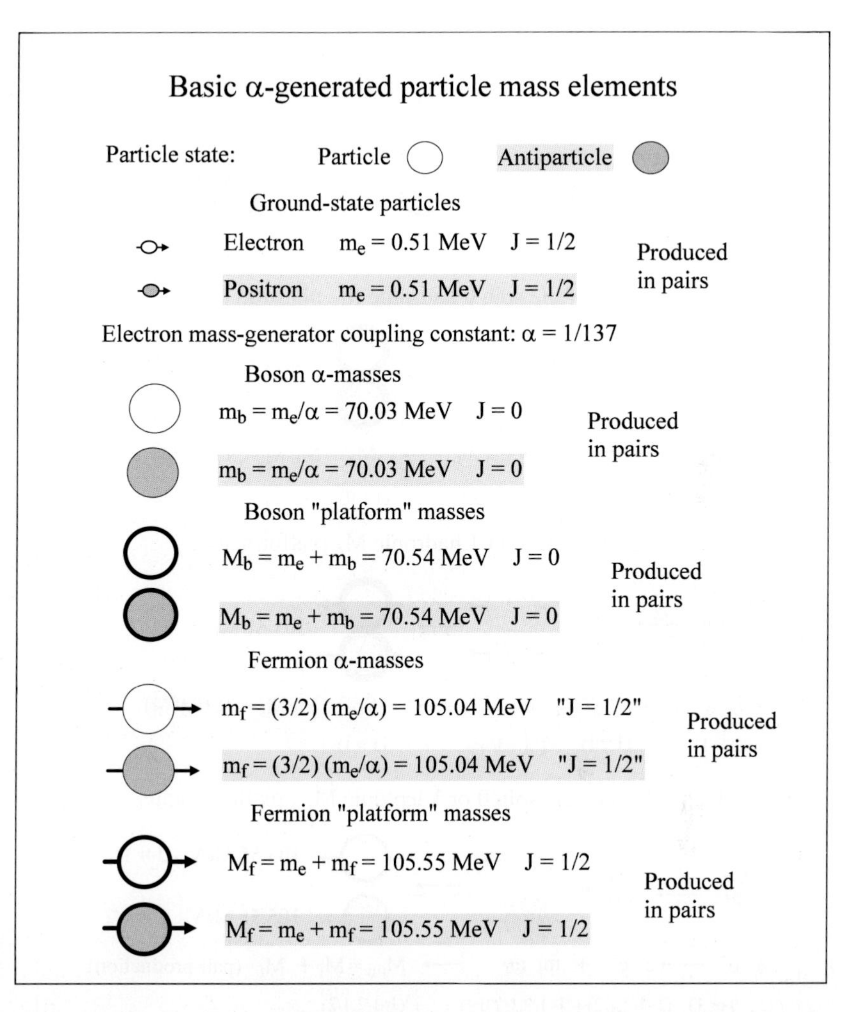

Fig. 0.2.15 Schematic depictions of the electron and positron elementary particle "ground states" (small circles); the boson and fermion m_b and m_f "α-masses" (large circles); and the boson and fermion M_b and M_f "platform" masses (large circles inside borders), which are obtained by adding the electron masses to the α-masses. The contribution of the electron mass to the platform mass is very small, but the quantum numbers carried by the electron are taken over and incorporated into the platform state. The mass m_f is the calculated spinning mass for a Compton-sized sphere: i.e., when a spin $J = 1/2$ electron mass m_e and a "spin-1/2" m_f mass are combined together, the result is a spinning sphere with total mass $M_f = m_e + m_f$ and spin $J = 1/2$. The sizes of the objects in Fig. 0.2.15 are suggestive of their relative mass values (in actuality, the radii of Compton-sized objects vary *inversely* with their masses).

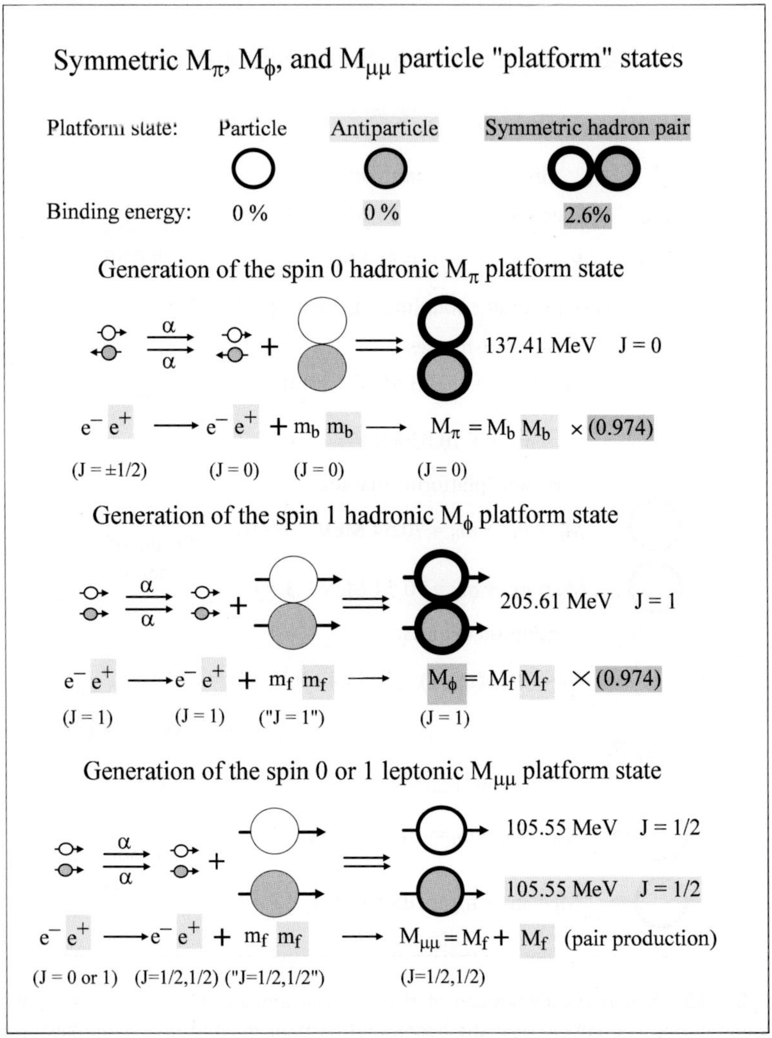

Symmetric M_π, M_ϕ, and $M_{\mu\mu}$ particle "platform" states

Platform state: Particle Antiparticle Symmetric hadron pair

Binding energy: 0 % 0 % 2.6%

Generation of the spin 0 hadronic M_π platform state

137.41 MeV J = 0

$e^- e^+ \longrightarrow e^- e^+ + m_b m_b \longrightarrow M_\pi = M_b M_b \times (0.974)$

(J = ±1/2) (J = 0) (J = 0) (J = 0)

Generation of the spin 1 hadronic M_ϕ platform state

205.61 MeV J = 1

$e^- e^+ \longrightarrow e^- e^+ + m_f m_f \longrightarrow M_\phi = M_f M_f \times (0.974)$

(J = 1) (J = 1) ("J = 1") (J = 1)

Generation of the spin 0 or 1 leptonic $M_{\mu\mu}$ platform state

105.55 MeV J = 1/2

105.55 MeV J = 1/2

$e^- e^+ \longrightarrow e^- e^+ + m_f m_f \longrightarrow M_{\mu\mu} = M_f + M_f$ (pair production)

(J = 0 or 1) (J=1/2,1/2) ("J=1/2,1/2") (J=1/2,1/2)

Fig. 0.2.16 Schematic depictions of the generation processes for the particle-antiparticle-symmetric M_π, M_ϕ and $M_{\mu\mu}$ platform states. The masses (relative sizes) and spin values are indicated, together with the particle (white) and antiparticle (gray) components. The quoted platform mass values for the hadronic M_π and M_ϕ platforms (which have thick black borders) include 2.6% hadronic binding energy (HBE) corrections. The particle and antiparticle components of the leptonic $M_{\mu\mu}$ platform (which have thin black borders) are essentially unbound and require no HBE corrections. The platforms are labeled in accordance with the lowest-mass particles that occupy them or their associated excitation towers. These excitation towers are displayed in Fig. 0.2.19.

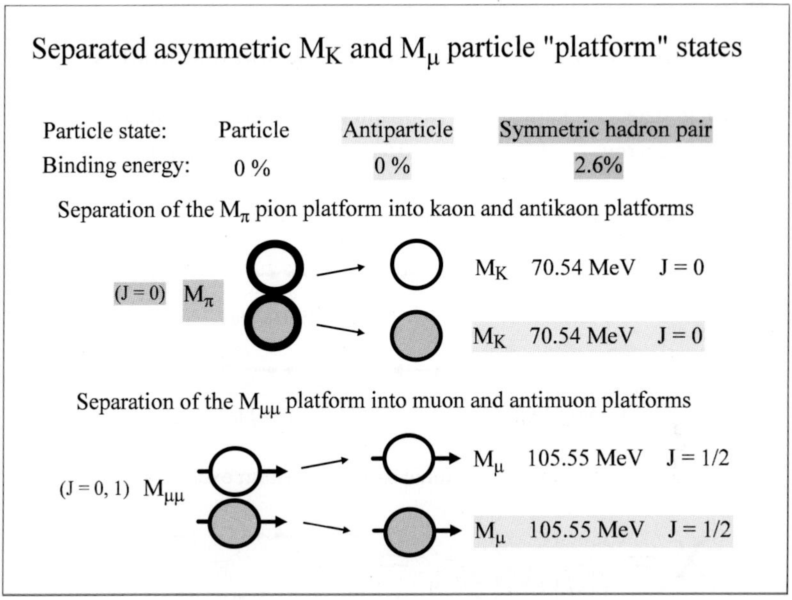

Separated asymmetric M_K and M_μ particle "platform" states

Particle state: Particle Antiparticle Symmetric hadron pair
Binding energy: 0 % 0 % 2.6%

Separation of the M_π pion platform into kaon and antikaon platforms

(J = 0) M_π
M_K 70.54 MeV J = 0
M_K 70.54 MeV J = 0

Separation of the $M_{\mu\mu}$ platform into muon and antimuon platforms

(J = 0, 1) $M_{\mu\mu}$
M_μ 105.55 MeV J = 1/2
M_μ 105.55 MeV J = 1/2

Fig. 0.2.17 Schematic depictions of the separation of the symmetric M_π platform into its asymmetric M_K and $\bar{M}_K$ components, and the separation of the symmetric $M_{\mu\mu}$ platform into its asymmetric M_μ and $\bar{M}_\mu$ components. The separated M_K and M_μ platforms do not require HBE corrections. The M_K platform is empty, but the K and K^* mesons occupy its excitation tower, as displayed in Fig. 0.2.20. The M_μ platform contains the muon, and the proton and τ lepton occupy its excitation tower, as displayed in Fig. 0.2.19.

the π, η, η', K, μ, p, τ, ϕ octet of M^X particles are reproduced to an average mass accuracy of 0.41%, which attests to both the usefulness of the constituent-quark representation and the relevance of the excitation quantum X.

Figure 0.2.20 shows the M^X excitations of Fig. 0.2.19 with the $K^*(894)$ and $K^*(1717)$ spin-1 kaons added in. These K^* mesons were displayed in Fig. 0.2.12 as elements in the pion Q_b excitation-doubling tower. Here they are shown as part of the X-quantized M^X representation. As can be seen, they fill out the symmetry of the M^X excitation towers. Also, by assigning them the same 2.6% hadronic binding energy as used for the other hadronically bound states in Fig. 0.2.20, we accurately reproduce their mass values, although their short lifetimes make these experimental mass values less precise. The occurrence of these spin-1 K^* mesons, with

The supersymmetric "platform" excitation quantum X

Particle state:	Particle	Antiparticle	Symmetric hadron pair
Binding energy:	0 %	0 %	2.6%

Isoergic but asymmetric α-mass basis state transformations

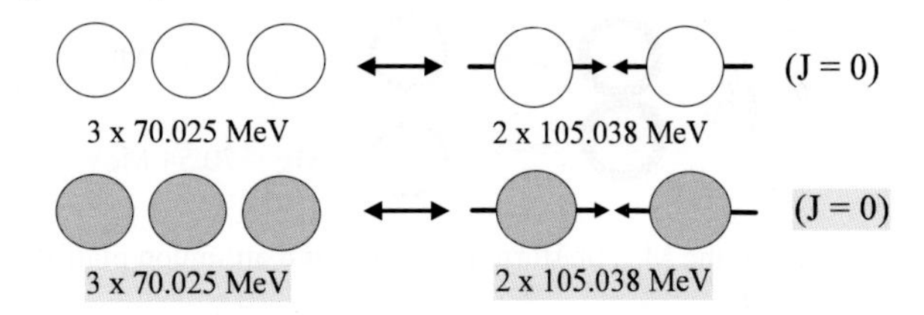

Isoergic and symmetric leptonic α-mass platform excitation

X = = (J = 0)

420.15 MeV 420.15 MeV

Isoergic and symmetric hadronic α-mass platform excitation

X = = (J = 0)

409.23 MeV 409.23 MeV

Fig. 0.2.18 The "supersymmetric" $X = 420$ MeV platform excitation quantum. The quantum X is the lowest-mass excitation unit that can have full particle-antiparticle symmetry in both the *pion* representation $X = Q_b\bar{Q}_b$, where $Q_b = m_b m_b m_b$ ($m_b = 70$ MeV), and in the *muon* representation $X = Q_f\bar{Q}_f$, where $Q_f = m_f m_f$ ($m_f = 105$ MeV). X also occurs in several forms in asymmetric representations. For example, the K meson masses all have the same generic form, $K = m_b X$, but the distinctive $K^{\pm}$, K_L^0 and K_S^0 lifetimes and decay modes are logically associated with different configurations for X. The isoergic nature of these configurations facilitates "crossover" transformations between them, as displayed at the top of Fig. 0.2.18. Transformations between boson and fermion mass quanta can be identified in a variety of particle decay modes.

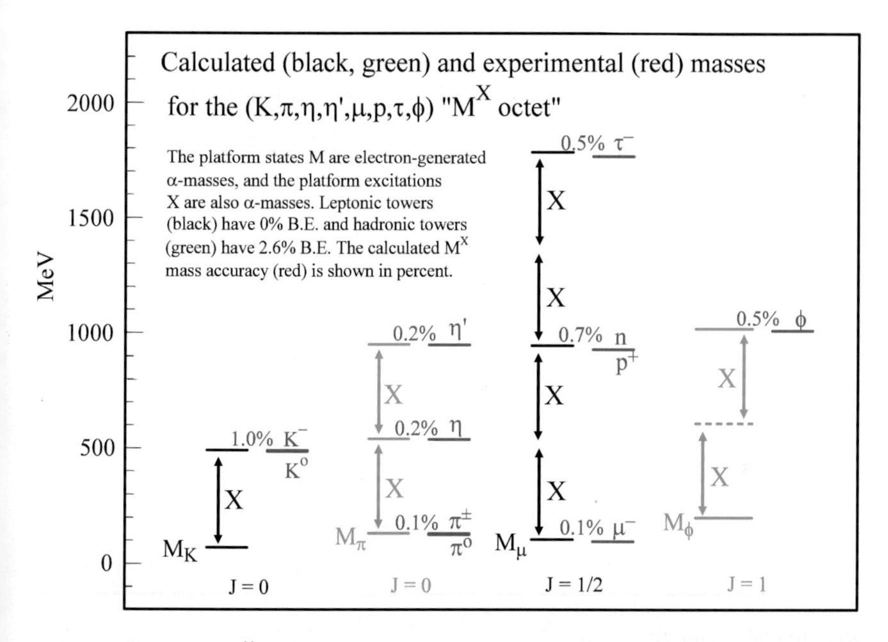

Fig. 0.2.19 The M^X excitation towers that contain the fundamental "M^X octet" of particles — π, η, η', K, μ, p, τ, ϕ. The towers are composed of X quanta (Fig. 0.2.18), and are superimposed on the M_K, M_π, M_μ and M_ϕ platform states. The mass values used for these X quanta and platform states are displayed in Figs. 0.2.16–0.2.18, and are more accurate than the approximate (but very close) mass values employed in Figs. 0.2.4–0.2.13, which were based on integer 70 and 105 MeV mass units with no contributions from the electron masses and no hadronic binding energy corrections. In Fig. 0.2.19, the hadronically bound particles (green) are assigned 2.6% binding energies, and the unbound particles (black) have zero binding energies. The mass values of the eight M^X particles in Fig. 0.2.19 are reproduced to an average mass accuracy of 0.41%, with no freely-adjustable parameters employed.

their q and s spin-1/2 quark substates, in the same M^K excitation tower as the spin-0 K meson in its various forms, illustrates the complexity of the α-quantized basis set, where both spinning and spinless mass quanta occur, and where isoergic transitions between them can take place in both the production and decay processes.

We now turn our attention to the supermassive W, Z isotopic-spin doublet and the top quark t. Are these a part of the same systematics that we have seen here in the particle states below 12 GeV, or do they constitute a completely different and independent set of particle entities? The experimental data furnish an answer to this question.

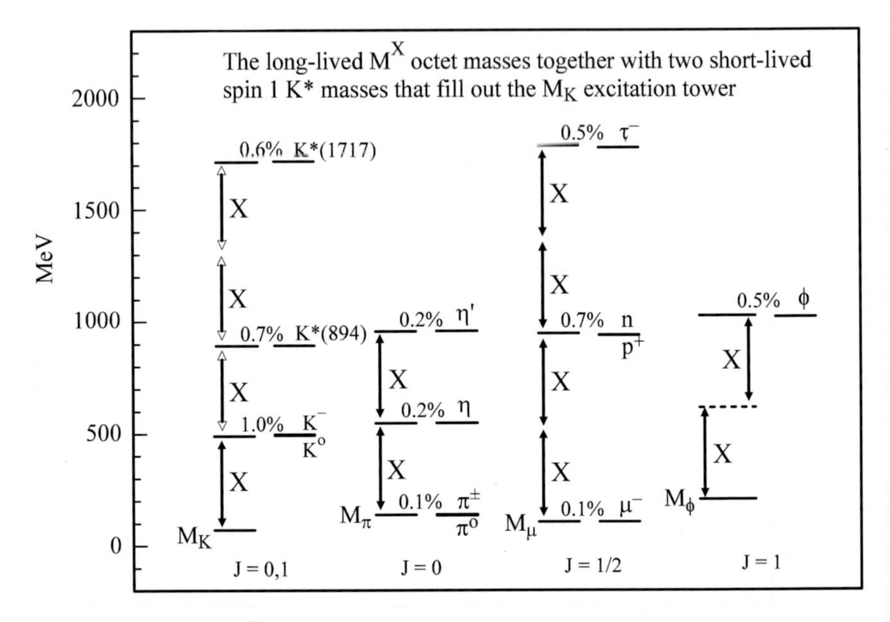

Fig. 0.2.20 The M^X excitation towers of Fig. 0.2.19 with the short-lived spin-1 $K^*(894)$ and $K^*(1717)$ kaon resonances added in. As can be seen, these K^* kaons fill out the X-quantized excitation columns in a symmetrical manner. The K mesons that occupy the $1X$ level of the M_K excitation column are spin 0 particles that appear in this α-quantized mass formulation with zero internal hadronic binding energies. The K^* mesons on the $2X$ and $4X$ levels of the M_K column each contain $q\bar{s}$ or $\bar{q}s$ hadronically bound spin 1 quark pairs. By its mass value, the $K^*(894)$ excitation also includes a 70 MeV m_b excitation quantum that reflects the available energy in the $2X$ mass generation process which seems to be driving this M_K platform excitation. The mass accuracies shown for the $K^*(894)$ and $K^*(1717)$ are the results obtained after 2.6% hadronic binding energies have been applied to their total unbound masses. The open arrow-points on the X excitations for these states indicate that they represent a different type of X-quantum structure, even though they fit in with the overall X-quantization systematics.

0.10 An α^{-2} Mass Leap: The $q^\alpha = 43{,}182$ MeV Basis Set for the W, Z Gauge Bosons and Top Quark t

Early elementary particle experiments extended up to energies around 2 GeV. The long-lived narrow-width mesons were at the lower mass values. As the accelerators pushed up to higher energies, particle physicists searched for elementary particles at ever-increasing mass values. The 1974 discovery of the narrow-width $J/\psi = c\bar{c}$ vector meson at 3 GeV came as a big surprise. Later, the even-narrower-width $\Upsilon = b\bar{b}$ vector meson showed up at 9.46 GeV, with excitations extending up to 11 GeV. After that came — nothing. Where was the missing $t\bar{t}$ vector meson that was needed to fill out the Standard Model quark family? The constituent-quark s, c and b mass values successively triple, so it seemed natural that a narrow mass-tripled $t\bar{t}$ resonance would show up around 28.5 GeV, but determined searches revealed nothing. Finally, guided by electroweak calculations (which have nothing to do with mass triplings), experimentalists searched in the region around 85 GeV (which corresponds to another quark–antiquark mass tripling), and they discovered the 80.4 GeV $W^{\pm}$ and 91.2 GeV Z^0 vector boson isotopic-spin doublet. These were the first two elementary particle states identified above 12 GeV. Evidence for a third and final elementary particle state, the long-sought-after top quark t, emerged a decade ago in the proton–antiproton colliding beam experiments at Fermilab. The t quark appeared at a mass of around 175 GeV. To date, these three particle states are the only ones to be pinned down above 12 Gev. Since the t was produced at the top of the energy range of the Fermilab experiments, higher-mass states will have to await the implementation of higher-energy accelerators.

With respect to theories of elementary particle masses, the W, Z and t particle states raise two immediate questions:

(1) *Do these states fit in with the mass systematics set in place at the lower energies, or are they completely different entities?*
(2) *Why is there a large gap between 12 and 80 GeV that contains no particle states?*

In the Standard Model classification of particles, the spin-1 W and Z are denoted as "gauge bosons" that mediate weak interactions. They are placed in the same category as the spin-1 gamma rays and gluons and the mysterious (and undiscovered) Higgs boson. The W and Z are not assigned any kind of quark substructure, and should bear no relationship to the t quark. The top quark t fills out the Standard Model u, d, s, c, b, t set of quarks, and thus might logically relate to the other quark masses, but it is so much higher in value that any direct connection is not easy to determine.

The first clue as to the nature of these supermassive states comes from their very-accurately-measured mass values. Figure 0.2.21 is a linear mass plot of the 36 threshold-state particles of Fig. 0.2.1, which extend up to the Υ resonances, together with the very massive W, Z and t particle states. This plot illustrates how large the energy jump is in going from the plethora

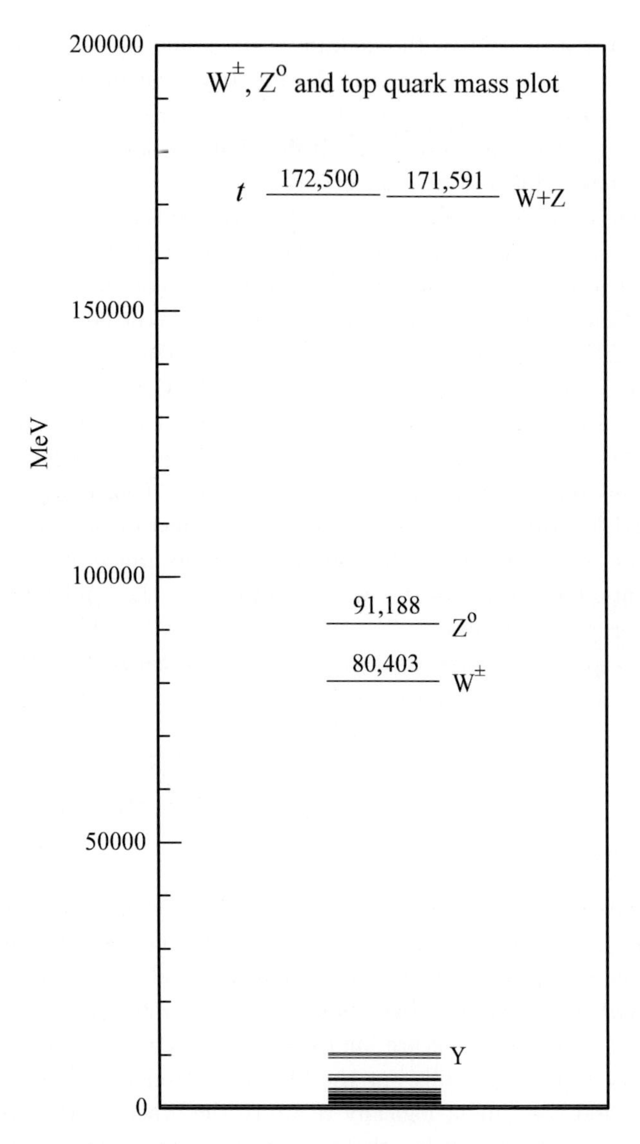

Fig. 0.2.21 The W and Z gauge bosons and the top quark t. This linear mass plot shows the 36 threshold-state mass levels arrayed at the bottom, where they overlap into a condensed continuum of states, plus the three observed particle states above 12 GeV. The W and Z gauge bosons have no known theoretical connection to the hadronic t quark, and yet their masses add up to accurately reproduce the mass of the t, as displayed in Fig. 0.2.21. A mass relationship between these states is established by the quark systematics discussed in Sec. 0.10 and displayed in Fig. 0.2.22.

of "low-energy" particle states to these three stratospheric masses: the mass of the t is about the same as the mass of a tungsten atom. But Fig. 0.2.21 contains another striking feature. The sum of the W and Z masses is also plotted, and it is almost precisely equal to the t mass! Not only was this result unexpected, but it does not seem to have been widely noticed. When a room full of particle physicists who had gathered recently at a colloquium devoted to the top quark were asked about this mass equality, none of them seemed aware that it even existed. And when it was pointed out to them, there was a short period of silence followed by the expressed opinion that it "must surely be a meaningless coincidence." Of course, according to the Standard Model assignments, comparing the W and Z masses to the t mass is like comparing apples to oranges, so there was no reason for them to even consider this comparison. This is especially true when we take into account the fact that the Standard Model deals with current-quark masses, which are only loosely defined, and not with constituent-quark masses, where much sharper definitions are usually offered.

The W, Z and t are the only particle states found above 12 GeV, so it seems advisable for us in the present work to assume as an *ansatz* that the relationship $m_W + m_Z = m_t$ may not be a coincidence, and then find out where it leads. Since there is some reason to connect the t mass to the u, d, s, c, and b masses, this ansatz implies that the W and Z masses may also be connected. But what is the nature of this connection? The clue to the answer may be supplied by the second question asked above — the reason for the large energy gap between states. We can find a somewhat similar gap if we start with the electron and go upward in mass. The electron has a mass of 0.5 MeV, and the muon, the next particle up the mass chain, has a mass of 105 MeV. The reason for this large mass gap — a factor of 207 — is that it requires a mass-generating α-leap to produce an excited state of the electron, as is displayed in Fig. 0.2.2. Maybe the reason for the large mass gap above 12 GeV is that a *second* α-leap is required in order to move up to the higher energies!

In the case of t-quark production, we can tell where this second α-leap logically originates. The proton–antiproton collisions in the Fermilab tevatron occur at such a high energy that the impacts are not between the whole proton and whole antiproton, but rather between the individual quarks within these particles. The production of the t quark requires a direct impact between a quark and an antiquark, which concentrates most of the available energy on this particular quark pair. A suitable event like this happens about once in every 10^{10} proton–antiproton collisions. Thus this second α-leap occurs as an excitation of a u or d quark in the proton, which is accompanied by a matching $\bar{u}$ or $\bar{d}$ excitation in the antiproton. This is in contrast to the excitation of an electron or positron in the first α-leap, where the entire electron or positron is involved, since these particles have no discernable substructures. In the first-order constituent-quark model employed here, where we deal only with isotopic-spin-averaged masses, we assign the same mass to both the u and d quarks, which we denote collectively as $q \equiv (u, d)$ quarks (Table 0.6.1). The calculated q mass is $q = 3m_f = 315.114$ MeV, where $m_f = 3m_e/2\alpha = 105.038$ MeV. (Technically, the electron mass should be added to these masses at the end of the calculation — see Fig. 0.2.3 — but it makes no appreciable difference at these high energies.) This q mass gives rise to the second-order α-leap mass $q^\alpha \equiv q/\alpha = 43,182$ MeV, where $q^\alpha = (u^\alpha, d^\alpha)$. We can refer to these massive second-order q^α quarks as "α-quarks."

We can denote the isotopic-spin-averaged mass of the $W^\pm$, Z^0 pair as $\overline{WZ} = 85,795$ MeV. Reproducing this $\overline{WZ}$ doublet as a $q^\alpha \bar{q}^\alpha$ α-quark pair gives a calculated $\overline{WZ}$ mass of 86,364 MeV, which agrees with the experimental mass to an accuracy of 0.66%. The $q^\alpha \bar{q}^\alpha$ pair can also generate both of the $W^\pm$ and Z^0 charge states.

If we assume that the $\overline{WZ}$ boson doublet is represented as a $q^\alpha \bar{q}^\alpha$ α-quark pair, this raises the question as to how to extend these results to encompass the top quark t. This question was answered empirically in Fig. 0.2.21, where it was demonstrated that $2m_{\overline{WZ}} \cong m_t$. For another answer, we study the excitation systematics of the first-order α-leap. Figure 0.2.22 displays three α-leap excitation towers. The left tower shows the first-order α-leap from an electron–positron pair to the pion platform, with subsequent excitations to the η and η' levels (see Figs. 0.2.11–0.2.13). The central tower shows the first-order α-leap from an electron–positron pair to the muon-pair platform, with subsequent excitations to the q-pair and s-pair levels (Fig. 0.2.4). The right tower shows an analogous second-order α-leap from a $q\bar{q}$ pair to the $q^\alpha \bar{q}^\alpha = \overline{WZ}$ platform, with subsequent excitations to the $T \equiv t\bar{t}$ level and a plausible T' level. The t quark mass is given by the equation

$$(m_t)_{\text{calc}} = 4m_{q^\alpha} = 4m_q/\alpha = 18m_e/\alpha^2 = 172.73 \text{ GeV},$$

as compared to the experimental top quark mass value

$$(m_t)_{\text{exp}} = 172.5 \pm 2.3 \text{ GeV}.$$

The close agreement between these two mass values furnishes *a posteriori* justification for the assumed excitation pattern of this second-order α-leap.

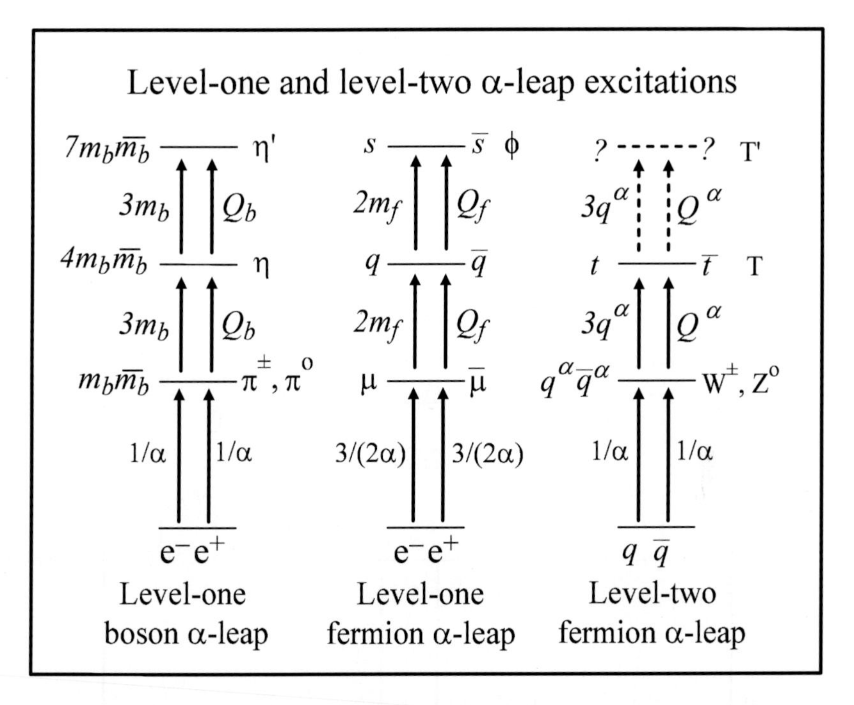

Fig. 0.2.22 Three α-leap excitation towers. The left tower shows the first-order α-leap from an electron–positron pair to the pion platform, and then the subsequent $Q_b \cong 210$ MeV platform excitations up each column of the tower to form the η and η' mesons. The center tower shows the first-order α-leap from an electron–positron pair to the muon-pair platform, and then the subsequent $Q_f \cong 210$ MeV platform excitations up each column to form the $q\bar{q}$ and $s\bar{s}$ quark pairs. The right tower shows the second-order α-leap from a $q\bar{q}$ quark pair to a $q^\alpha\bar{q}^\alpha$ pair, where $q^\alpha \equiv q/\alpha$, and where the $q^\alpha\bar{q}^\alpha$ pair represents the WZ platform. It then displays the subsequent $Q^\alpha = 129{,}546$ MeV platform excitations up each column to form the $t\bar{t}$ quark pair and another possible higher-mass quark pair.

These second-order α-leap mass values together with some basic first-order α-leap mass values are plotted logarithmically in Fig. 0.2.23, where they are displayed together in the form of a "muon mass tree" that reveals that excitation systematics. The "branches" of this tree contain all types of particle mass excitations — quarks, leptons, baryons, mesons and gauge bosons — which combine together to create a distinctive excitation pattern. The tree energy levels are formed entirely from two additive fermion mass quanta: the "single-alpha-leap" mass unit $m_f \equiv (3/2)(m_e/\alpha) \cong 105$ MeV for states below 12 GeV, and the "double-alpha-leap" mass unit $m_f^\alpha \equiv$

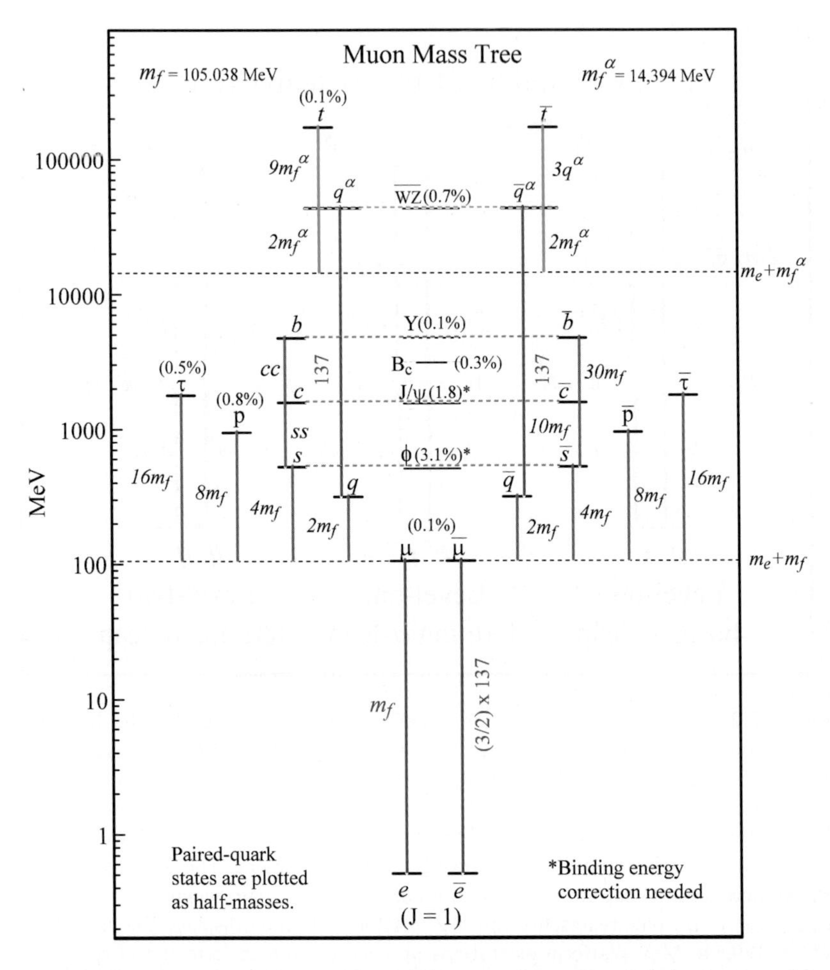

Fig. 0.2.23 The "Muon Mass Tree." This constitutes one of the most striking phenomenological results to emerge from the experimental data. The quark and particle states are composed of α-generated and muon-related $m_f \cong 105$ MeV and $m_f^\alpha \cong 14,394$ MeV spin-1/2 fermion mass quanta, which add together linearly as excited levels above the electron-positron ground state. The tree occupation levels form a Mendeleev-like mass diagram that encompasses both the Standard Model quarks and basic elementary particle states: the mass-averaged $q \equiv (u,d)$ quarks and s, c, b, t quarks; the μ and τ leptons; the proton; the ϕ, J/ψ_{1S} and Υ_{1S} vector mesons; the B_c meson; and the mass-averaged W and Z gauge bosons. The mass accuracies are 1% or better, except for the paired-quark ϕ and J/ψ mesons, which require 2–3% hadronic binding energies (that are not applied here). The HBE's vanish at higher energies (Sec. 3.15). The levels below and above 12 GeV correspond to single and double α-leaps, respectively, from the electron-pair ground state. The systematics displayed here is summarized in Table 0.10.1.

$m_f/\alpha \cong 14.394$ GeV for states above 12 GeV. The unpaired-quark levels require no hadronic binding energy (HBE) corrections. The paired-quark levels are plotted in Fig. 0.2.23 with half-masses. The levels above roughly 6 GeV have negligible HBE's (a consequence of "asymptotic freedom"), whereas the lower-energy paired-quark levels require 2–3% binding energy corrections (Figs. 3.15.1 and 3.15.2), which are *not* applied in the mass calculations of Fig. 0.2.23. As can be seen, the unpaired-quark lepton and proton masses are reproduced as multiples of m_f to an accuracy level of better than 1%. The higher-mass B_c and Υ_{1S} mesons, which have particularly clear-cut quark substructures, are also reproduced to better than 1% accuracy. Hence the isotopic-spin-averaged $q \equiv (u, d)$ constituent-quark mass and the s, c and b constituent-quark masses, whose "excitation-doubling" and "mass tripling" excitation modes are displayed in Fig. 0.2.23, are necessarily reproduced at this same level of mass accuracy. When we move above 12 GeV to the supermassive W and Z vector mesons and top quark t, the basic mass unit becomes m_f^α, which is produced in direct $q \equiv (u, d)$ quark–antiquark collisions at the Tevatron in the form of the "α-quarks" $q^\alpha = 3m_f^\alpha$. The isotopic-spin-averaged W, Z mass doublet $\overline{WZ}$ and the quark t both occur as multiples of q^α, and again at an accuracy level of better than 1%. The comprehensiveness and overall accuracy of these results speak for themselves. Two conclusions we can draw from Fig. 0.2.23 are that (1) the muon mass is indeed a universal basis state for both leptons and hadrons, and (2) this *muon mass tree* does *not* accommodate the low-mass pseudoscalar meson nonet, which requires its own *pion mass tree*.

The "pion mass tree" is displayed in Fig. 0.2.24. It has the same general form as the muon mass tree of Fig. 0.2.23, but is formed from additive boson mass quanta $m_b \equiv m_e/\alpha \cong 70$ MeV. It is much more limited in scope, containing just the "generic pion quarks" q_π, q_η, q_K and the pseudoscalar (PS) meson nonet, which consists of the mass-averaged π and K isotopic spin multiplets and the η and η' isotopic spin singlets. The α-quantized excitation systematics of the pion mass tree is similar to the excitation systematics of the muon mass tree. The unpaired K mesons are reproduced to a mass accuracy of 1%, but the paired-quark π, η and η' mesons require a 2–3% hadronic binding energy that is not applied in Fig. 0.2.24.

The excitation systematics and mass calculations of Figs. 0.2.23 and 0.2.24 are summarized in Table 0.10.1, which contains the 9 measured quark and particle masses of the muon mass tree and the 4 measured pseudoscalar masses of the pion mass tree. If we phenomenologically impose a uniform hadronic binding energy of 2.6% on each of the 5 (isotopic-spin averaged)

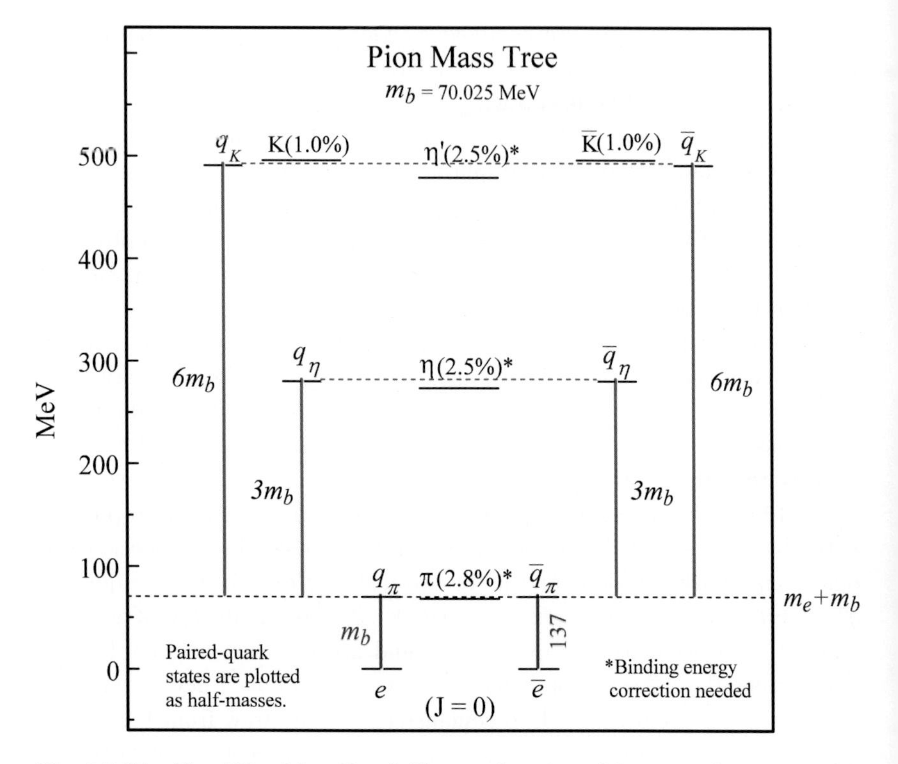

Fig. 0.2.24 The "Pion Mass Tree." The quark and particle states that occupy this pion excitation tree, which is complementary to the muon tree of Fig. 0.2.23, are composed entirely of α-generated $m_b \cong 70$ MeV spin-0 mass quanta which combine together linearly. The occupation levels of this pion tree contain a set of "generic pion quarks" — q_π, q_η, q_K — whose main properties are their mass and spin values, together with the pseudo-scalar nonet of low-mass, spin-0 mesons — the mass-averaged $\pi \equiv (\pi^\pm, \pi^0)$ and $K \equiv (K^\pm, (K^0, \bar{K}^0))$ isospin multiplets and the η and η' isospin singlets. The unpaired-quark K mesons, which do not involve hadronic binding energies, are reproduced at the 1% mass accuracy level. The paired-quark π, η and η' mesons require 2–3% HBE energy correction (Figs. 3.15.1 and 3.15.2), which are not applied here, to bring them into the 1% mass accuracy range. The spin-0 m_b mass quantum of the pion tree is the spinless counterpart of the spin-1/2 m_f mass quantum of the muon tree. In this constituent-quark mass formalism, the same quark set cannot be used to reproduce both the low-mass pseudoscalar mesons and the much-higher-mass baryons and hyperons. Pion mass quanta m_b and muon mass quanta m_f often occur combined together in the short-lived excited states that correspond to these longer-lived particle "ground states." The pion mass tree systematics displayed here is summarized in Table 0.10.1.

Table 0.10.1. α-quantized mass values for particles and quarks. The basis-state mass structures are displayed, and are grouped into excitation types. The mass values are additive, and calculated mass accuracies are shown.

Input data: $m_e = 0.510998918$ MeV, $\alpha = 1/137.03599911$
First-order and second-order α-generated mass quanta:
$m_b \equiv m_e/\alpha = 70.025$ MeV ($J = 0$ boson)
$m_f = (3/2)m_b = 105.038$ MeV ($J = 1/2$ fermion)
$m_f^\alpha \equiv m_f/\alpha = 14,394.0$ MeV ($J = 1/2$ fermion)

Particle state	Mass structure	Observed state	Mass accuracy
	Excitation doubling		
μ lepton	$m_e + m_f$	muon	(0.10%)
$q \equiv (u, d)$ quarks	$\mu + 2m_f$		
s quark	$\mu + 4m_f$		
p nucleon	$\mu + 8m_f$	proton	(0.81%)
τ lepton	$\mu + 16m_f$	tauon	(0.52%)
	Mass triples		
q quarks*	$m_e + 3m_f$		
q^α "α-quarks"*	$m_e + 3m_f^\alpha$		
	Mass triplings		
s quark	$m_e + 5m_f$	$s\bar{s} = \phi$	(3.13%)**
c quark	$m_e + 15m_f$	$c\bar{c} = J/\psi$	(1.78%)**
b quark	$m_e + 45m_f$	$b\bar{b} = \Upsilon_{1S}$	(0.06%)
		$c\bar{b} = B_c$	(0.28%)
	Second-order α-quark states		
W, Z gauge bosons*	$q^\alpha \bar{q}^\alpha$	$m_{q^\alpha \bar{q}^\alpha} = m_{WZ}$*	(0.66%)
t quark	(experimental mass equation)	$m_t = m_W + m_Z$	(0.53%)
t quark	$m_e + 12m_f^\alpha$	top quark mass	(0.13%)
	Pseudoscalar mesons		
π	$m_b\bar{m}_b$	pion*	(2.77%)**
η	$4m_b\bar{m}_b$	eta	(2.51%)**
η'	$7m_b\bar{m}_b$	eta prime	(2.46%)**
K	$7m_b$	kaon*	(1.00%)

*Isotopic-spin-averaged mass value.
**Paired-quark masses below 4 GeV require a hadronic binding energy correction of 2–3% (Sec. 3.15).

paired-quark states that have masses below 5 GeV (see Sec. 3.15), then the mass accuracies of the 13 particle states displayed in Figs. 0.2.23 and Figs. 0.2.24, and listed in Table 0.10.1, are all in the 1% range, with no adjustable parameters employed. The completeness of this set of basic particle states, together with the wide range of included masses and the high accuracy of the calculated mass values, constitutes a compelling empirical verification of these complementary sets of muon and pion quarks — and of the muon and pion mass trees.

The use of a second-order α-leap to encompass the W and Z bosons and t quark is important in that it brings these particle states into the constituent-quark formalism, but it is also significant in fortifying the entire concept of mass α-quantization. We started on this phenomenological journey by demonstrating how the elementary particle lifetimes of the long-lived threshold states are quantized in powers of α, where this scaling extends over 11 powers of α. Then we studied the masses of these same threshold states and discovered that they are quantized over a single power of α^{-1}. To now go on to the higher-mass states and reproduce them as a generation process involving the second power α^{-2} reinforces the entire concept of reciprocal lifetime and mass α-quantization. To move on to a third power α^{-3} in particle masses would involve the design of particle accelerators for much higher energies, which is not likely to happen any time soon. However, the scaling might extend in the other direction — towards the small but evidently finite masses of the electron, muon and tau neutrinos. The subject of second-order mass α-quantization is discussed in more detail in Sec. 3.20.

We have now completed our journey through the landscape of α-quantized elementary particle lifetimes and masses. Our goal was to provide an overview of the experimental systematics that reveal the operation of the fine structure constant α as an elementary particle mass generator. Leptons and hadrons are intertwined together in this landscape, and both are required in order to complete the picture. The hadrons furnish most of the landscape, and provide most of the interactions, but the leptons supply the ground state for the masses — the humble electron. Solutions to problems that have long-defied explanation, such as the proton-to-electron, muon-to-electron and tau-to-electron mass ratios, emerge from the experimental mass α-quantization in a manner that requires essentially no theory at all. In the main body of the book we start this journey all over again, and we fill in the details. Chapter 1 gives the rationale for the search, Chapter 2 is the journey through particle lifetimes, and Chapter 3 is the study of the corresponding particle masses. But the story does not end there. Guided by the results of these phenomenological lifetime and mass investigations, we embark on a brief theoretical journey into the spectroscopic landscape of the particles themselves. This journey does not involve large amounts of experimental data, but only a few spectroscopic numbers. Our goal here is to provide mathematical models that can correlate these numbers by using well-established physical laws. We are progressing from the experimental phenomenology of particle lifetimes and masses to the mathematical phenomenology — mathology — of the particles themselves. This final journey is outlined in the last section of the Introduction.

0.11 Mathological Studies of Elementary Particle Spectroscopy

The journey through the phenomenological world of elementary particle experimental lifetimes and masses has provided little direct information about the particles themselves, other than the fact that they seem to exist in an understandable and well-ordered physical universe. In an attempt to understand these particles, we invoke a different type of phenomenology, a *mathematical phenomenology* that consists of constructing mathematical models which can reproduce the spectroscopic features of the particles. This *mathology* is discussed in the final chapters of the book, where we consider the electron (Chapter 4), the electron wave and the photon (Chapter 5), and the fine structure constant α and electric charge e (Chapter 6). A few ramifications are listed in Chapter 7.

The salient spectroscopic features of the electron are its mass of 0.511 MeV, its spin $J = 1/2\,\hbar$, and its magnetic moment $\mu = e\hbar/2mc$. Other electron features are its anomalous magnetic moment $\mu_{\text{anom}} \simeq \mu\alpha/2\pi$, its vanishing electric quadrupole moment, and its point-like electromagnetic interaction length. The mathological challenge posed by the electron is to account numerically for these features within the framework of accepted physical theories. This challenge is customarily presented to physicists in the following form: given the mass, spin and magnetic moment of the electron, correlate these properties in an object whose radius appears to be point-like ($\sim 10^{-17}$ cm), and in any event is no larger in magnitude than the classical electron radius $R = e^2/mc^2 = 2.8 \times 10^{-13}$ cm. The answer to this challenge is that there is no possible classical explanation using established electromagnetic and mechanical laws. This result has precluded any progress along these lines. However, if we alter this challenge slightly by substituting the words "electric charge" for the word "radius", then the challenge is solvable mathologically, as we demonstrate in Chapter 4. It should be kept in mind that the electron interacts only electromagnetically — through the operation of its electric charge, so when the *size* of the electron is being measured, it is actually just the size of the charge on the electron.

A first-order mathological challenge with respect to the electron is to account for its gyromagnetic ratio, g = magnetic moment/spin. The gyromagnetic ratio for atomic electron orbitals is $g = e/2mc$, whereas the electron itself has $g = e/mc$. Although it is frequently asserted that no classical explanation exists for this result, an explanation emerges from the relativistically spinning sphere (RSS) model described in Chapter 4. Several other pertinent results emerge from this model — in particular, the 3/2 mass ratio between relativistically spinning and spinless spherical masses that ties together the $J = 1/2$, $m_f = 105$ MeV and $J = 0$, $m_b = 70$ MeV α-masses (Fig. 0.2.14).

The mathological challenge for the photon is even more formidable than for the electron: account for the spin $J = \hbar$ of a massless particle, as well as its electromagnetic features. This is the problem that Albert Einstein famously pondered for 50 years without arriving at a solution. However, we can solve this problem mathologically and obtain a "generic" photon model if we change the word "massless" to the words "zero mass" in the challenge. These are not the same concepts. A zero mass object can be obtained by employing a "positive mass" state together with the "negative mass hole state" that results when the positive mass is removed from a uniform mass substrate. Of course, then we have to postulate the existence of this substrate, but that is a worthwhile price to pay if it enables us to explain how a circularly polarized photon can mechanically rotate a quarter-wave plate. This topic and others are mathologically explored in Chapter 5.

In Chapter 6 we turn finally to the coupling constant α that plays such a pervasive role in all of these interactions. Its role in QED is well known. Its role in particle lifetimes and masses is displayed in Chapters 2 and 3. Its relevance to the electron emerges in the equation for the anomalous magnetic moment, $\mu_{\rm anom} \simeq \mu\alpha/2\pi$. And α appears in the photon model in the form of the $1/\alpha^2$ scaling factor between the electrostatic energy $e^2/2r = \hbar\omega$ of the photon and the masses m^+ and m^- of the rotating particles in the particle-hole and antiparticle–antihole pairs that make up the generic photon. We can trace the operation of the coupling constant

α through four different "phase transformations" as it couples electrostatic field energy to the masses of the elementary particles. The electric charge e emerges from this mathology as an electrostatic charge quantum that can fragment but never be broken apart, and which can assume various spatial configurations, as follows: a variable-size finite-energy "charge ballon" in the electrostatic field; a zero-energy "point charge" in the leptons; and a "rubber-band-like cloud" with "fractionally charged lumps" in the particle quarks.

Chapter 7 contains brief discussions of ramifications in cosmology, in the concept of "mechanical" mass, and in our non-Standard-Model interpretation of the pseudoscalar mesons. Hopefully the insights suggested by these Introductory phenomenological and mathological tours of the elementary particle microworld will furnish the motivation to tarry longer and enjoy the wonderful panorama of elementary particles that make up the universe — including ourselves.

Chapter 1 Lifetime and Mass α-Quantization: Physics Beyond the Paradigm

1.1 The Missing Elementary Particle Ground State and Its Mass Generator

This book is about the fine structure coupling constant $\alpha = e^2/\hbar c \cong 1/137$ and about the ground state on which it operates. We present evidence which indicates that the coupling constant α is the most powerful — and most misunderstood — operator in elementary particle physics. More specifically, this book contains analyses of two sets of elementary particle data that, independently of theory, exhibit clear-cut periodicities in powers of the numerical factor 137 or its reciprocal value 1/137. These α-based periodicities occur in the long-lived, low-mass, threshold-state particles, which serve as the foundation states for the higher-mass excitations, and which are the most revealing in their structure. The first set of data — particle lifetimes — is analyzed in Chapter 2. The second set of data — particle masses — is analyzed in Chapter 3. The accuracy and completeness of these α-quantized data patterns furnishes strong phenomenological evidence for the conclusion that the fine structure constant α, which we know serves as the coupling constant in quantum electrodynamics (QED), also serves as a coupling constant in quantum chromodynamics (QCD). One practical benefit we obtain from this approach is that we can finally handle the oldest unsolved problem in elementary particle physics — the proton-to-electron mass ratio. Guided by the phenomenology of α-dependent threshold-state lifetimes and masses, we obtain a reasonably accurate value for this important ratio without having to resort to the use of adjustable parameters.

In the current paradigm of elementary particle physics, as embodied in the Standard Model, the leptons of QED (the weakly-interacting particles) and the hadrons of QCD (the strongly-interacting particles) are treated theoretically as separate entities. However, the particle *lifetime* data displays mentioned above show leptons and hadrons combined together on a common α-spaced lifetime grid. Even more compellingly, the particle *mass* data displays reveal that the leptons and the proton occupy a common α-dependent mass excitation tower. The lepton and proton masses are interleaved together, and the overall pattern can only be understood by considering all of these masses in one unified picture. What does this tell us about the similarities and differences between leptons and hadrons? The answer, broadly stated, is that leptons and hadrons share a common set of mass building blocks, and the essential difference between leptons and hadrons arises mainly from the fact that the electric charges on leptons are not fractionated, whereas the electric charges on hadrons are fractionated and spread among the quark substates. This charge fractionization process produces the gluon fields and the distinctive hadron interactions. The forces between elementary particles are created almost exclusively by the charges. The masses are passive entities which carry the bulk energy content and inertial properties of the particles, and also the particle–antiparticle and lepton family quantum numbers, but otherwise are essentially non-interacting. These results follow not only from the lifetime and mass systematics of the elementary particles, which are presented in Chapters 2 and 3, but also from information we can extract from the spectroscopic properties of these particles [1], which are discussed in Chapter 4.

What are the implications of this extensive particle α-quantization with respect to the well-established systematics of the Standard Model? The essential point here, which is in line with the above discussion about leptons and hadrons, is that the distinctive features of QCD — the accurate predictions of isotopic spins, the unique asymptotically-free quark binding energies, the calculations of quark and gluon jets — all involve the *charges* on the elementary particles. The Standard Model handling of elementary particle *masses* has not been nearly as successful. The lifetime of an elementary particle is essentially a measure of the stability of its mass structure, and the mass of the particle reflects the structure itself. Thus the present α-quantization studies pertain to the masses, and have very little to do (at least on this phenomenological level) with the charge states. The α-spaced lifetime groups we show here accurately reflect the Standard Model quark structure, and thus reinforce the reality of the quarks in these particles. But the α-quantized masses we obtain, which are constituent-quark masses, are quite different from the current-quark masses of the Standard

Model, especially in the lower-mass states. QCD is primarily a theory of elementary particle charges, not masses. This conclusion was also set forth by Gottfried and Weisskopf in their classic two-volume *Concepts of Particle Physics* [2], where they commented as follows:

> *Unfortunately, QCD has nothing whatsoever to say about the quark mass spectrum, nor, for that matter, does any other existing theory.* [3]

The determination of the particle mass spectrum is thus thrown back to the experimental data themselves, which fortunately tell their own story when properly arrayed.

In the remainder of Chapter 1 we briefly discuss some of the problems facing particle physics. Then we move on in Chapter 2 to the lifetime systematics, and in Chapter 3 to the mass systematics. These two chapters involve the phenomenology of the experimental data. In Chapter 4 we introduce some spectroscopic information, which involves mathematical phenomenology (mathology). Finally, in Chapters 5–7 we discuss some of the ramifications that ensue from this pursuit of the coupling constant α.

1.2 The Particle Mass Mystery: Physics from the Higgs Down or the Bottom Up?

The main reason we study elementary particles is to learn something about the make-up of our universe. Leon Lederman expressed this viewpoint very succinctly in his book *The God Particle* [4]:

> *This book is devoted to one problem, a problem that has confounded science since antiquity. What are the ultimate building blocks of matter?* [5]

The present book is focused on this same problem, although with a different approach, and with a different conclusion. An *elementary particle*, by implicit definition, is a fundamental building block out of which everything else is constructed. If we start with a macroscopic object and then divide it into smaller and smaller pieces, we presumably finally reach bed-rock — the smallest pieces there are. These are the elementary particles. And perhaps their most fundamental attribute is their mass — their total energy content. Thus what we want most to learn about these particles is their mass structure. However, this has turned out to be unexpectedly difficult to accomplish. The total mass of a particle can be accurately measured, but that does not explain why it has this particular value. In the area of atomic physics, by way of contrast, the mass structure of the basic elements is relatively simple. The mass of an atom is mainly in its nucleus, and nuclei are composed of protons and neutrons, which have about the same mass. So the mass of an atom is accurately obtained by simply counting up the number of protons and neutrons in the nucleus. But this simplicity does not carry over to the observed elementary particle states. A good example is provided by the massive *leptons*, of which there are three: the electron, muon, and tau, together with their antiparticles. Using mass units of MeV (million electron volts), with the velocity of light c set equal to unity, as is customary in particle physics, we have:

electron (0.51 MeV), muon (105.66 MeV), tau (1776.99 MeV).

These lepton mass values do not seem to form any kind of logical progression. The leptons are weakly-interacting particles, whose interactions are purely electromagnetic. The strongly-interacting particles, denoted as *hadrons*, also have mysterious masses. Hadrons occur as *mesons* (with integral spins) and *baryons* (with half-integral spins). The lightest meson, the neutral pion, has a mass of 135 MeV, and the lightest baryon, the proton, has a mass of 938 MeV. None of these mass values emerge directly from the existing elementary particle theories.

The entire problem of elementary particle mass generation has remained obscure. In his classic book *QED* [6], Richard Feynman devoted the first three chapters to the interactions of electrons and photons. Then in the fourth and final chapter he took this information over and discussed how much of it applies to hadron interactions. At the very end of the book, in his final observation, Feynman commented on the mass problem as follows:

> *Throughout this entire story there remains one especially unsatisfactory feature: the observed masses of the particles, m. There is no theory that adequately explains these numbers. We use the numbers in all our theories, but we don't understand them — what they are, or where they come from. I believe that from a fundamental point of view, this is a very interesting and serious problem.* [7]

Present-day theories attribute elementary particle masses to the operation of the hypothetical "Higgs particle," which seems required in order to have a renormalizable particle theory [4]. In most fields of physics, masses are determined by the lightest members of the field, whose masses then compound or excite to form the higher-mass states. But the Higgs particle, by way of contrast, is a very heavy particle — so heavy that it has not yet been found. In his book *Facts and Mysteries in Elementary Particle Plysics* [8], Martinus Veltman describes its salient features in the following way:

> *In short, it [the Higgs] must be coupled to* ***any*** *particle having a mass. Moreover, the coupling must always be proportional to the mass of the particle to which it is coupled.*
>
> *To date the Higgs particle has not been observed experimentally. Unfortunately the theory has nothing to say about its mass, except that it should not be too high (less than, say, 1000 GeV) ... .*
>
> *Because this Higgs particle seems so intimately connected to the masses of all elementary particles, it is tempting to think that somehow the Higgs particle is responsible for these masses. Up to now we have no clue as to where masses come from: they are just free parameters fixed by experiment.*
>
> *There is clearly so much that we do not know! ...* [9]

A basic difficulty in trying to understand elementary particle masses is that we need to know more than just the total mass of the particle. We need to know the substructure of the mass. We have discovered so many elementary particles that we now believe they are not truly *elementary.* It seems logical that these elementary particles must be composed of even-more-elementary substates. Before the age of high-energy accelerators, only a handful of elementary particles were known to exist. However, we now have about 200 measured particle states, which suggests that some of these must be more "elementary" than others. A significant step towards simplifying this plethora of particles was the recognition that they can be constructed from a much smaller set of "quark" substates. The so-called Standard Model (SM) of physics, the theoretical model that dominates present-day particle physics, features six spin-1/2 quarks (d, u, s, c, b, t), together with their corresponding antiquarks [10]. The names for these quarks are *down, up, strange, charm, bottom* and *top.* These names are denoted collectively as the quark "flavors." The quarks carry the fractional charge states $(-1/3, +2/3, -1/3, +2/3, -1/3, +2/3)\, e$, respectively, with opposite signs for the antiquarks. The down and up quarks d and u reproduce the proton and neutron (which are referred to collectively as "nucleons"). The d and u quarks also appear as components in most of the other hadron states. The s, c, b and t quarks carry the strangeness, charm, bottom and top flavor quantum numbers, respectively. The quark charge assignments, together with the specified quark combinations, accurately reproduce the measured isotopic spins (charge states) of the various particles. This has been one of the major successes of the Standard Model, and has led to many predictive accomplishments.

The difficulties with the quark model arise when attempts are made to separate a particle into its quark substates. We logically expect an integrally charged particle in a high energy collision to break apart into its fractionally charged u, d, s, c, b quark and antiquark components, so that we can measure the masses of the individual quarks. But experimentally this does not occur. (The top quark t is a special case, and is discussed separately in Sec. 3.20.) Specifically, what has been established is that particle break-ups or decays into *fractionally-charged* particles do not occur. If particles break apart into quarks, the quark fragments combine with other created quark states to form combinations that carry integer electric charges. The fragmented *quark charges*, which seem to exist as 1/3 and 2/3 entities on the quarks inside of a particle, cannot be extracted as fractional charges outside of the particle: only integrally-charged particles are found in the decay products. Thus if the electric charges do actually separate into 1/3 and 2/3 fractions on the quark states, as is indicated by the isotopic spin rules, they are evidently still tied together by unbreakable confining forces. The question then remains as to whether the *quark masses*, apart from the quark charges, can be separated and dislodged in collision or decay processes. Empirically, the masses of the final-state particles are not those of the initial-state u, d, s, c, b quarks. Thus Standard Model quarks are not observable as isolated entities, so that it is difficult to ascertain their intrinsic masses experimentally.

There are two different reasons we can advance to explain why single-quark u, d, s, c, b masses are not observed in particle annihilations: either (1) the quark *masses* are bound so *strongly* that quarks cannot be pulled apart, or (2) the quarks have *lightly-bound* mass *substructures*, so that when a particle is shattered or decays, the quarks also shatter into their mass substates (but not with the fractional quark charges). The first reason (which seems to apply to the quark *charges*) requires a rubber-band type of mass binding energy that increases with separation distance, and is generally the favored explanation. This type of binding is denoted in the Standard Model as "asymptotic freedom." However, the second reason is the one that seems to emerge from the present studies, which indicate that the Standard Model quarks have weakly-bound (a few percent) mass substructures. In either case, we have no direct way to determine the intrinsic u, d, s, c, b quark masses (but magnetic moments [11] provide interesting clues). Without intrinsic quark masses, we cannot deduce the masses of the pion and proton, for example, without actually measuring them.

In his analysis of *Facts and Mysteries in Elementary Particle Physics* [8], Veltman expresses his feelings about the mass problem as follows:

> *Here is another major problem of elementary particle physics. Where do all these masses come from?* [12]

In the current paradigm of elementary particle physics [4, 8, 10], the assumption is made that the masses are created from the Higgs downward. However, the phenomenology of α-quantized lifetimes and masses suggests that the coupling constant α acting on the electron creates these masses from the bottom up. In the present book, experimental evidence is displayed to substantiate this "power of α" viewpoint.

1.3 The Double Mystery of the Fine Structure Constant $\alpha = e^2/\hbar c$

In addition to the mystery of the origin of elementary particle masses, there is the mystery of the dimensionless fine structure constant $\alpha = e^2/\hbar c \cong 1/137$. In his book *Elementary Particles: Building Blocks of Matter* [13], Harald Fritzsch describes α as follows:

> *Let us emphasize that this number is the most prominent number in all of the natural sciences. Ever since its first introduction it has caused a lot of speculation. After all, α gives the strength of the electromagnetic interaction, which gives it fundamental importance for all the natural sciences and for all technology.* [14]

The mystery about α is actually a double mystery. The first mystery — the origin of its numerical value $\alpha \cong 1/137$ — has been recognized and discussed for decades. The second mystery — the range of its domain — is generally unrecognized. One of the most important successes in modern physics has been the theory of QED. This is the theory of the interactions of electrons and muons with photons [6]. The QED electron–photon interaction strength is expressed in terms of the dimensionless coupling constant

$\alpha = e^2/\hbar c$. In the QED calculation of the anomalous magnetic moment of the electron, for example, a perturbation expansion is made in powers of α, and the calculated value matches the experimental value to an accuracy of one part in 10^{10}! A calculation of comparable accuracy occurs for the anomalous magnetic moment of the muon, another lepton. The theoretical framework for the much stronger hadron interactions is QCD, [4, 8, 10], which was patterned after QED, and which has its own coupling constant. The leptons, including the fine structure constant α, have not been incorporated into QCD.

In QED the coupling constant α gives the exact interaction strength for an electron to produce a photon, including all of the various pathways along which the electron can travel in generating the photon [15]. The calculation of these pathways to high order in perturbation theory has taken years of work. The physical basis for the numerical value of this coupling constant, $\alpha \cong 1/137$ is unknown. Feynman has summarized this situation very vividly in his book *QED*, where he uses the symbol e to stand for α:

> *There is a most profound and beautiful question associated with the observed coupling constant, e — the amplitude for a real electron to emit or absorb a real photon. It is a simple number that has been experimentally determined to be close to −0.08542455. (My physicist friends won't recognize this number, because they like to remember it as the inverse of its square: about 137.03597 with an uncertainty of about 2 in the last decimal place. It has been a mystery ever since it was discovered more than 50 years ago, and all good theoretical physicists put this number up on their wall and worry about it.) Immediately you would like to know where this number for a coupling comes from: is it related to pi, or perhaps to the base of natural logarithms? Nobody knows. It's one of the* ***greatest*** *damn mysteries of physics: a* ***magic number*** *that comes to us with no understanding by man. You might say the 'hand of God' wrote that number, and 'we don't know how He pushed His pencil.' We know what kind of a dance to do experimentally to measure this number very accurately, but we don't know what kind of a dance to do on a computer to make this number come out — without putting it in secretly!*
>
> *A good theory would say that e is 1 over 2 pi times the square root of 3, or something. There have been, from time to time, suggestions as to what e is, but none of them has been useful.* [16]

The latest value for the numerical value of α is $1/137.03599911$ [10]. In spite of its extreme accuracy, this numerical value does not convey much physical insight as to its phenomenological significance.

Feynman was not the only physicist fascinated by the number 137. Lederman points out [4] that many physicists have pondered where it came from. As he states:

> *Werner Heisenberg once proclaimed that all the quandaries of quantum mechanics would shrivel up when 137 was finally explained.* [17]

Also:

> *Pauli was in fact obsessed with 137, and spent countless hours pondering its significance.* [18]

Lederman himself, while he was director at Fermilab, lived in a 150-year-old farm house situated on Eola Road on the laboratory grounds, to which he assigned the postal address 137 Eola Road.

The puzzlement over the derivation of the "pure number" 137, or $1/137$, has persisted for more than half a century. Eddington's attempts to explain it are legendary [16]. In the present studies we argue the case that, in addition to its numerical value, there is in fact a second mystery involved with the constant α, which is the question:

> *What is the extent of the domain in which α operates as a coupling constant?*

According to the Standard Model and its associated Grand Unified Theories, the domain of α is restricted to the leptonic particles that are considered in QED, where for example it generates photons in electron interactions. But the experimental data seem to point to a different conclusion. As we will demonstrate in detail in Chapter 2, the long-lived threshold-state elementary particles have lifetimes which exhibit a clear-cut periodicity in powers of α. Due to the conjugate nature of lifetimes and mass widths, this experimental α-dependence also applies to the resonance widths of these threshold-states. Since it applies to the "mass widths," then it should affect the "masses" themselves. Specifically, there must logically be some kind of "α-structure" in the masses that accounts for the observed width and lifetime α-quantizations, since masses are "primary" physics quantities, whereas widths and lifetimes merely express the stability of the masses, and hence are in a sense "secondary" physics quantities. Following this line of reasoning to its conclusion, we arrive at the result that the electron and its coupling constant α generate not only the photon, but also the spectrum of leptons and hadrons. Thus the domain of the fine structure constant $\alpha = e^2/\hbar c$ seems phenomenologically to be larger than currently believed. This viewpoint has in fact long been suggested by the experimental data, and was pointed out in the early literature [19].

1.4 The Dichotomy of Leptons and Hadrons: Interactive Charges and Passive Masses

In the Standard Model classification of particles [10], the *massive* elementary particles are divided into two major classifications — leptons and hadrons. The massive particles we consider here start with the electron at 0.5 MeV and extend up through the Upsilon mesons in the 9 to 11 GeV (billion electron volt) range. The superheavy W and Z vector mesons at 80 and 91 GeV are placed in a different SM category, that of gauge bosons. (The mass systematics developed in the present work shows an interesting mass extrapolation between the W and Z vector bosons and the much lighter ϕ, J/ψ and Υ vector mesons, as we discuss in Sec. 3.19.)

There are only three massive *leptons* — the electron (e^-), muon (μ^-), and tau (τ^-) — and three massive antileptons — the positron (e^+), antimuon (μ^+), and antitau (τ^+). The leptons have negative integer electric charges, and the antileptons have positive integer electric charges. Leptons are weakly-interacting particles, whose interactions with other particles and with each other are purely electromagnetic. By way of contrast, there are about 200 massive *hadrons* (plus associated antihadrons), which are divided into two general categories according to spin — the integral-spin (boson) *mesons* and the half-integral-spin (fermion) *baryons*. Hadrons occur in a variety of integer charge states. They are strongly-interacting particles (Greek *hadros* = "stout"), and their interactions with other hadrons are dominated by the short-range "strong force," which is much greater than the long-range electromagnetic force.

A striking difference between leptons and hadrons is their measured sizes. Since lepton interactions such as Møller (e^-, e^-) and Bhabha (e^-, e^+) scattering [20] are known to be accurately electromagnetic, measurements of their "sizes" are in actuality determinations of their "charge sizes," and these measurements show *no size at all.* To be more precise, the experiments show no measurable sizes down to the experimental limit of about 10^{-17} cm. By way of contrast, comparable measurements on hadrons show roughly the Compton sizes, $R_{\text{hadron}} \sim \hbar/mc$ that we expect to find on the basis of their masses m and their spins and magnetic moments [21]. This brings up an interesting question:

> *Is the electron a point particle, or is just the* **charge** *on the electron point-like?*

This is in fact a crucial question to answer, and the prevailing opinion seems to be that the electron is truly point-like. However, if this is so, then both the spin angular momentum and the magnetic moment of the electron are entities that we cannot explain on the basis of any known physics involving mechanics and electrodynamics, and the electron is consigned to the unreachable depths of string theory. The thrust in the present studies is to establish the fact that particle lifetimes and masses are in fact in accord with known physical principles, and we would expect the electron to fall into this same category. This matter is discussed further in Chapter 4. But even if the *mass* of the electron is of finite extent, as seems required for its spin angular momentum, the *charge* on the electron *must* effectively be a point. Otherwise the accurate agreement between theory and experiment for Møller and Bhabha scattering would not be obtained.

Another striking difference between leptons and hadrons involves their charge distributions. All of these particles have integer electric charges in units of the fundamental electron charge e. But the leptons have no discernable substructure to the charge, whereas the hadrons have a quark charge substructure that is indicated both by the isotopic spins (charge states) of the particles, and also by the results of deep inelastic scattering experiments. It appears as if the integer charge e on a hadron is broken into $\frac{1}{3}e$ and $\frac{2}{3}e$ fractions, and these fractions are located on individual *quarks* within the hadron. Unfortunately this is still a theoretical concept, because very determined experimental efforts have failed to produce any evidence of fractional charge states in the decay products of hadrons that have been shattered in high-energy collisions (or from any other decay modes). By assigning fractional charge states to quarks it is possible to reproduce the observed integer charge states of the families of hadrons so accurately that the reality (or at least the usefulness) of this procedure seems firmly established. The assumption of fractional quark charges has led to many predictive successes, and is a core feature of the Standard Model [10].

In addition to the isotopic spin successes, confirmatory evidence for a quark substructure in hadrons comes from deep inelastic scattering experiments [22]. At the low energies of early experiments, electron scattering off protons followed the results expected from simple Coulomb scattering. However, at the much higher energies later reached at the Stanford Linear Accelerator Center (SLAC), a different kind of scattering was observed. Bjorken and Feynman pointed out that this scattering could be explained by assuming that the charge inside the proton is not smoothly distributed, which would lead to Coulomb-like scattering, but instead is in the form of localized charge centers on each of three quarks inside the nucleon, with fractional charges on the quarks. At very high energies the quarks scatter essentially independently and in a quasi-free manner (they are "asymptotically free"). These experimental results furnished the first direct evidence of an actual quark structure inside the proton. As a result of the fact that the electric charge is spread out over three quarks, measurements of the overall electromagnetic size of the proton (and the neutron) give a finite result [21].

Leptons have integer point-like electric charges that show no evidence of substructure. Hadrons have integer charges that are separated into 1/3 and 2/3 fractions on quarks within the hadron. Leptons interact only weakly (electromagnetically). Hadrons interact both weakly (electromagnetically) and strongly (with the short-ranged strong force). Thus it seems logical to conclude that the origin of the strong force resides in the fractionization of the electric charge.

The measured size of the charge e on the electron is less than 10^{-17} cm. The electrostatic self-energy of this charge, if evaluated in conventional electromagnetic theory, is much larger than the observed mass of the electron. Since this cannot be correct, conventional theory does not apply, and we must assume that the point electron charge e does not have a conventional self-energy. Spectroscopic calculations [23] in fact indicate that the charge self-energy must be assumed to be zero. The magnetic energy that corresponds to the magnetic moment of the electron can be estimated, and it is roughly 0.1% of the total electron energy [24]. Thus the electron must possess a "mechanical" mass that represents the remaining 99.9% of the energy of the electron [25]. Since the electron interactions are purely electromagnetic, it is apparent that this "mechanical" mass is essentially non-interacting. If we were to strip the charge off an electron, we would be left with a non-interacting spin-1/2 mechanical mass that carries the electron lepton quantum number; this is the property we now ascribe to the electron neutrino (albeit at a vastly smaller mass value). If we carry this result over and apply it to hadrons, so that the hadron strong force comes from its (fractured) charge and associated gluon field, then the residual "mechanical" mass of the hadron is plausibly also more-or-less non-interacting. Thus we have the result that particle charges are interactive, and particle mechanical masses are passive.

This discussion of leptons and hadrons is aimed at establishing one main point: the distinctions we can draw about the differences between leptons and hadrons center mainly on their different charge structures. We actually know very little about their mass structures, apart from the fact that lepton masses show no evidence of a substructure whereas hadron masses indicate a division into quarks. This raises two questions.

(a) *Do leptons and/or hadronic quarks have fundamental mass "building blocks"?*

(b) *If so, do the lepton and hadron "building blocks" have anything in common?*

The assumption that is implicitly made in the Standard Model is that the answers to both of these questions would be *no*. Thus the SM treatments of the lepton and hadron masses are not related, and the proton-to-electron mass ratio, which straddles this divide, remains unexplained. Even within the separate domains of leptons and hadrons the masses remain a mystery. The muon-to-electron mass ratio of 206.8 and the tau-to-muon mass ratio of 16.8 do not lend themselves to ready explanations, and the pion mass, which is the lowest hadron state, is unaccounted-for. One way out of this impasse is to examine elementary particle *lifetimes*, which represent the stability of the mass structures, and see if the lifetimes convey any information about the mass structures themselves. As we will demonstrate in our analyses in Chapters 2 and 3, this turns out to be a fruitful line of endeavor.

There is one final question we can ask about the dichotomy of leptons and hadrons, which follows from the questions raised above.

If leptons and hadrons are in fact two completely separate types of particles, then do they share a common ground state, or do they each have a separate ground state?

In the case of leptons, it makes sense to denote the electron and positron as the particle and antiparticle ground states, respectively. They are the lowest-mass states, and they are stable. Electron–positron pairs can be excited into muon and tau particle–antiparticle pairs, which eventually decay back down to electrons and positrons. But hadrons have the proton and antiproton as the only stable states, and these are not the lowest-mass hadron states. This leads to our final question here.

Where is the hadron ground state?

The spinless pion is the lowest-mass hadron, but it is unstable, and it has balanced particle–antiparticle symmetry. Several other mesons have lower masses than the proton, but they are also unstable. If we do not have an obvious candidate for the hadron "ground state," which we can define as *the state that is used to generate all of the other hadron states*, then we are free to look anywhere, including at very high masses. This makes the ultramassive Higgs particle a viable candidate. But the Higgs would be an unprecedented ground state in the annals of physics. This ground-state situation is examined in detail in Chapter 3.

1.5 Experiment, Phenomenology, Theory: The Three Steps to Success

The three fundamental steps in the development of theories in physics, and in science in general, are to see *what* is there (experiment), to determine *how* it is arranged (phenomenology), and finally to explain *why* it happens that way (theory). As historians of science point out, our theories are in general forced on us step by step by the experimental data and accompanying phenomenology, and they are not usually theories that we would plausibly arrive at in the absence of data.

One danger in constructing theories is that, given the experimental database, we accidentally overlook some of the phenomenology, and thus end up developing a theory which cannot account for that phenomenology. The resulting theory may not be comprehensive enough to answer questions that we really need to know about. Historically, the need for an elementary particle theory dates back to the discovery of the electron in 1897 and the delineation of the geometry of the proton in 1911. These two particles carry precisely the same magnitude of electric charge with opposite polarities, but they have radically different masses. The main challenge to particle theorists for the past century has been to account for this surprising electron-to-proton mass ratio. Unfortunately, the Standard Model, the currently-prevailing elementary particle theory [10], puts electrons and protons into two apparently-unrelated categories — leptons and hadrons, as we discussed in Sec. 1.4, and thus provides no clues as to the nature of this mass ratio.

In searching for a way out of this dilemma, we need to re-examine the elementary particle data base to see if there are any regularities that have not been recognized, and which can be used to confront the particle theorists. One area that leaps to the forefront is the collection of elementary particle lifetimes. The long lifetimes of the low-mass threshold states, right from the early days of accelerator particle physics, have exhibited a lifetime separation into distinct groups, with the groups separated by fairly accurate factors of 1/137, where $\alpha = e^2/\hbar c \cong 1/137$ is the fine structure constant. These results were first published in 1970 [26], using an early Particle Data Group compilation [27]. The α-spaced lifetime groups contained 13 measured lifetimes (the Σ^0, η and η' lifetimes, for example, had not yet been measured), and the calculations for the lifetime table were performed on a Marchand mechanical desk calculator. Over the past 35 years an additional 23 long-lived particle lifetimes have been measured and documented (Sec. 2.11), and have accurately fitted into this α-spaced systematics, as described in Chapter 2. However, the Review of Particle Physics [10], which summarizes not only the elementary particle data, but also the *physics* that presumably applies to these data, makes no mention of this phenomenology, which has not been incorporated into the Standard Model. The significance of the α-dependence of the lifetimes lies not so much in the lifetimes themselves, although that is certainly of interest, but rather in the fact that since they represent the stability of the masses, the observed α-dependence logically carries over to the masses themselves.

In Sec. 1.6 we describe the elementary particle database. Then in Sec. 1.7 we discuss the nature of the linkage between lifetimes and masses. In Sec. 1.8 we display the numerics of the proton-to-electron mass ratio. After these preliminaries, we move on to Chapters 2 and 3, the pivotal chapters in the book, where we use a series of graphical data displays to lay out the overall systematics of elementary particle lifetimes and masses. Since this is unfamiliar territory, we need to have a good idea of the forest before we spend a lot of time examining the trees. This global overview was already provided in the Introduction, where we used color to bring out some of the systematics more clearly. What we discovered in the Introduction was that threshold-state particle lifetimes reveal an experimental

α-structure which is independent of theory, and this α-structure carries over to the masses in a way that we could not untangle without the empirical guidance of a comprehensive collection of experimental data [10]. Interesting, the global views of the α-spaced lifetime groups in the Introduction also demonstrate how clearly these lifetime groupings reflect their Standard Model s, c, b quark flavor structures, which were deduced quite independently of the lifetime data displays.

Benoit Mandelbrot, known for his pioneering book *The Fractal Geometry of Nature* [28], utilized the power of the computer to produce graphic displays of fractal patterns that would otherwise be impossible to display. Many self-similar fractal patterns closely resemble chaotic shapes that we observe in nature, and they are an aid in developing mathematical descriptions for these shapes. As Mandelbrot commented [29]:

> *Graphics is wonderful for matching models with reality.*

Later in the book he applied this comment to a specific experimental situation [30]:

> *Feynman 1979 reports that fractal trees made it possible for him to visualize and model the 'jets' that arise when particles collide head on at very high energy.*

In the present studies we are not dealing with fractal shapes, but we are using graphics as the best way to describe unfamiliar experimental results. These pictorial representations display the *phenomenology* in a way that we hope will help to link *experiment* and *theory* together in a comprehensive triad of results.

1.6 The Review of Particle Physics (RPP) Elementary Particle Data Base

Research in elementary particle physics has led to the identification of roughly 200 different massive particle states (particles with measured rest masses) that can be produced in energetic particle collisions. Most of these states are very short-lived, persisting for no more than a couple of orders of magnitude longer than the collision transit time. However, a few are long-lived metastable threshold states that signal the onset of quark excitations within the particles. In the past half century, after the advent of high-energy accelerators, the proliferation of these particle states, together with the expensive equipment required to produce and analyze them, has led to worldwide collaborations in experimental work and in the compilation of the experimental data and its systematic analysis. The Particle Data Group (PDG) summarizes these results biennially in the Review of Particle Physics (RPP).

The precursor to the RPP was a data compilation put together annually at UC Berkeley by Arthur Rosenfeld, and known informally among particle physicists as the "Rosenfeld Tables." Figure 1.6.1 shows Table VI of a Rosenfeld compilation dated September 1962. This single-page table, which was thumb-tacked on the present author's bulletin board for a long time, summarizes the data on the strongly interacting particles (hadrons). It lists 20 particle states (some with multiple isotopic spins — different charge states) that are still recognized today. This is roughly 10% of the present members of the hadron "elementary particle zoo." The mass values shown in Table VI are quite accurate, but the lifetime (mass width) measurements are more rudimentary, with all resonance widths less than 15 MeV listed either as upper limits or as effectively zero. The great improvements in the measurements of particle lifetimes or resonance widths represent one of the most striking achievements of particle physics in the past half century, and they reflect the increasing innovation and sophistication in the design of massive particle detectors.

The database for the present work is RPP2006 [10]. Of the approximately 200 elementary particles and resonances listed in RPP2006, 157 have reasonably-well-established experimental lifetime values. These constitute the lifetime database for the present studies, which is listed in Appendix A. In this database, a few charge multiplets with very similar and rather short lifetimes are characterized as single states. Also, some very-broad-width S-state resonances with poorly-determined widths have been excluded, as well as a few narrow-width states whose lifetimes are presently listed in RPP2006 only as upper limits. The corresponding mass database for these same particles is listed in Appendix B.

In addition to the elementary particle data on spectroscopic properties and interactions, RPP2006 also provides summaries of the relationships between these particles, and it describes how they fit in with the general formalism of the Standard Model.

Table VI.

TENTATIVE DATA ON STRONGLY INTERACTING PARTICLES Sept. 1962 A. H. Rosenfeld

Particle	Established Quantum No. $I(J^{PG})$	Possible Assignment: Quantum No. $I(J^{PG})$	Possible Assignment: Regge[1] Trajectory	Mass (MeV)	Γ[2] (MeV)	Mass^2 (BeV)^2	Dominant Decays: Mode	%	Q[4] (MeV)	p or p_{max} (MeV/c)
Vacuum ?	-	$0(2^{++})$	$+\omega_\alpha$	-	-	-	(even no. pions) $\bar{K}K(K_1K_1$ etc.) [5]			
η	$0(0^{-+})$		$+\omega_\beta$	548	< 10	.30	neutrals [3] $\pi^+\pi^-\pi^0$	75 25±4	- 136	- 175
ω	$0(1^{--})$		$-\omega_\gamma$	782	< 15	.62	$\pi^+\pi^-\pi^0$ [3, 5] $\pi^0\gamma$	86 14±4	368 647	326 379
π $\{\pi^0, \pi^\pm$	$1(0^{--})$		$-\pi_\beta$	π^0 135 $\pi^\pm$ 140	0 0	.018 .02	$\pi^0\to\gamma\gamma$ [6] $\pi^\pm\to\mu\nu$	100 58	135 34	67 30
ρ	$1(1^{-+})$		$+\pi_\gamma$	750	100	.56	$\pi\pi$[3] (p-wave)	100	471	348
ζ (?)	1(?)	$1(0^{+-})$	$-\pi_\alpha$	560	< 15	.31	$\pi\pi$	?	290	245
K $\{K^0, K^\pm$	$\frac{1}{2}(0^-)$		κ_β	K^0 498 $K^\pm$ 494	0 0	.24	$K_1^0\to\pi^+\pi^-$ [6] $K^\pm\to\mu\nu$	2/3K_1 58	219 388	206 236
$K^*_{1/2}(888)$	$\frac{1}{2}(1^-)$		κ_γ	888	50	.78	$K\pi$(p-wave)	100	251($K^0\pi^-$)	283
$K^*_{1/2}(730)$?	$\frac{1}{2}$ (?)	?	?	730	< 20	.53	$K\pi$	?	101($K^-\pi^0$)	161
N $\{n, p$	$\frac{1}{2}(\frac{1}{2}+)$		N_α	n 940 p 938	0	.88	$e^-\bar{\nu}p$ [6] -	100 -	.78 -	1.2 -
$N^*_{1/2}(1688)$="900MeV πp"		$\frac{1}{2}(\frac{5}{2}+)$	N_α	1688	100	2.84	$N\pi$(f-wave)	?	610	572
$N^*_{1/2}(1512)$="600MeV πp"		$\frac{1}{2}(\frac{3}{2}-)$	N_γ	1512	150	2.28	$N\pi$(d-wave)	?	434(π^-p)	450
$N^*_{3/2}(1238)$= "Isobar"	$\frac{3}{2}(\frac{3}{2}+)$		Δ_δ	1238	100	1.53	$N\pi$(p-wave)	100	160(π^-p)	233
$N^*_{3/2}(1920)$	$\frac{3}{2}(J\geq\frac{5}{2})$	$\frac{3}{2}(\frac{7}{2}+)$	Δ_δ	1920	200	3.69	$N\pi$ + other	?	842(π^-p)	722
Λ	$0(\frac{1}{2}+)$		Λ_α	1115	0	1.24	π^-p [6]	67	38	100
$Y^*_0(1815)$	$0(J\geq\frac{5}{2})$	$0(\frac{5}{2}+)$	Λ_α	1815	120	3.29	(KN+other)	?	383(K^-p)	541
$Y^*_0(1405)$	0(?)	$0(\frac{1}{2}-)$	Λ_β	1405	50[5]	1.97	$\Sigma\pi$ $\Lambda 2\pi$	{100}	69($\Sigma^-\pi^+$) 10($\Lambda\pi^+\pi^-$)	144 69
$Y^*_0(1520)$	$0(\frac{3}{2}-)$		Λ_γ	1520	15	2.31	$\Sigma\pi$(d-wave) $\bar{K}N$(d-wave) $\Lambda 2\pi$	60 30 10	194($\Sigma^0\pi^0$) 88(K^-p) 125($\Lambda\pi^+\pi^-$)	267 244 253
Σ $\{\Sigma^+, \Sigma^0, \Sigma^-$	$1(\frac{1}{2}+)$		Σ_α	1189 1191 1196	0 0 0	1.42 1.42 1.42	$n\pi^+$ [6] $\Lambda\gamma$ $n\pi^-$	50 100 100	110 76 117	185 74 192
$Y^*_1(1385)$	$1(J\geq\frac{3}{2})$	$1(\frac{3}{2}+)$	Σ_δ	1385	50	1.92	$\Lambda\pi$ $\Sigma\pi$	98 2±2	135($\Lambda\pi^0$) 49($\Sigma^-\pi^+$)	210 119
$Y^*_1(1685)$?	1(?)	?	?	1685?	?	2.85?	($\Lambda\pi$+others)	?	435	459
Ξ $\{\Xi^0, \Xi^-$	$\frac{1}{2}$(?)	$\frac{1}{2}(\frac{1}{2}+)$	Ξ_α	1311 1321	0	1.72	$\Lambda\pi^0$ [6] $\Lambda\pi^-$	- -	61 66	131 138
$\Xi^*(1530)$	$\frac{1}{2}$(?)	?	?	1530	<7	2.34	$\Xi\pi$	100	74($\Xi^-\pi^0$)	148

Fig. 1.6.1 The "Rosenfeld Tables," the predecessor to the Review of Particle Physics. Shown here is Table VI of a data compilation by Arthur Rosenfeld of UC Berkeley in September 1962, which is labeled as "tentative." It summarizes the experimentally identified hadron resonances in a single page. The modern RPP2006 [10] labels for these states, in the order listed in Table VI (below the vacuum state) are: η, ω, π, ρ, (out), K, $K^*(892)$, (out), N, $N(1680)F_{15}$, $N(1520)D_{13}$, $\Delta(1232)P_{33}$, $\Delta(1950)F_{37}$, Λ, $\Lambda(1820)F_{05}$, $\Lambda(1405)S_{01}$, $\Lambda(1520)D_{03}$, Σ, $\Sigma(1385)P_{13}$, $\Sigma(1670)D_{13}$, Ξ, $\Xi(1530)P_{13}$. The mass values shown in Table VI agree well with the modern values. However, the lifetime resonance widths in Table VI that are narrower than 15 MeV are either listed as upper limits or are assigned "zero" widths.

1.7 The Linkage Between Particle Lifetimes/Widths (Stability) and Particle Masses (Structure)

A direct method for calculating elementary particle masses has thus far eluded physicists, as we have described above. However, there is an indirect way of approaching elementary particle masses which has not been exploited in devising theories of these particles. This involves making use of the following two facts.

(1) *Lifetimes and mass widths are conjugate quantum mechanical variables which have a reciprocal relationship to one another via the Heisenberg uncertainty principle.*
(2) *Mass widths, which represent the stability in the mass structure of the particle, are plausibly linked to the mass structure itself, since the explanations for observable patterns in particle lifetimes must reside in "reciprocal patterns" in particle mass structures.*

In the case of atomic physics, the use of conjugate coordinates was instrumental in unraveling the secrets of the atom. One of the major challenges in physics during the *first half* of the twentieth century was to understand the structure of the atom. A key to this understanding was the discovery of the conjugate relationship between *coordinates* and *momenta*, a discovery which has become a cornerstone of quantum mechanics, and is embodied in Heisenberg's equations. This led to a comprehensive quantum mechanical model of the atom. A major challenge during the *last half* of the twentieth century was to understand the structure of the elementary particle. A key to this understanding, which has yet to be fully implemented, is the conjugate relationship between *lifetimes* and *mass widths*, with a subsequent extension to include the *masses*.

Quantum mechanics is central to both of these endeavors, and its use of conjugate coordinates is important in each. Historically, the effort to understand atomic structure was a crucial factor in the development of quantum mechanics, and it led to the discovery of the role played by noncommuting observables. The first decisive step in unraveling atomic structure was the demonstration by Rutherford that the positive charge of the atom is concentrated at the center. The next step was the development of the Bohr orbitals, with electrons placed in planetary orbits around the positively-charged nucleus. These orbits were quantized in units of angular momentum, so that velocities combined with spatial positions (to give angular momenta) became the variables of interest. The subsequent discovery of the de Broglie wavelength of the electron explained the orbital angular momentum quantization in terms of the coordinate wavelength quantization, which led directly to the creation of wave mechanics. Heisenberg clarified the noncommuting nature of conjugate coordinate (x) and momentum (p) variables (operators) in studies that led to the formulation of the Uncertainty Principle, $\Delta x \cdot \Delta p \geq \hbar$. Thus the knowledge obtained initially from the *angular momentum* spectra in the atom was used to obtain important information about the conjugate *position coordinates* that describe the geometry of the atom.

It is interesting to note that Planck's constant h, which was initially introduced in the quantization of black body radiation, subsequently appeared in the quantization of atomic orbitals and in the quantification of the uncertainty that arises from the use of conjugate variables. The constant h also appears in the fine structure constant $\alpha = e^2/\hbar c$. Perhaps most strikingly, h occurs as a direct factor in the macroscopic force that operates in the Casimir effect [31].

In atomic physics the conjugate nature of *coordinate* and *momentum* representations is utilized. Relevant conjugate quantities in elementary particle physics are *mass widths* and *lifetimes* (energy uncertainty and time uncertainty), which are also represented by noncommuting variables. Specifically, the uncertainty in the lifetime of a particle (Δt), which is expressed in the form of its *mean life* $\tau \equiv \Delta t$ [10], is related to the uncertainly in the mass of the particle (Δm), which is conventionally expressed as the full width Γ [10] of the measured resonant state, and these are quantitatively tied together by the Heisenberg uncertainty principle

$$\Delta t \cdot \Delta mc^2 \geq \hbar\,, \tag{1.7.1}$$

where the numerical factor in front of $\hbar$ depends on the choice of variables (e.g., full width [10] or half width at half maximum for Gaussian distributions), and where the equality that represents the lower limit of the uncertainty in Eq. (1.7.1) is used to define the conjugate transformation of variables. In the present paper we use the appellations *mean life* and *lifetime* interchangeably. Experimentally, the lifetimes τ of the long-lived particles are measured directly, but the lifetimes of the short-lived particles, which do not have measurable flight paths, have to be deduced from the resonance (full) widths Δm, using the equality in Eq. (1.7.1), which has the form

$$\tau \cdot \Gamma = \hbar\,. \tag{1.7.2}$$

This result is used [10] to transform from mass resonance widths to lifetimes. It thus seems logical that we can invert this conjugate mass width and time relationship and, by linking the mass widths to the masses, use lifetimes as a tool to investigate particle masses. Specifically, if there is a discernable periodicity $f(\tau \equiv \Delta t)$ in the particle lifetime structure, then the explanation for this periodicity must involve a (reciprocal) periodicity $g(\Delta m)$ in the corresponding mass width structure, and plausibly also in the mass structure. Mathematically, we have $f(\tau) \cdot g(\Delta m) = \hbar$ as the limiting equality in Eq. (1.7.1), so that any periodicity f in τ calls for a reciprocal conjugate periodicity $g \propto \hbar/f$ in Δm, and possibly also in m. After making this hypothesis, we can test it by studying the systematics of the particle mass structure and ascertaining whether a periodicity $g(m)$ actually exists.

Lifetime structures are in some respects easier to measure than mass structures, since elementary particles appear to have a (quark) mass substructure that is not amenable to direct observation (single quarks are not observed in collision or decay events). Thus to investigate the "periodicity" g in the quark mass structure, we must determine the observable periodicity f in the global lifetime structure, and then use $f(\tau) \cdot g(\Delta m) = \hbar$ to deduce the mass quantization $g(\Delta m)$ and look for $g(m)$. In particular, if $f = f(\alpha)$, then we expect to find $g = g(\alpha^{-1})$. In Chapter 2 we demonstrate in a series of graphical data plots that an α-quantized periodicity $f(\tau) = f(\alpha)$ is observed in the lifetimes of the long-lived threshold-state elementary particles, and in Chapter 3 we show that a corresponding periodicity $g(m) = g(\alpha^{-1})$ is in evidence in the masses of these same elementary particles. This α-dependence draws lepton and hadron masses together in one unified picture that reveals important contributions from each of these particle families.

1.8 The Numerical Challenge of the Proton-to-Electron Mass Ratio

The field of elementary particle physics was launched with the discovery of the electron by J. J. Thomson in 1897, and with the identification of the point-like nature of the proton by E. Rutherford in 1911. These two particles have electric charges that are identical in magnitude but of opposite signs. This result is in agreement with the general properties of Maxwellian electromagnetism, which features charge-symmetric behavior for positive and negative electric charges. Thus the electron and proton emerged as the logical negative-charge and positive-charge entities for Maxwellian electrodynamics. However, the masses of the electron and proton turned out to be radically different: the proton is 1836 times as massive as the electron. This result was not anticipated in classical electromagnetic theory, and it posed a problem for elementary particle physicists:

Explain the 1/1836 electron-to-proton mass ratio.

This problem has never been solved. The field of elementary particles sustained a slow steady growth during the first half of the 20th century, and then an explosive growth during the second half, fueled by the development of high energy accelerators for producing these particles. Starting with just the electron and proton, there are now roughly 200 identified elementary particle states, and comprehensive theories have been devised to account for very detailed properties of their interactions and decay modes. However, the challenge of explaining the electron-to-proton mass ratio has never been met. Almost a century later it still remains a mystery. This suggests that something is lacking in our theoretical program. Harald Fritzsch, in his book *Elementary Particles* [13], summarizes this situation very succinctly:

> *We know the Standard Model does not address a number of decisive problems, most urgently the question for the origin of particle masses. The most telling of these, as far as the structure of matter is concerned, are the electron mass of 0.511 MeV and the proton mass of 938 MeV.* [32]

When studying an unfamiliar area in physics, it is helpful to obtain an overall picture of where you are headed, and why it is useful for you to even go there. In order to provide motivation for a phenomenological examination of elementary particle lifetimes and masses, we set ourselves the goal of obtaining a reasonable estimate of the electron-to-proton mass ratio directly from the observed patterns in the experimental data, without recourse to specific elementary particle models, and without employing adjustable parameters. In order to yield a meaningful result, this procedure should also yield information about the masses of the particles in the neighborhood of the electron and proton, which it does in a comprehensive and somewhat surprising manner.

Well before any elementary particles had been identified, Maxwell electromagnetic theory was developed to account for the observed electric and magnetic fields, and to serve as a framework for charged particle interactions. The Maxwell equations feature the negative and positive electric charges that had long ago been identified by Benjamin Franklin, and they use these charges in a completely symmetric manner. When the negatively-charged electron was discovered by Thomson in 1897, it was naturally identified with the negative charges in the Maxwell fields. The only question was whether the electron just carries the negative electric charge, or whether the charge itself *is* the whole electron. This question still persists to some extent at the present time, fueled by the fact that the electron seemingly has no measurable size. The concept of a point electron rules out much of what we know about Compton-sized particles, and it opens the door to speculations that are not amenable to direct experiment, and thus not bounded by experiment.

The discovery of localized protons at the center of the atom provided a carrier for the positive charge that appears in Maxwell's equations. The absolute values of the charges on the proton and electron are equal to within experimental error:

$$\begin{aligned}\text{Proton} &: +4.80320441 \times 10^{-10}\ \text{esu}\,,\\ \text{Electron} &: -4.80320441 \times 10^{-10}\ \text{esu}\,.\end{aligned}$$

The most accurate measurement of this charge ratio is actually between protons and antiprotons, where we have [10]

$$\text{Charge ratio: } \bar{p}/p = 0.99999999991 \pm 0.00000000009\,.$$

However, this symmetry between the proton and electron absolute electric charges does not carry over to the masses. The proton and electron masses are [10]

$$\begin{aligned}\text{Proton} &: 938.272029\ \text{MeV}\,,\\ \text{Electron} &: 0.510998918\ \text{MeV}\,.\end{aligned}$$

Thus their mass ratio is

$$\text{Mass ratio: } p/e = 1836.1527\,.$$

A challenge to physicists is to be able to deduce this number, or something close to it, from the systematics of the elementary particle data. Unfortunately, our current elementary particle theories [10] place protons and electrons into two unrelated categories of particles — hadrons and leptons, respectively — and hence offer no way of calculating their mass ratio. In lieu of a theory, particle physicists have come up with a handy mnemonic:

$$6\pi^5 = 1836.1181\,.$$

This illustrates a fact that was noted by Feynman in commenting on attempts to "explain" the numerical value 137 that appears in the coupling constant α:

> *Every once in a while, someone notices that a certain combination of pi's and e's (the base of the natural logarithms), and 2's and 5's produces the mysterious coupling constant, but it is a fact not fully appreciated by people who play with arithmetic that you would be surprised how* ***many*** *numbers you can make out of pi's and e's and so on.* [16]

In the present studies we attempt to move beyond this mnemonic. A major accomplishment of elementary particle physicists during the past century has been the accumulation of a comprehensive body of highly accurate experimental data. One would think, or at least hope, that these data would tell their own story, and in fact they seem to do just that, as we demonstrate in the chapters ahead. The reader may judge the results of our search.

Chapter 2 The Phenomenology of α-Quantized Particle Lifetimes and Mass-Widths

2.1 The Zeptosecond Boundary between Threshold-State and Excited-State Lifetimes

One of the most significant properties of an elementary particle is its mean life τ [10]. Since most accelerator-produced particles have very short mean lives, their flight paths are short, so that the measurement of a path length (which together with the velocity gives the transit time) requires high spatial resolution in the detector. A great deal of experimental effort has been expended along these lines, with very impressive results. However, the shortest-lived states have no accurately-measurable flight paths, and the mean lives must be deduced from the mass widths of the resonances, using the fact that mass widths and mean lives are conjugate variables which are tied together by the uncertainty principle (Eq. (1.7.1)). In the present work we commonly use the term "lifetime" to denote "mean life," which is the Particle Data Group designation [10].

The lifetime measurements are obtained at high energy physics laboratories around the world, and are compiled together biennially in the Review of Particle Physics. The latest compilation of elementary particle data, RPP2006 [10], is 1232 pages long. This includes data compilations and summaries, and also theoretical fits to the data, using various facets of the Standard Model (SM). A substantial fraction of RPP2006 is devoted to individual elementary particle decays, where many interesting results have been achieved. However, little attention has been paid to the overall patterns that are exhibited by these lifetimes when they are plotted together in a global manner. These global lifetime patterns have been analyzed and published for the past 36 years (Sec. 2.11). In the present chapter we make a detailed examination of these observed lifetime regularities.

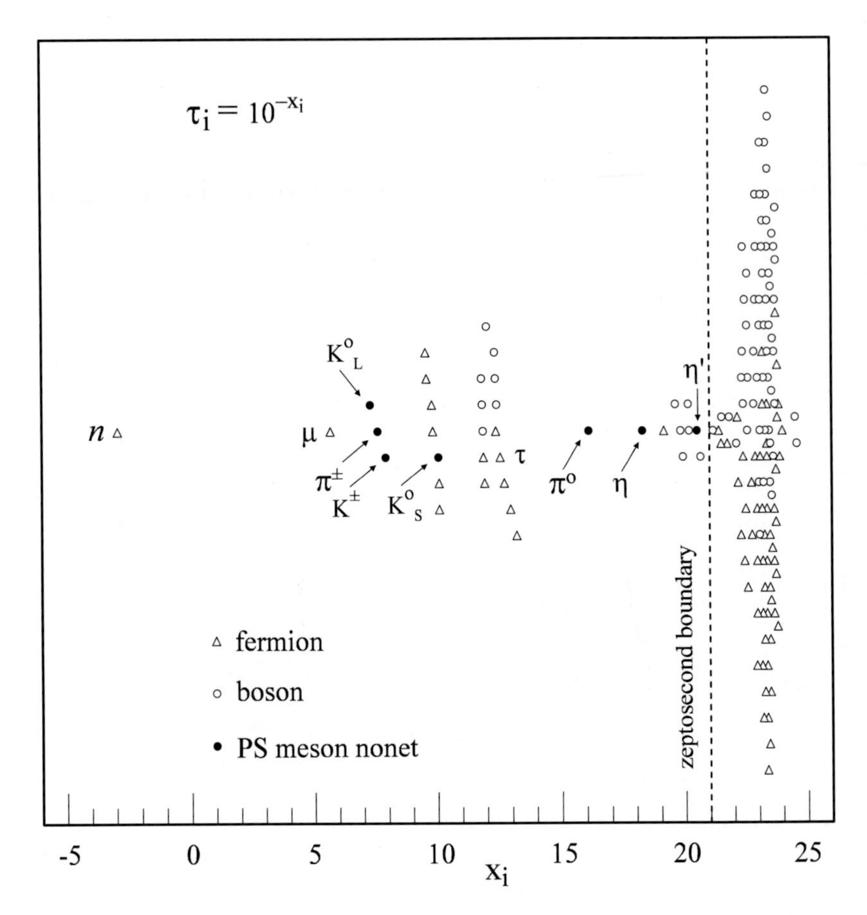

Fig. 2.1.1 The 157 well-measured elementary particle lifetimes τ in RPP2006 [10], plotted here as (negative) exponents x_i in powers of 10. The half-integer-spin fermions are shown as open triangles, the integer-spin bosons as open circles, and the zero-spin low-mass pseudoscalar (PS) mesons as filled circles (with arrows and labels). The dotted zeptosecond boundary line at the lifetime value $\tau = 10^{-21}$ s serves as the visual dividing line between the 121 short-lived *excited-state* particles on the right, which exhibit a *continuum of lifetime values*, and the 36 lower-mass, long-lived *threshold-state* particles on the left, which exhibit widely-spaced *discrete lifetime groups*.

The experimental elementary particle lifetime data set used in the present studies was briefly described in Sec. 1.6, and is listed in Appendix A. This data set, which is based on RPP2006 [10], consists of 157 particles of various types that have well-measured lifetimes. These lifetimes are displayed in Fig. 2.1.1, where they are plotted as exponents $-x_i$ to the base

10 in the equation

$$\tau_i = 10^{-x_i} \,. \tag{2.1.1}$$

The longest-lived unstable elementary particle shown here is the neutron, with a lifetime of 886 s. The shortest is the Z^0, with a lifetime of 2.64×10^{-25} s. This is a span of 28 orders of magnitude, and it necessitates a logarithmic representation. A vertical boundary line is drawn at $\tau = 10^{-21}$ s. A total of 121 short-lived particles have lifetimes $\tau < 10^{-21}$ s, and 36 much-longer-lived particles have lifetimes $\tau > 10^{-21}$ s. As can be seen in Fig. 2.1.1, the 121 short-lived particle lifetimes occur with essentially a *continuum* of values that begin to the right of the boundary line, expand rapidly at 10^{-22} s, and terminate at about the transit time for an interaction event in a high energy accelerator. The 36 long-lived particle lifetimes to the left of the boundary line occur in *widely separated groups.* As we will demonstrate, these widely spaced lifetime groups correspond to thresholds where the various types of quark excitations first occur. If we did not know about the existence of the s, c, and b quarks, we could deduce them from this lifetime spectrum. And this, of course, is essentially what happened in the discoveries of the c and b quarks. It was the large mass and very narrow width (long lifetime) of the J/ψ vector meson that signaled the presence of a new c-quark type of excitation. This result was repeated a few years later with the b quark, in the discovery of the even-narrower-width Upsilon vector meson at three times the mass of the J/ψ.

A time interval of 10^{-21} s is denoted here as a "zeptosecond" (zs). This is analogous to the use of the *zeptogram* (10^{-21} gm) as a basic unit for measuring atomic masses in the most sensitive state-of-the-art microbalances. The 36 long-lived threshold-state elementary particles have lifetimes $\tau > 1$ zs, and the 121 short-lived excited-state particles have lifetimes $\tau < 1$ zs. We can picture the 36 long-lived threshold-state particles as "zepto-plus" states, and the 121 short-lived excited-state particles as "zepto-minus" states, thus indicating on which side of the zeptosecond boundary line they occur.

It is interesting to note that a statistical analysis of the scaling of elementary particle lifetimes in powers of α, with the muon used as the reference lifetime, was carried out by David Akers [33], who then combined this experimental information with theoretical muon decay calculations and arrived at a "fundamental unit of time" equal to 2.43 zs, which closely matches the phenomenological boundary of 1 zs established here on the basis of the lifetimes displayed in Fig. 2.1.1.

The lifetimes in Fig. 2.1.1 are plotted on a logarithmic abscissa in order to compress them into a single picture, while at the same time keeping the shorter lifetimes separated enough to be visible as distinct entities. In order to get a feel for the great range of lifetimes displayed in Fig. 2.1.1, it is interesting to write down a few lifetimes τ in units of zeptoseconds, which we show in Table 2.1.1. To have this great a range of values in one set of physical entities is remarkable, and to be able to measure them [10] over this range is an impressive achievement. In fact, two different methods are used to obtain the lifetimes or mean lives. The six longest lifetimes τ in Table 2.1.1 are determined directly from path lengths and known velocities, but the three shorter lifetimes have to be deduced from the mass widths Γ of the resonances. To show how this works out, we take the nine lifetimes of Table 2.1.1 and replot them in terms of their mass widths, using the mass unit $\Gamma_{zs} = \hbar/1$ zs $= 0.6582$ MeV. This gives the mass width Γ values shown in Table 2.1.2. The three bottom particles in Table 2.1.2 have widths that are wide enough to be easily measured, but the six top particles do not, so direct lifetime measurements are used. The main purpose here in displaying Tables 2.1.1 and 2.1.2 is to illustrate numerically the wide range of values we have to consider when we study the global patterning of elementary particle lifetimes. These global patterns are important, because only by putting all of the particles together in the same plot can we see just what is happening over this vast range of particle lifetimes.

Table 2.1.1 Some representative elementary particle lifetimes [10] expressed in units of zeptoseconds, where 1 zs $= 10^{-1}$ s — a sextillionth of a second. Particles lifetimes greater than 1 zs occur in widely-spaced groups, whereas the shorter lifetimes occur with essentially a continuum of values.

neutron	$\tau =$	885, 700, 000, 000, 000, 000, 000, 000	zs
muon	$\tau =$	2, 197, 000, 000, 000, 000	zs
$\pi^{\pm}$	$\tau =$	26, 033, 000, 000, 000	zs
Ξ^{-}	$\tau =$	163, 900, 000, 000	zs
$B^{\pm}$	$\tau =$	1, 671, 000, 000	zs
π^{0}	$\tau =$	84, 000	zs
Υ_{1S}	$\tau =$	121.9	zs
J/ψ_{1S}	$\tau =$	7.047	zs
Z	$\tau =$	0.00026379	zs

Table 2.1.2 The elementary particle lifetimes of Table 2.1.1 expressed in units of "Γ_{zs}", where 1 $\Gamma_{zs} \equiv \hbar/(1\ \text{zs}) = 0.6582$ MeV. RPP2006 [10] lists 36 threshold-state particles with widths $\Gamma < 1\ \Gamma_{zs}$, and 121 excited-state particles with widths $\Gamma > 1\ \Gamma_{zs}$.

neutron	$\Gamma =$	0.000000000000000000000001129	Γ_{zs}
muon	$\Gamma =$	0.0000000000000004552	Γ_{zs}
$\pi^{\pm}$	$\Gamma =$	0.000000000000038412	Γ_{zs}
Ξ^{-}	$\Gamma =$	0.000000000006101	Γ_{zs}
$B^{\pm}$	$\Gamma =$	0.0000000005984	Γ_{zs}
π^{0}	$\Gamma =$	0.0000119	Γ_{zs}
Υ_{1S}	$\Gamma =$	0.08207	Γ_{zs}
J/ψ_{1S}	$\Gamma =$	0.1419	Γ_{zs}
Z	$\Gamma = 3,790.9$		Γ_{zs}

The question that now arises is whether there are common factors that tie these lifetimes together in some kind of a comprehensive lifetime grid, or whether they are for the most part just randomly spaced phenomena. The 121 excited states in Fig. 2.1.1 that have lifetimes $\tau < 1$ zs are clumped so closely together that no patterns emerge, at least on the time scales considered here. However, a close inspection of the 36 particles that have lifetimes $\tau > 1$ zs reveals five interesting features, which we discuss in detail in the present chapter.

(1) The lifetimes occur singly or in groups at spacing intervals S, where S (or $1/S$) is the lifetime ratio between groups. Visual inspection of Fig. 2.1.1 indicates that S is somewhat larger than a factor of 100 (see Figs. 0.1.3 and 0.1.4). In Secs. 2.2 and 2.6 we carry out detailed linear and quadratic statistical fits to the lifetime data in order to obtain an experimental value for S.
(2) Three of the lifetime groups have partial slopes (as plotted here in an arbitrary spreading of lifetimes along the y-axis) which are roughly the same. As we demonstrate in Sec. 2.5, these slopes correspond to a fairly accurate factor-of-2 lifetime hyperfine (HF) structure.
(3) In Sec. 2.5 we also examine the lifetime ratios of the hyperon pair Λ–Ω and charmed hyperon pair Λ_c–Ω_c. The Λ and Λ_c hyperons each contain one flavored s or c quark, and the Ω and Ω_c hyperons each contain three flavored s or c quarks. Both hyperon pairs have lifetime ratios of about

3 to 1, which suggests that the flavored s and c quarks in these hyperons independently trigger the decays. Hence the threshold-state particles also have a factor-of-3 HF component.

(4) Unpaired-quark decays (which are triggered by the decay of an unpaired particle or antiparticle subquark mass quantum) are characteristically slower by a factor of α^{-4} than paired-quark decays (which have balanced particle–antiparticle symmetry). Also, radiative decays (which do not necessitate a charge transfer between quarks during the decay process) are characteristically faster by a factor of α^{-4} than charged-particle decays (which involve interquark charge transfers). These results are displayed in Sec. 2.8.

(5) The two lifetime groups near $\tau_i = 10^{-12}$ s in Fig. 2.1.1 form parallel vertical lines (apart from the factor-of-2 and factor-of-3 HF structures). One line corresponds to unpaired b quark states, and the other to unpaired c quark states (Sec. 2.9). These two groups are separated by a factor of about three in lifetimes (Sec. 2.9), and also, as we will see, by a factor of about three in mass values (Sec. 3.17), with the higher-mass b states having the longer lifetimes.

Our main interest in Chapter 2 is to pin down feature (1), the overall group spacing interval S, both to ascertain its numerical value and also to see how far it extends in fitting these lifetimes. But in order to do this, it is important to empirically evaluate the effect of features (2) and (3) — the factor-of-2 and factor-of-3 HF lifetime structure — which logically represent quite different lifetime effects than the overall S spacing. This HF hyperfine structure is in fact of considerable interest in its own right, because it is a direct indication of a common quark mass substructure in these particles. To test the effect of HF corrections on the spacing interval S, we start in Sec. 2.2 with the basic low-mass $\pi^{\pm}$, π^0, η and η' pseudoscalar mesons, which involve no HF corrections, and we carry out two precision determinations of the scaling factor S. Then, after discussing the HF corrections in Sec. 2.5, we go on in Sec. 2.6 to incorporate the other threshold-state particles into the S analysis. We demonstrate in Sec. 2.6 that applying phenomenological HF corrections enhances the sharpness and accuracy of the determination of S. By enhancing the accuracy of the value for S, we mean that HF corrections bring it closer to the value $\alpha^{-1} \cong 137$, which is the value that has theoretical justification.

In evaluating these results, it should be kept in mind that Fig. 2.1.1 contains *all* of the experimentally well-determined lifetimes [10]. The *comprehensiveness* of this lifetime systematics is as important as its accuracy and span of values.

2.2 The Nonstrange π^+, π^-, π^0, η, η' PS Meson Quintet: The "Crown Jewels" of Lifetime α-Quantization

The lowest-mass, and logically the simplest, hadronic particles are the pseudoscalar (PS) spin 0 mesons π, η, η' and K. From the standpoint of lifetime systematics, the "strange" K^+, K^+, K_L^0 and K_S^0 mesons — the strange "PS meson quartet" — involve factor-of-2 HF corrections, as we discuss in Sec. 2.5. This leaves the "nonstrange" π^+, π^-, π^0, η and η' mesons — the nonstrange "PS meson quintet" — as the most straightforward ones to analyze from the standpoint of the overall lifetime scaling factor S. These are the ones we study first. The lifetimes of the PS quintet particles in zs are shown in the Table 2.2.1. These lifetimes are spread out over more than 12 orders of magnitude, and hence should furnish an accurate evaluation of the large-order scaling. Since they cover almost the entire lifetime range of the threshold particles that have lifetimes $\tau > 1$ zs, the global scaling grid they establish should also apply to the other threshold states. Figure 2.2.1 shows the lifetime plot of the PS quintet, which is the same logarithmic plot as that of Fig. 2.1.1, but shifted to the $\pi^\pm$ reference point. As can be seen, the lifetimes τ are spread out over almost 13 orders of magnitude, and they extend down almost to the zeptosecond boundary at 10^{-21} s. Figure 2.2.2 is the same plot as that of Fig. 2.2.1, but is in terms of the "full width" Γ rather than the "mean life" τ, where we employ the conventions of the PDG in RPP2006 [10]. Figure 2.2.2 is included here to demonstrate the fact that the graphical lifetime analysis is the same whether we use lifetimes or mass widths for the particles: one scaling is the reciprocal of the other, and the transformation from one to the other is accomplished by changing the sign of the lifetime logarithm x_i.

Table 2.2.1 The lifetimes of the nonstrange PS meson quintet, expressed here in units of zeptoseconds.

$\pi^\pm$	$\tau =$ 26,033,000,000,000	zs
π^0	$\tau =$ 84,000	zs
η	$\tau =$ 506	zs
η'	$\tau =$ 3.26	zs

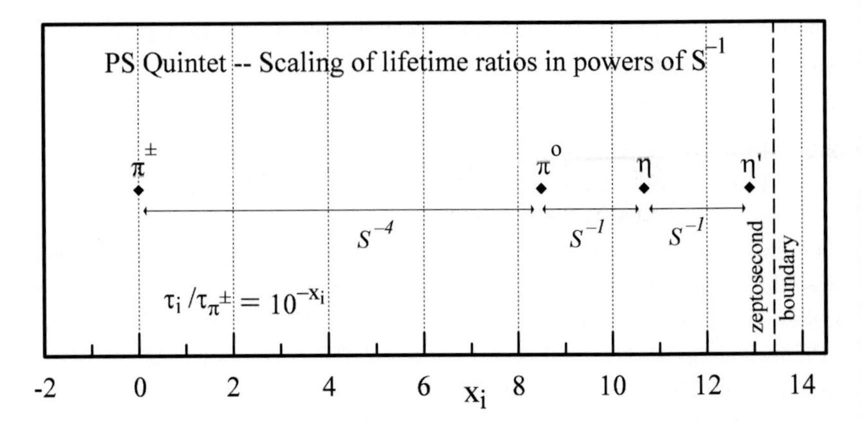

Fig. 2.2.1 The ($\pi^\pm$, π^0, η, η') PS quintet. These are the nonstrange PS mesons from Fig. 2.1.1, with their lifetimes now expressed as ratios to the $\pi^\pm$ lifetime and plotted as logarithms $-x_i$ to the base 10. The arrows indicate the visual scaling in factors of S^{-1}, where $S \gtrsim 100$ is a common scaling factor. The dashed vertical line denotes the zeptosecond lifetime boundary at $\tau = 10^{-21}$ s.

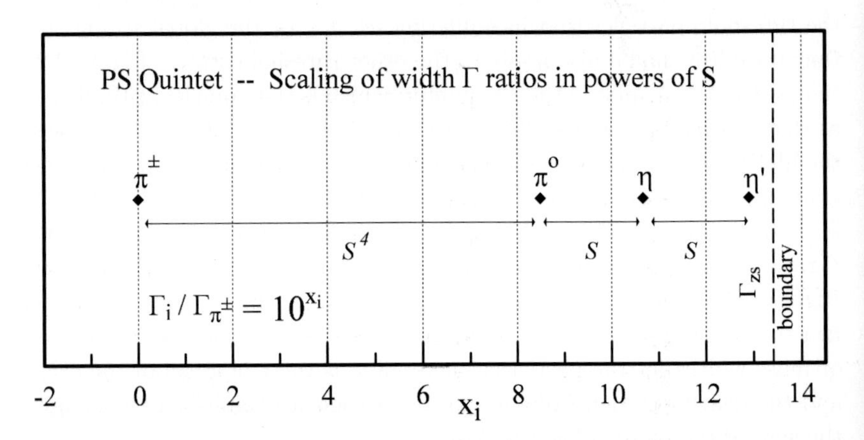

Fig. 2.2.2 This is the same plot as Fig. 2.2.1, but for resonance mass widths Γ instead of lifetimes τ. The relative data points are the same in both figures, but the scaling here is in factors of S instead of S^{-1}. The dashed vertical line denotes the zeptosecond boundary value $\Gamma_{zs} = 0.6582$ MeV.

The search for an α-quantization in the PS quintet lifetimes is straightforward. We calculate lifetime *ratios* with respect to a standard reference lifetime, and we then see if these ratios are multiples of $1/137$. Our first task is to select a basic reference lifetime. This we take to be the $\pi^\pm$ lifetime, since the pion is the lowest-energy unstable hadronic elementary particle, and $\pi^\pm$ is its longest-lived and most accurately measured lifetime state. Using the $\pi^\pm$ as a reference, we write the lifetimes in the form

$$\tau_i/\tau_{\pi^\pm} = 10^{-x_i} \,. \tag{2.2.1}$$

It seems apparent from the x_i values displayed in Fig. 2.2.1 that lifetimes do not scale in powers of 10^{-1}: the π^0, η and η' lifetimes do not lie on the grid lines established by the $\pi^\pm$ lifetime. In order to find a more suitable scaling parameter, we choose an arbitrary scaling factor $S > 1$ and then use its inverse S^{-1} to scale the lifetimes. We can tell from a visual inspection of Fig. 2.2.1 that the η/π^0 and η'/η lifetime ratios are approximately equal to one another, so they are assigned the same scaling factor, S^{-1}. Visual inspection also shows that the $\pi^0/\pi^\pm$ lifetime ratio is roughly S^{-4} on this logarithmic scale. This scaling systematics is indicated by the horizontal arrows in Fig. 2.2.1. Thus the scaling equations for π^0, η and η' are

$$\tau_i/\tau_{\pi^\pm} = S^{-x_i} \,, \quad x_{\pi^0} \cong 4 \,, \quad x_\eta \cong 5 \,, \quad x_{\eta'} \cong 6 \,. \tag{2.2.2}$$

If we work with mass widths Γ instead of lifetimes τ, we have the same scaling systematics, but with the sign of the exponent x_i changed, as shown in Fig. 2.2.2. By comparing the horizontal arrows to the abscissa in Fig. 2.2.1, we can see that the actual scaling factor S for the PS quintet is somewhat larger than 100 (see Figs. 0.1.3 and 0.1.4).

In order to deduce the optimum phenomenological value for S as a scaling factor, we make direct fits to the experimental data. We make these fits using two different methods: a *linear* fit of the lifetime values to an S-spaced lifetime grid, and a *quadratic* fit in a χ^2 sum of the lifetime values weighted by the uncertainties in these values. We first discuss the linear fit, which is carried out with the scaling factors displayed in Eq. (2.2.2). The accuracy of this fit to the S-grid is determined by examining the exponents x_i that are obtained by matching the experimental lifetime ratios. If the S scaling is accurate, the x_i will have almost integer values. Hence the "correct" scaling factor S for the PS quintet lifetimes is the one that most nearly gives integral exponent values x_i in Eq. (2.2.2); that is, the proper S *minimizes* the deviations Δx_i of the exponents x_i from integer values. In order to determine S, we first define the average *Absolute Deviation from an Integer* (ADI) of the exponents x_i as

$$\mathrm{ADI} = \frac{1}{N}\sum_{i=1}^{N} |\Delta x_i| \,, \quad \Delta x_i \equiv x_i - I_n \,, \quad I_n = \text{nearest integer} \,. \tag{2.2.3}$$

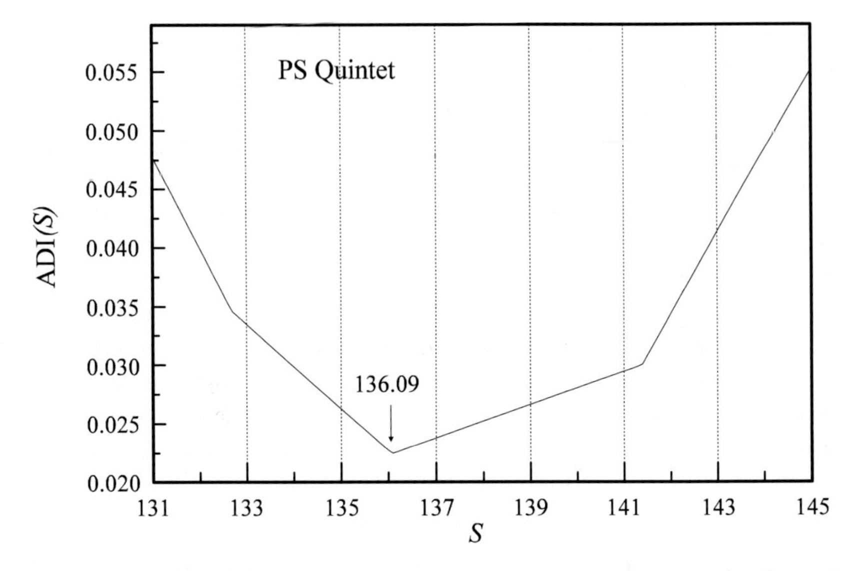

Fig. 2.2.3 The *Absolute Deviation from an Integer* ADI(S) for the ($\pi^{\pm}$, π^0, η, η') PS quintet of Fig. 2.2.1. The average ADI value is defined in Eqs. (2.2.2) and (2.2.3). As can be seen in Fig. 2.2.3, ADI(S) has a minimum at $S_{\text{min}} = 136.09$, which is close to the α-quantized value $\alpha^{-1} = 137.04$. The ADI value at this minimum is 0.0225, which is more than a factor of 10 smaller than the random-fit value of 0.25. This indicates mathematically that these PS quintet lifetimes are accurately reproduced with a single scaling factor $S \cong \alpha^{-1}$.

Then we carry out calculations over a range of S values and plot out the function ADI(S). Since the maximum absolute deviation $|\Delta x_i|$ of an individual lifetime exponent x_i from an integer value is 0.5, a *random* distribution of x_i's from 0 to 0.5 will yield ADI $\cong$ 0.25. But if the lifetimes being tested correspond to a definite S value, then the ADI should show a dip at this value. A calculation of ADI(S) as a function of S was carried out using Eqs. (2.2.2) and (2.2.3). Due to the small number of lifetimes in the fit, and the change in sign as an integer value for x_i is passed, the slope of the ADI(S) curve is not continuous. Figure 2.2.3 shows the ADI(S) curve over the range $131 \leq S \leq 145$. As can be seen, a minimum at $S = 136.09$ is clearly in evidence. The ADI value at this minimum is 0.0225, which is much smaller than the value ADI $\cong$ 0.25 that corresponds to a random fit. This shows that the exponents x_i in Eq. (2.2.2) for the π^0, η and η' lifetime ratios with respect to that of the $\pi^{\pm}$ are very nearly integers, which verifies that S is the common scaling factor S for these PS lifetimes. Furthermore, the value $S_{\min} = 136.09$ is within 1% of the value $\alpha^{-1} \cong 137.04$, which does not seem to be an accident. As we tried to convey in Table 2.2.1, the lifetime leap from the $\pi^{\pm}$ to the π^0 is enormous — more than eight orders of magnitude. To have this leap be almost precisely equal to α^4, and then to have it followed by additional leaps of α for the two higher excited states, is direct experimental evidence for the relevance of α in this domain. These low-mass spin 0 hadrons have the simplest mass structures and most clear-cut lifetime scaling factors.

The second method used here to determine S is to make a *quadratic* χ^2 fit to the lifetime *values* as weighted by the experimental *errors* on the lifetime measurements [10]. That is, we write down mathematical expressions for the lifetimes $\tau_i(S)$ as ratios to the $\pi^{\pm}$ lifetime, and then find the value of S that minimizes the chi-squared sum [34]

$$\chi^2(S) = \sum_i \left(\frac{\tau_i(S) - \tau_i(\exp)}{\Delta\tau_i(\exp)} \right)^2 . \qquad (2.2.4)$$

This method assigns the most importance to the most accurately measured data points. It only works well in a statistical sense if all of the data have roughly comparable error limits. Otherwise, the χ^2 sum is dominated by the fits to the data that have very small error bars, which may not be intrinsically of more importance than the other data points. (The neutron and muon, for example, have very accurately determined lifetimes, and do not fit in naturally with the other threshold particles in this type of analysis.) The chi-squared analysis serves as a comparison to the ADI analysis in trying to determine if the lifetime scaling factor S is in fact α^{-1} for these hadronic particles.

The χ^2 analysis was carried out in the following manner. The pseudoscalar $\pi^{\pm}$ meson was selected as the reference lifetime, as in the ADI analysis. The π^0, η and η' lifetimes τ were written in the form $\tau_i(S) = \tau_{\pi^{\pm}} S^{-x_i}$, $x_i = 4, 5, 6$, respectively, for insertion into Eq. (2.2.4). Then S was varied so as to obtain a minimum value for χ^2. The results are displayed in Fig. 2.2.4. As can be seen, the χ^2 minimum occurs at $S = 138.96$, in close agreement with the ADI value of 136.09, and also with the α^{-1} value of 137.04.

If we consider the ADI and χ^2 minimizations to be two independent ways of obtaining a numerical value for S in the lifetime scaling of the PS quintet, then we can average these minima, which gives $S = 137.53$. Since $\alpha^{-1} = 137.04$, we have agreement at an accuracy level of better than 0.5%, although the scattering of the individual data points means

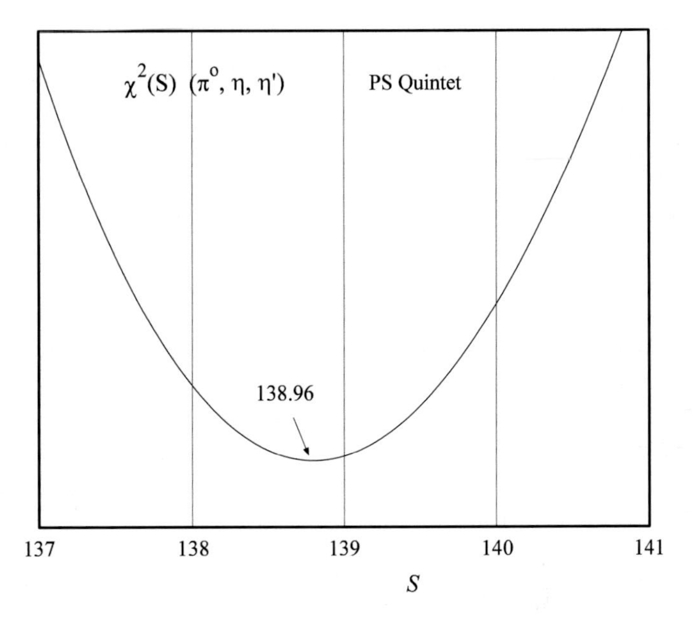

Fig. 2.2.4 The *chi-squared sum* $\chi^2(S)$ for the ($\pi^{\pm}$, π^0, η, η') PS quintet [34]. The π^0, η and η' lifetimes are calculated theoretically as S^{-4}, S^{-5} and S^{-6} ratios, respectively, to the $\pi^{\pm}$ reference lifetime. These calculated lifetimes are entered into the $\chi^2(S)$ sum of Eq. (2.2.4), together with the experimental lifetimes and errors from Ref. [10]. As shown in Fig. 2.2.4, the $\chi^2(S)$ sum has a minimum at $S_{\text{min}} = 138.96$. This *quadratic* value for S_{min} and the *linear* ADI value $S_{\text{min}} = 136.09$ (Fig. 2.2.3) bracket the value $\alpha^{-1} = 137.04$, and thus furnish direct evidence that the fine structure constant α plays a dominant role in these hadronic decays.

that we should not take this precision too seriously. However, the broad conclusion we reach is that α or α^{-1} is in fact the scaling factor that ties these lifetimes together. We can buttress this conclusion by extending these results to include the other threshold-state lifetimes, which we do in the remainder of Chapter 2. We can also make use of the fact that the conjugate threshold-state particle *mass widths*, which are linked to the *lifetimes* by the uncertainty principle, Eq. (1.7.1), exhibit a reciprocal dependence on α. This *mass-width* α-dependence is logically related to an α-dependence in the *masses* themselves. Thus we should examine the threshold-state particle masses to see if they also exhibit an α-dependence. If they do, then the case for the extension of α into the hadronic domain becomes much more compelling. This task is carried out in Chapter 3, where we demonstrate in Sec. 3.7 that the *mass* α-quantization for the PS meson quintet is fully as accurate as the *lifetime* α-quantization we have demonstrated here.

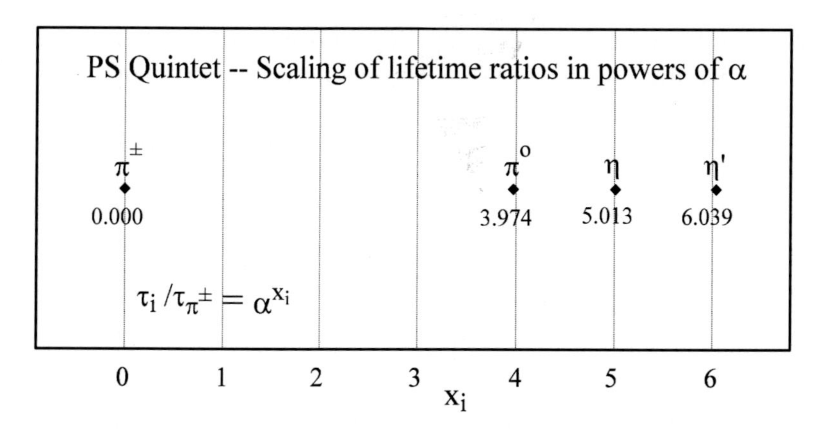

Fig. 2.2.5 The ($\pi^{\pm}$, π^0, η, η') PS quintet lifetimes plotted as logarithms x_i to the base α, with $\pi^{\pm}$ as the reference lifetime (Eq. (2.2.5)). The accuracy of this lifetime α-quantization over almost 13 orders of magnitude is visually apparent. The numerical values of the exponents x_i are shown below the data points. It should be noted that these are *all* of the nonstrange pseudoscalar mesons. The completeness of this α-grid lifetime mapping is as important as its accuracy over a wide range of values.

The goal we have been working toward in these ADI(S) and $\chi^2(S)$ PS meson quintet minimization studies is to establish the fact that their lifetimes are in fact functions of the fine structure constant α. The best way to demonstrate this result visually is to replot these lifetimes on an α-spaced lifetime grid, using the equation

$$\tau_i/\tau_{\pi^\pm} = \alpha^{x_i}\,, \tag{2.2.5}$$

with the $\pi^{\pm}$ meson as the reference lifetime. These results are displayed in Fig. 2.2.5, where the accuracy of the α-spacing is apparent over a wide range of lifetimes — six powers of α. (Also see Fig. 0.1.4.) The lifetime logarithms x_i are displayed under the data points. As can be seen, the x_i are very close to integer values. In addition to this numerical accuracy, the other point to be taken into consideration is the fact that all of the nonstrange pseudoscalar mesons are included here. There are no "rogue" counterexamples. These PS meson quintet lifetimes furnish direct experimental verification of the relevance of the fine structure constant α to elementary particle decays, irrespective of any theoretical considerations. This is an important fact to emphasize, because it must be kept in mind that we are now working outside of the paradigm of current elementary particle grand unified theories (GUTs), which have no room for this empirical α-dependence in hadronic systems.

2.3 The Strange K^+, K^-, K_L^0, K_S^0 Meson Quartet: α-Scaling and Factor-of-2 Hyperfine (HF) Structure

The spin 0 pseudoscalar mesons (the PS meson nonet) were singled out and labeled in the lifetime plot of Fig. 2.1.1. The *nonstrange* π^+, π^-, π^0, η and η' "PS quintet" mesons were placed on a common central ordinate, and the *strange* K^+, K^-, K_L^0 and K_S^0 "PS quartet" mesons were arbitrarily displaced above and below this ordinate. The quintet mesons are particle–antiparticle symmetric: they are their own antiparticles. The quartet mesons are a mixed bag. The K^- and K^+ mesons are not particle–antiparticle symmetric: they are kaon and antikaon states, respectively. The K_L^0 and K_S^0 mesons are considered to be linear combinations of the kaon and antikaon states K^0 and $\bar{K}^0$, and thus have mixed symmetries. In Sec. 2.2 we analyzed the lifetime scaling factor S^{-1} of the PS quintet mesons, and found it to be accurately equal to the fine structure constant α, as summarized in Figs. 2.2.3–2.2.5. In the present section we examine the PS quartet mesons to see how they fit in with the lifetime α-quantization of the quintet mesons.

Figure 2.3.1 is a plot of the PS quartet mesons on an expanded portion of the same α-spaced lifetime grid that was used for the quintet mesons in

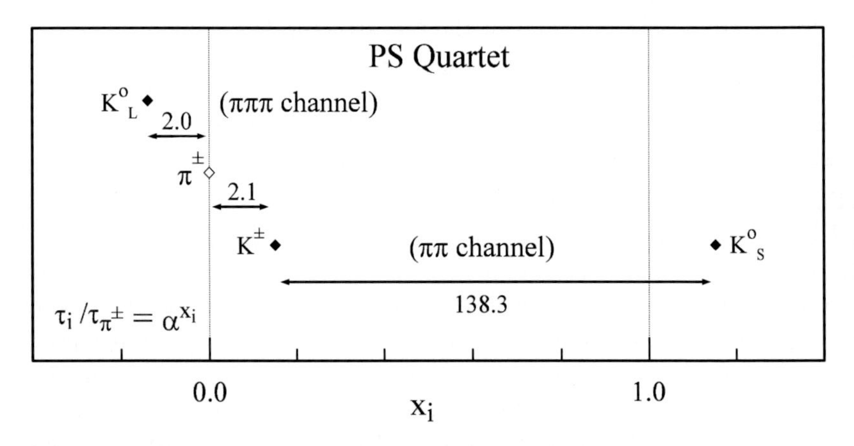

Fig. 2.3.1 The ($K^\pm$, K_L^0, K_S^0) PS quartet lifetimes, shown with the reference $\pi^\pm$ lifetime on an α-spaced lifetime grid. The $K^\pm$ and K_S^0 mesons, which are grouped together in the $\pi\pi$ decay channel, have a lifetime ratio of almost exactly α^{-1}. The $K_L^0/\pi^\pm$ and $\pi^\pm/K^\pm$ lifetime ratios are each a factor of about 2. This characteristic factor-of-2 "hyperfine" (HF) lifetime structure is also observed in other hadron isotopic spin multiplets (Sec. 2.5).

Fig. 2.2.5. As can be seen, the K_L^0 meson has a lifetime that is a factor of 2 longer than the $\pi^\pm$ reference lifetime ($x_i = 0$), and the $K^\pm$ has a lifetime that is a factor of 2.1 shorter than the $\pi^\pm$. These are the "crown jewels" of the factor-of-2 hyperfine (HF) lifetime structure that we discuss in Sec. 2.5. The systematics of elementary particle lifetimes is revealed most clearly in these low-mass hadron excitations with their clear-cut mass substructures. The $K_L^0/K^\pm$ lifetime ratio is 4.2. The K_S^0 lifetime is considerably shorter than either the K_L^0 or $K^\pm$, and is shifted over to the $x_i \cong 1$ α-spaced grid line. Since the K_L^0 decays into three negative-parity pions, whereas the K_S^0 and $K^\pm$ both decay into two pions (and thus have opposite parity to the K_L^0), it seems natural to associate the K_S^0 with the $K^\pm$, as we have done in Fig. 2.3.1. With this assignment, we can see visually that these two lifetimes differ by just about one power of α. Experimentally, the $K_S^0/K^\pm$ lifetime ratio [10] is 1/138.33, where we use the $K^\pm$ lifetime together with the K_S^0 lifetime that is obtained by assuming CPT invariance (Appendix A). This lifetime ratio closely resembles the $\alpha = 1/137.04$ lifetime scaling factor that we deduced from the PS quintet lifetimes. This gives us another determination of the lifetime scaling factor S. It also gives us another example of the factor-of-2 lifetime hyperstructure HF, since the K_S^0 is evidently shifted off the α-spaced $x_i = 1$ grid line by the same amount that the $K^\pm$ is shifted off the $x_i = 0$ grid line. These low-mass spin 0 meson states are logically the simplest meson excitations, which gives added significance to their observed α-dependence and factor-of-2 HF structure.

Table 2.3.1 summarizes the three determinations of the pseudoscalar meson lifetime scaling factor S that we have obtained from the present analyses. The purpose in showing these numerical comparisons is not to stress the precise accuracy of the results, but simply to establish the fact that the fine structure constant α really does apply to these lifetime ratios,

Table 2.3.1 Pseudoscalar meson determinations of the lifetime scaling factor S.

PS quintet: ADI(S) minimum at $S = 136.09$ (Fig. 2.2.3)

PS quintet: $\chi^2(S)$ minimum at $S = 138.96$ (Fig. 2.2.4)

PS quartet: $K^\pm/K_S^0$ lifetime ratio $= 138.33$ (Appendix A and Ref. 10)

Average of these three determinations: $S = 137.79$

Reciprocal of fine structure constant: $\alpha^{-1} = 137.04$

Accuracy of this result: 0.5%.

since this is not the viewpoint that is currently prevalent in the community of elementary particle physicists.

The other feature of the PS quartet that is of interest here is the factor of 4.2 ratio between the K_L^0 and $K^{\pm}$ lifetimes. What makes this lifetime ratio particularly intriguing is the fact that the $\pi^{\pm}$ lifetime is the geometric mean of these two kaon lifetimes. It is not clear *a priori* why these π and K lifetimes should bear any relationship to one another. As Fig. 2.3.1 shows, the $\pi^{\pm}$ lifetime is about twice as long as the $K^{\pm}$ lifetime, and the K_L^0 lifetime is twice as long as the $\pi^{\pm}$ lifetime. Thus this triad of seemingly unrelated particles exhibits a quite accurate factor-of-2 hyperfine (HF) structure. This result does not seem to be accidental, because the same factor-of-2 HF structure is also observed in several other threshold-state lifetime groupings, as we discuss in Sec. 2.5, and have displayed in Figs. 0.1.7 and 0.1.8. We will demonstrate in Sec. 2.6 that this HF structure is an independent lifetime feature that is superimposed on the global scaling in powers of α. Since particle lifetimes represent in essence the stability of particle masses, this factor-of-2 HF lifetime structure suggests a corresponding mass substructure within these particles, with the mass substructure, together with the charge states, dictating the number of "decay triggers" for each particle. We will see evidence in Chapter 3 for this mass substructure, although a comprehensive theory of "decay triggers" is beyond the scope of the present phenomenological analyses.

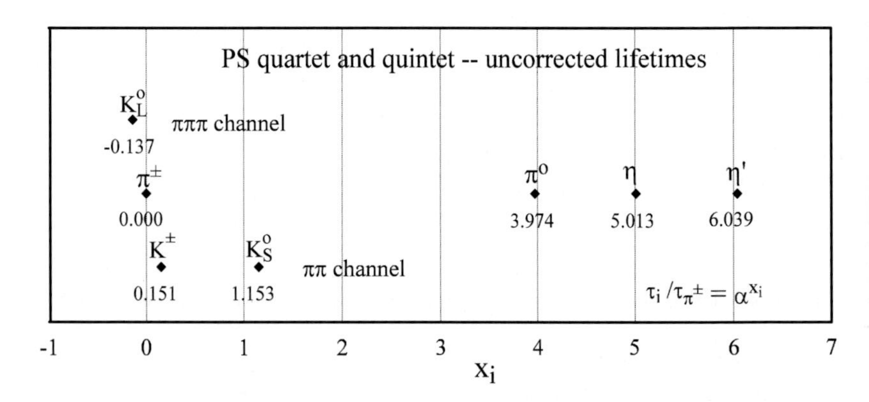

Fig. 2.3.2 The pseudoscalar meson quartet and quintet lifetimes plotted on an α-spaced lifetime grid that is centered on the $\pi^{\pm}$. The quintet lifetimes fall on the α-grid (Fig. 2.2.5), but the quartet lifetimes are shifted off the grid by factors of $+2$ or -2 (Fig. 2.3.1). The quartet and quintet mesons make up the PS meson nonet.

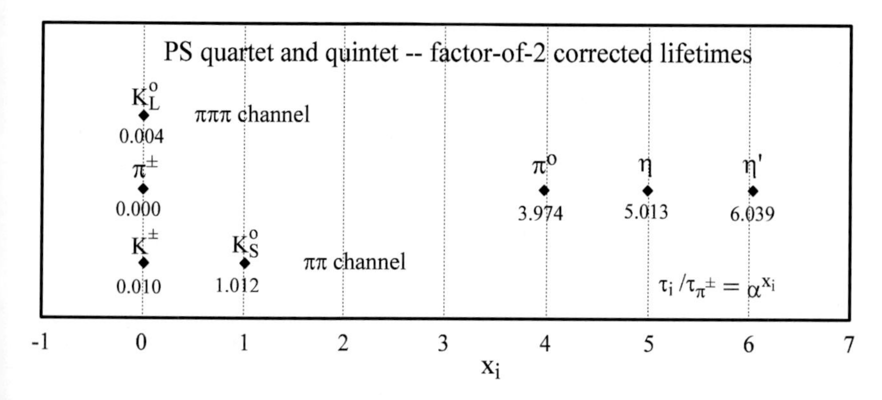

Fig. 2.3.3 The pseudoscalar meson quartet and quintet lifetimes of Fig. 2.3.2, but now with correction factors of exactly 2 applied to the quartet lifetimes to bring them into correspondence with the lifetime α-grid. With these factor-of-2 HF corrections (see Sec. 2.5), all of the PS mesons have lifetime logarithms x_i that are very close to integer values. This α-scaling of the PS meson lifetimes seems to be too accurate, too comprehensive, and too far-ranging to be regarded as accidental.

We initially grouped the K^0_S meson with the $K^\pm$ meson in Fig. 2.3.1 because of their similar $\pi\pi$ decay modes, as contrasted to the $\pi\pi\pi$ decay mode of the K^0_L meson. Having done this, we now see from their fits to the universal lifetime α-grid that they are both similarly displaced from the grid. Thus they seem to share common substructural elements. Figure 2.3.2 shows all nine of the PS meson lifetimes displayed on this α-grid, with the numerical values of the lifetime logarithms x_i displayed below the data points. Figure 2.3.3 is the same plot, but with exact factor-of-2 HF lifetime "corrections" applied to bring the kaon lifetimes into line with the α-scaling of the pion and eta lifetimes. As can be seen in Fig. 2.3.3, when these arbitrary HF corrections are applied, all nine of the PS mesons have lifetime logarithms x_i that are very nearly integers. The comprehensiveness, great span of lifetimes encompassed, and accuracy of these results speak for themselves.

2.4 The PS Meson Lifetime Nonet: Physics Outside of the Standard Model

The subject of the pseudoscalar meson nonet has been accorded this separate section in Chapter 2 because we want to make a simple but important statement about the PS mesons, as follows:

> *From the point-of-view of experimental lifetimes, the η and η' mesons are on a completely equivalent footing.*

If we think in nuclear physics terms, the η and η' mesons appear as the first and second excited states, respectively, above the ground-state π^0 meson. This concept is reinforced by the observed decay modes $\eta \rightarrow \pi^0 + \pi\pi$ and $\eta' \rightarrow \eta + \pi\pi$. In the formulation of Gell-Mann's famous "eightfold way" of the Standard Model, the pseudoscalar meson octet has precisely *eight* unique particle states [35]. The problem which arises here is that the PS meson quintet of Sec. 2.2 and the PS meson quartet of Sec. 2.3 make a total of *nine* closely-related pseudoscalar mesons that are eligible to occupy the *eight* states in the PS octet. The three pions and two charged kaons accurately fill five of the slots in the PS meson octet, where the accurate reproduction of their isotopic spins and strangeness quantum numbers represents a major triumph of the eightfold way. The K^0 and $\bar{K}^0$ neutral kaons that logically occupy two more slots in the PS octet are not observed directly, but (presumed) linear combinations of these states are observed as the hybrid K_L^0 and K_S^0 kaons. This leaves only one octet slot available for the η and η' mesons. At the time the eightfold way was being formulated, the properties of the η' were not well known, and it was originally denoted as the X^0 particle [36]. However, as its properties became clarified, it was recognized as a bona fide pseudoscalar meson. This difficulty was resolved theoretically by employing a *singlet* configuration associated with the *octet* to provide the required additional slot. Since the η and η' have the same quantum numbers, they can presumably "mix" together, and the assumption was made that linear combinations of these two particles occupy the slots in the octet and the singlet. But this means that the η and η' mesons are not being treated on an equal footing, as the experimental data seem to mandate.

From the standpoint of *lifetime* systematics, the equal experimental values for the π^0/η and η/η' lifetime ratios have not posed a problem for Standard Model physicists, because Standard Model physics does not presently contain a global lifetime formalism. However, when we go over to *mass* systematics, where the η and η' again appear on an equal footing, the problems raised with respect to the Standard Model become more incisive. As we discuss in Sec. 3.21, the phenomenological problems that arise with respect to the PS meson masses suggest a needed reformulation of the

Standard Model quarks. This reformulation retains the charge symmetries of QCD, but not the treatment of masses within the current paradigm of the Standard Model.

The very accurate symmetries in the π^0, η and η' lifetime spacings, as illustrated in Fig. 2.2.5, are the forerunners of the equally accurate symmetries in the π^0, η and η' mass spacings, as illustrated in Figs. 0.2.11 and 3.7.2. These lifetime and mass symmetries are at the 1% level of accuracy, and Standard Model physics does not account for them. All theories are, in the final analysis, dependent on the results of experiments, and theorists who overlook results that do not seem to fit in with their preconceptions may end up wasting a lot of time and effort, as well as a lot of journal space in the libraries.

2.5 Hyperfine (HF) Factor-of-2 and Factor-of-3 Lifetime Structure

The topic of a pervasive factor-of-2 hyperfine (HF) structure in the lifetimes of the long-lived threshold-state elementary particles is phenomenologically simple, but is significant from the standpoint of elementary particle theories. The experimental information about this HF structure is straightforward:

> *almost half of the 36 particles that have lifetimes longer than 1 zs occur as members of factor-of-2 (or factor-of-3 for Λ–Ω pairs) lifetime multiplets.*

This HF structure is observed in the pseudoscalar mesons and in the s and c unpaired-quark excitations, but not yet in the b unpaired-quark excitations. The significance of this effect appears when we try to account theoretically for this factor-of-2 granularity, which seems to require some kind of particle substructure. This topic has been addressed by Harald Fritsch in his book *Elementary Particles* as follows.

> *It may well be that, in the future, we will have to give up the notion of a point-like mass. It may, after all, be that the masses of leptons and quarks are a consequence of a hitherto unobserved substructure of these particles, just as the proton mass is due to its substructure. It cannot be ruled out at this time that substructure effects and hints at new building blocks inside the leptons and quarks will be gleaned from powerful new accelerator experiments.* [37]

The evidence for quark substructure may not in fact require powerful new accelerators, but may already be in evidence in the threshold-state lifetimes. If we picture each threshold-state particle as being composed of a small number of mass substates, some of which are annihilated in the particle decay process, then the factors of two can be accounted for in terms of the number of available substate "decay triggers." But if we picture the masses of the particles as arising mainly from the amorphous "gluon fields" of the quarks, as is assumed in the Standard Model, then it is not immediately evident how such a clear-cut HF granularity can be explained.

The first example of HF structure is shown in Fig. 2.5.1, where the K^0_L, $\pi^\pm$, and $K^\pm$ lifetimes are displayed as exponents x_i on an α-spaced lifetime grid, with the $\pi^\pm$ lifetime as the $x_i = 0$ anchor for the α-grid. The $K^0_L/K^\pm$ lifetime ratio is 4.2, which is of interest in itself (Fig. 0.1.6), but the item of real interest is the fact that the $\pi^\pm$ lifetime is the geometric mean of these lifetimes, as shown in Fig. 2.5.1. In terms of decay "triggers," we can account for these lifetime ratios by ascribing one, two, and four triggers to the K^0_L, $\pi^\pm$, and $K^\pm$ mesons, respectively. The $\pi^\pm$ meson is a particle–antiparticle-symmetric hadron that is composed of matching 70 MeV particle and antiparticle substates (Sec. 3.3), only one of which is annihilated in the $\pi^\pm \rightarrow \mu^\pm$ decay process (Sec. 4.3), so the assignment of

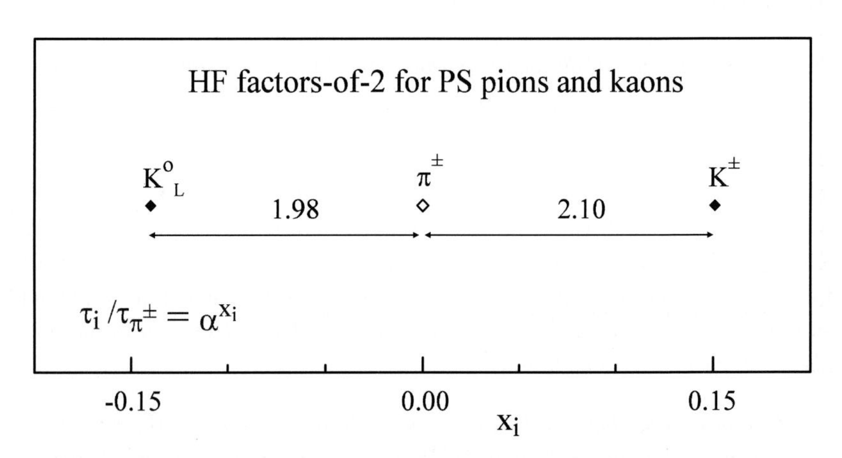

Fig. 2.5.1 Quantitative factor-of-2 lifetime ratios for the pseudoscalar mesons. The $K^0_L/K^\pm$ lifetime ratio is 4.13. The $\pi^\pm$ lifetime, which bears no obvious relationship to these two kaon lifetimes, is almost at their geometric mean, so that the K^0_L–$\pi^\pm$–$K^\pm$ triad consists of two factor-of-2 lifetime steps. This suggests that these particles are composed of related mass sub-elements which generate one, two, and four decay "triggers," respectively, and thus create this HF hyperfine lifetime structure.

two decay triggers for it seems to be a natural procedure: either 70 MeV quantum can initiate the decay. The kaons each contain seven 70 MeV mass quanta. In the $K_L^0 \to \pi\pi\pi$ decay, a single 70 MeV quantum is annihilated, which is consistent with the assignment of one decay trigger for the K_L^0. The $K^\pm \to \pi\pi$ decay involves the annihilation of three 70 MeV quanta, so the assignment of four decay triggers for it is not so transparent.

If K_L^0, $\pi^\pm$, $K^\pm$ were the only example of a factor-of-2 HF lifetime multiplet, it might be regarded as accidental, but when we move over to $x_i \cong 1$, the next level on the lifetime α-grid, we find this situation repeated, as displayed in Fig. 2.5.2, which shows the $x_i \cong 1$ α-grid lifetime plot for the s-quark baryons. The Ξ^- and Σ^- lifetimes are clustered close to the $x_i = 1$ grid line, in the α-grid position that corresponds to the $\pi^\pm$ lifetime in Fig. 2.5.1. The Ξ^0 lifetime is displaced almost a factor of 2 to the left,

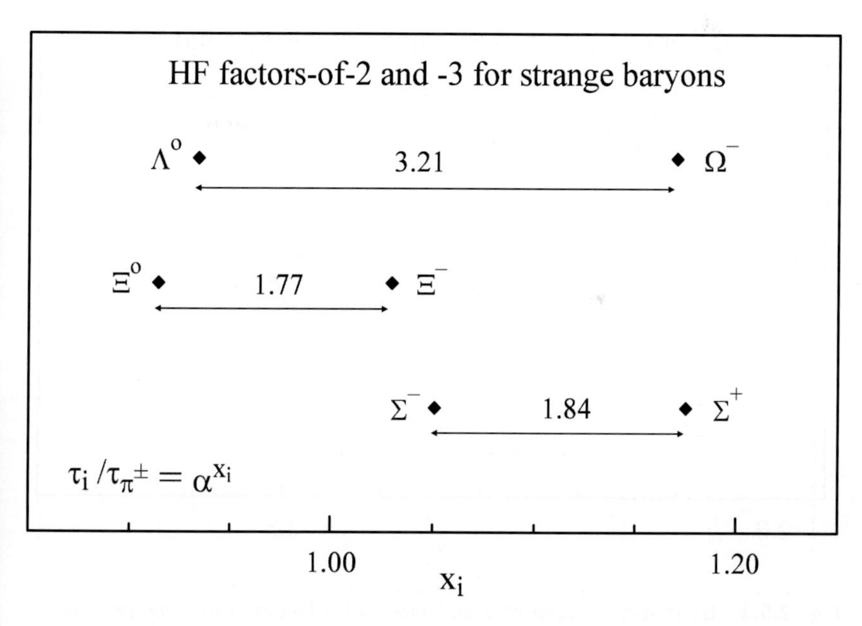

Fig. 2.5.2 Quantitative factor-of-2 and factor-of-3 lifetime ratios for the hyperons (strange baryons). The Σ and Ξ hyperons each feature an approximate factor-of-2 lifetime ratio, and the combination of these four lifetimes (which are plotted to scale) forms a factor-of-2 lifetime triad that closely resembles the K_L^0–$\pi^\pm$–$K^\pm$ lifetime triad displayed in Fig. 2.5.1. The $\Lambda = qqs$ and $\Omega = sss$ hyperons at the top of Fig. 2.5.2 have a lifetime ratio of approximately 3, which seems related to the fact that they contain one and three strange quarks, respectively. All of these hyperons contribute to the HF hyperfine structure of the threshold-state particles.

and the Σ^+ lifetime is almost a factor of 2 to the right, thus echoing the K_L^0 and $K^\pm$ lifetimes in Fig. 2.5.1. The Λ^0 and Ω^- lifetimes are also in roughly the same positions as the Ξ^0 and Σ^+, respectively, but the observed lifetime ratio between those two particles is 3.2. The Λ^0 contains one s quark, and the Ω^- contains three s quarks, which suggests that the s quarks are the decay triggers for these states, and this type of decay triggering may overrule the factor-of-2 decay triggering. This conclusion is reinforced by the similar lifetime ratio for the charmed Λ_c^+ and Ω_c^0 quarks, as displayed in Fig. 2.5.3 and Table 2.5.1. Thus, without a comprehensive theory for all of these HF lifetime multiplets, we cannot tell exactly what corrections should be applied to "remove" the HF perturbations of the overall α-spaced lifetime grid.

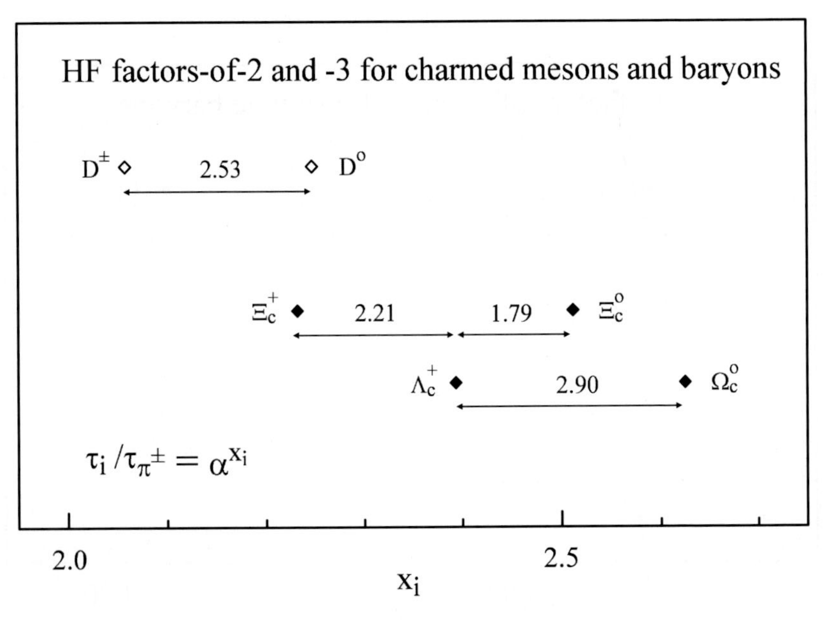

Fig. 2.5.3 Quantitative factor-of-2 and factor-of-3 lifetime ratios for mesons and baryons that contain a single charmed quark. The D mesons at the top of Fig. 2.5.3 have a lifetime ratio which is somewhat larger than a factor of 2 (see the ratio displayed in Fig. 2.5.4). The Ξ_c^+ and Ξ_c^0 hyperons have a lifetime ratio of 3.95 that echoes the K_L^0 and $K^\pm$ lifetime ratio (see Fig. 0.1.6). The Λ_c^+ lifetime falls at about the geometric mean of the Ξ_c lifetimes, thus forming a Ξ_c^+–Λ_c– Ξ_c^0 factor-of-2 lifetime triad similar to those displayed in Figs. 2.5.1 and 2.5.2 (see Fig. 0.1.8). The factor-of-3 Λ_c/Ω_c lifetime ratio is similar to the Λ/Ω lifetime ratio of Fig. 2.5.2 (see Fig. 0.1.9).

Table 2.5.1 Experimental lifetime factors-of-2, -3, and -4 in the low-mass, long-lived, threshold-state elementary particles (see Figs. 2.5.1–2.5.3).

Factors-of-2	Factors-of-3	Factors-of-4
$K_L^0/\pi^\pm = 1.96$	$\Lambda^0/\Omega^- = 3.21$	$K_L^0/K^\pm = 4.13$
$\pi^\pm/K^\pm = 2.10$	$\Lambda_c^+/\Omega_c^0 = 2.90$	$\Xi_c^+/\Xi_c^0 = 3.95$
$\Xi^0/\Xi^- = 1.77$	average $= 3.06$	average $= 4.04$
$\Sigma^-/\Sigma^+ = 1.84$		
$D^\pm/D^0 = 2.53$		
$\Xi_c^+/\Lambda_c^+ = 2.21$		
$\Lambda_c^+/\Xi_c^0 = 1.79$		
average $= 2.03$		

The final examples of these unpaired-quark HF lifetime multiplets are the c-quark particle states displayed in Fig. 2.5.3, which have lifetimes that are clustered slightly above the $x_i = 2$ level on the α-grid. As we demonstrate in Sec. 2.9, the unpaired b-quark particle states cluster right on the $x_i = 2$ α-grid line, and the unpaired c-quark states are systematically a factor of three shorter in lifetimes. As can be seen in Fig. 2.5.3, the $D^\pm$ and D^0 lifetimes are separated by a factor of 2.53. The Ξ_c^+ charmed baryon and D^0 charmed meson lifetimes are roughly equal (this relationship is reinforced in Fig. 2.5.4). The Ξ_c^+ and Ξ_c^0 lifetimes are separated by a factor of 3.95 (see Table 2.5.1), and the Λ_c^+ lifetime is located almost at their geometric mean, thus creating two more factor-of-2 spacings (these Ξ-Λ-Ξ spacings are analogous to the K-π-K factor-of-2 spacings shown in Fig. 2.5.1). The Λ_c^+ and Ω_c^0 lifetime spacing is 2.9, which compares to the Λ^0 and Ω^- spacing of 3.2. Hence we see the same general HF systematics at the $x_i \cong 2$ α-grid level that we observed at the $x_i \cong 0$ and $x_i \cong 1$ levels. It is important to note that these three α-grid levels contain all of the s, c and b unpaired-quark particle states. There are no "rogue" particles that do not fit in with this α-quantized and HF-quantized lifetime systematics.

The HF systematics of the charmed mesons and baryons is clarified in Fig. 2.5.4, where we add in the $D_s^\pm$ and $B_c^\pm$ charmed meson states. The $B_c^\pm$ meson actually contains both c and b quark and antiquark states, but as we demonstrate in Sec. 2.10, the c quark is dominant from the standpoint of lifetime systematics (whereas the b quark dominates the mass systematics). The D^0, $D_s^\pm$, $B_c^\pm$ and Ξ_c^+ charmed meson states all have essentially the

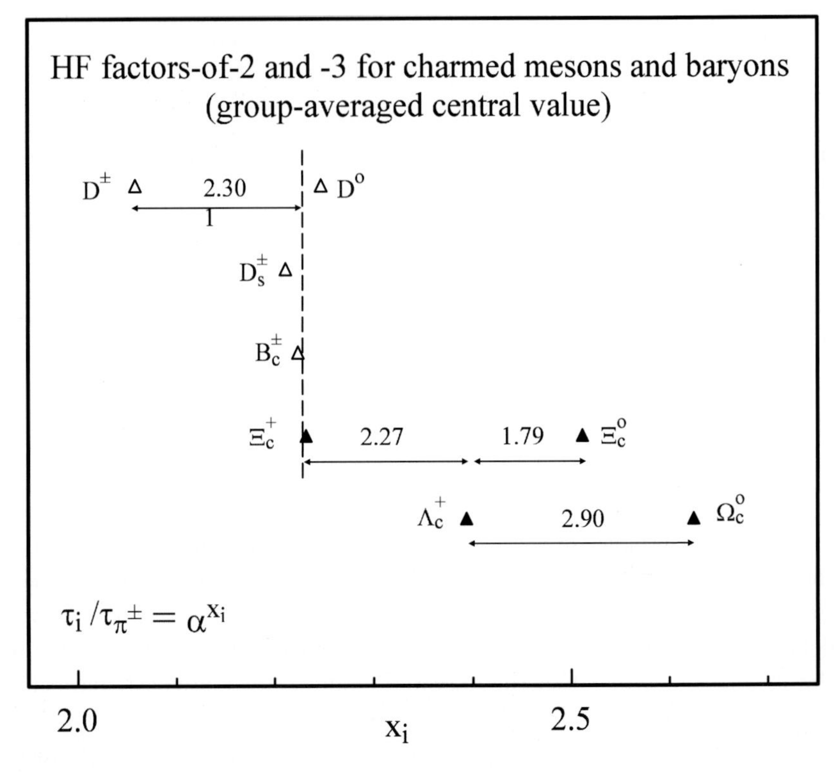

Fig. 2.5.4 These are the same charmed lifetimes shown in Fig. 2.5.3, but with the strange $D_s^{\pm}$ and bottom $B_c^{\pm}$ lifetimes added in, and with a four-particle-average used to define the "central" lifetime of the group (the vertical dotted line at $x_i = 2.229$). The $D^0 = c\bar{u}$, $D_s^+ = c\bar{s}$, $B_c^+ = c\bar{b}$ and $\Xi_c^+ = cud$ charmed particles that form this "central group" have in common only the fact that they each contain an unpaired c quark, which clearly dominates the decay rates for all of them.

same lifetime, and the vertical dotted line in Fig. 2.5.4 represents their average value. This line represents the "central lifetime" of the $x_i \cong 2$ HF lifetime multiplet. The $\pi^{\pm}$ lifetime is the "central" value for the $x_i \cong 0$ multiplet, and the Ξ^- and Σ^- lifetimes represent the "central" value for the $x_i \cong 1$ multiplet. These are all in some sense "two-trigger" excitations. The $D^{\pm}$ is the "one-trigger" excitation for this HF multiplet, and the Λ_c^+, Ξ_c^0 and Ω_c^0 are higher-trigger states for these more complex structures.

Figure 2.5.5 shows the lifetimes for all of the particles displayed in Figs. 2.5.1–2.5.4, plotted on the lifetime α-grid with no HF corrections applied. Figure 2.5.6 displays these same lifetimes after factor-of-2 and

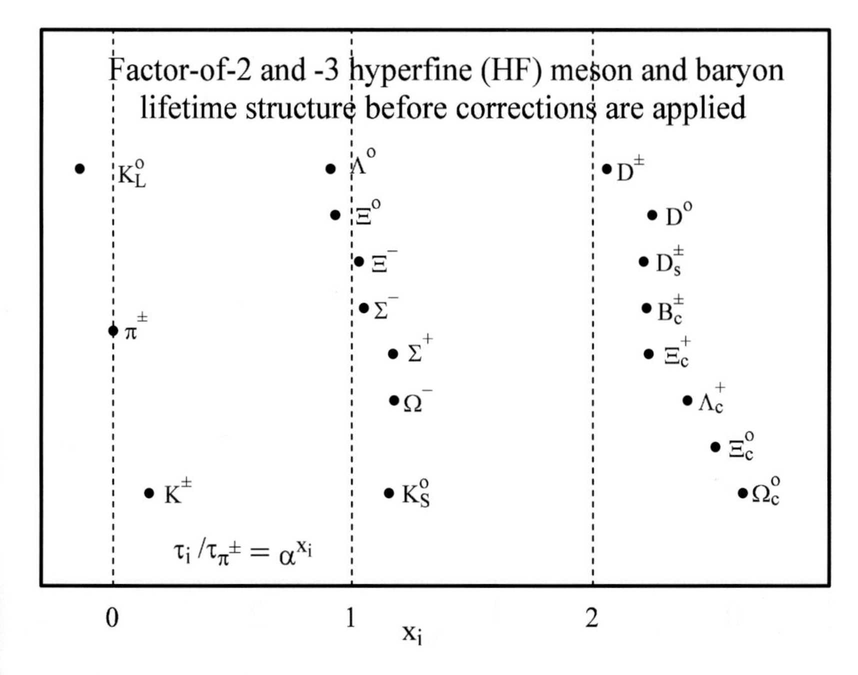

Fig. 2.5.5 A plot of the uncorrected lifetimes for all of the threshold-state particles whose lifetimes exhibit a factor-of-2 or factor-of-3 HF hyperfine structure. All of these particles feature unpaired-quark decays, and they include all of the particles in this region of the global α-spaced lifetime grid: there are no "rogue" particles in evidence. Since these 18 particles are half of the total of 36 threshold-state particles (particles with lifetimes $\tau > 1$ zs), it can be seen that the HF structure is a pervasive phenomenon. It is our clearest phenomenological evidence for a common mass substructure in all of these quark states.

factor-of-3 HF "corrections" have been applied to move all of the particles in a HF multiplet into the "central" position in each lifetime α-level, using integer values of 2 and 3 for the correction factors. The HF corrections that are required to accomplish this are shown in parentheses by each particle. These phenomenological corrections clarify the delineation of the α-spaced groups. Finally, in Fig. 2.5.7 we add in the unpaired b-quark states, which all fall accurately on the $x_i \cong 2$ α-grid line. Figure 2.5.7 includes all of the hadronic elementary particles in this lifetime range.

The discussion in Sec. 2.5 has centered on the experimental evidence for a pervasive lifetime HF structure that is superimposed on the overall lifetime scaling in powers of α. In addition to displaying these factor-

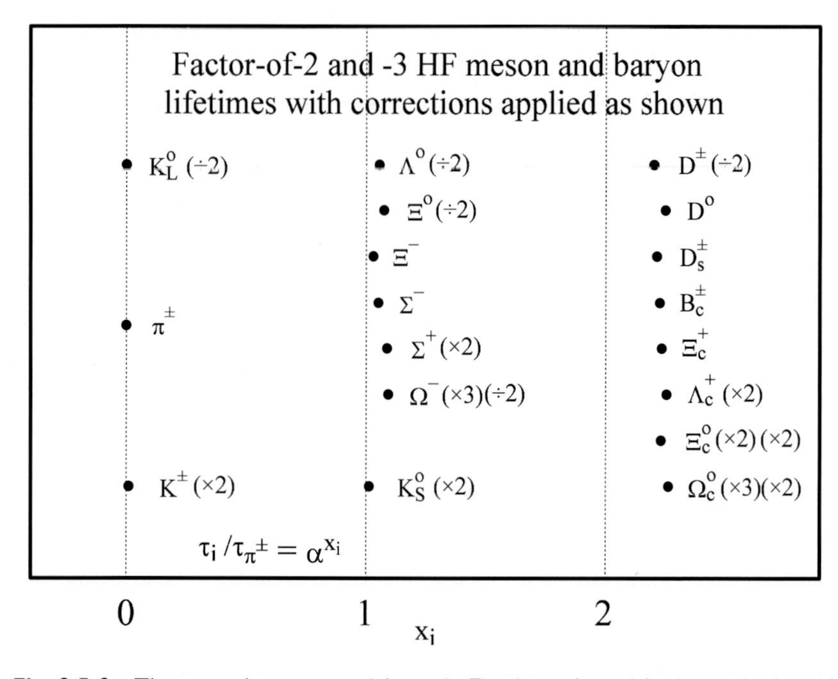

Fig. 2.5.6 These are the same particles as in Fig. 2.5.5, but with phenomenological factor-of-2 and factor-of-3 lifetime "corrections" applied as indicated in parentheses to move all of the particles in each α-grid group into its "central region" (see Fig. 2.5.4 for the c-quark central region).

of-2 and factor-of-3 HF multiplets, we also showed how the lifetime data appear after phenomenological corrections have been arbitrarily applied to remove the HF effect. The question that naturally arises is whether this HF correction process "improves" the quality of the lifetime data set, or whether it actually skews it in one direction or another. Fortunately, if we make the *ansatz* that the lifetimes of the long-lived ($\tau > 1$ zs) threshold-state particles occur on an α-spaced grid, we can directly test the effect of the HF corrections. To accomplish this, we make up two corresponding sets of lifetime data, one with HF corrections and one without, and then we apply the linear ADI(S) minimization analysis and the quadratic $\chi^2(S)$ minimization analysis that were described in Sec. 2.2 to the two data sets in question, where S is the scaling factor for the observed lifetime grouping. If the value of $S_{\min}$ gets closer to the value $\alpha^{-1} \cong 137$ with HF corrections than without them, then within the context of our ansatz the corrections represent an improvement. This is the topic of the next section.

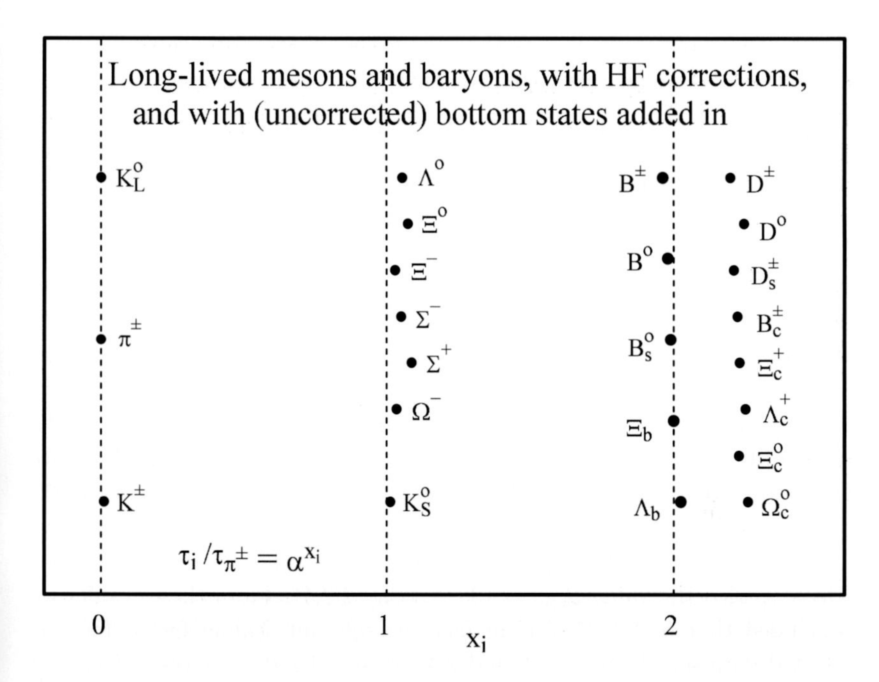

Fig. 2.5.7 This is the same plot as Fig. 2.5.6, but with the uncorrected bottom *b*-quark meson and baryon lifetimes added in. As can be seen, these *b*-quark lifetimes are all approximately equal, and they fall right on the $x_i = 2$ α-spaced lifetime grid, which was established long before the discovery of the *b*-quark states. Once the HF structure has been removed (corrected-for) in the *c*-quark states, the clear-cut factor-of-3 separation between the *b*-quark and *c*-quark lifetimes becomes apparent (Sec. 2.9). This factor-of-3 lifetime separation echoes a factor-of-3 separation in their mass values (Fig. 2.9.2).

2.6 The α-Quantization of the 36 Long-Lived Threshold-State Particle Lifetimes

The goal in Chapter 2, which is a key objective in this book, is to evaluate the experimental evidence that points to the fine structure constant $\alpha = e^2/\hbar \cong 1/137$ as a relevant scaling factor for threshold-state elementary particle lifetimes. If this viewpoint can be established, then the lifetime α-dependence suggests a reciprocal α-dependence in the masses of these particles. Whether or not this mass α-dependence actually occurs is a question that can be answered by a direct examination of the mass systematics of these same threshold-state particles.

The first step towards this goal was taken in Sec. 2.1, where we identified the 157 elementary particles that have well-measured lifetimes τ (Appendix A). We demonstrated in Fig. 2.1.1 how these lifetimes divide into a continuum of 121 short-lived *excited states* and a widely-spaced set of 36 long-lived *threshold states*, with the lifetime value $\tau = 10^{-21}$ s $\equiv$ 1 zs serving as the boundary line between these two domains.

The second step was taken in Sec. 2.2, where we selected the low-mass nonstrange *PS quintet* mesons π^+, π^0, π^-, η and η' as a trial basis set, and plotted their lifetimes as (negative) logarithmic powers x_i to the base S, where S is an experimentally-determined lifetime scaling factor (Eq. (2.2.2) and Figs. 2.2.1 and 2.2.2). These PS quintet lifetimes span a lifetime range of more than 12 orders of magnitude, and they encompass most of the range of the long-lived 36 particle data set. We then applied an "Absolute Deviation from an Integer" (ADI) *linear* minimization analysis to the PS quintet, which yielded the value $S_{\min} = 136.09$ (Fig. 2.2.3). Finally we used the "chi-squared" sum in a $\chi^2(S)$ *quadratic* minimization analysis, which yielded the value $S_{\min} = 138.96$ (Fig. 2.2.4). From these results we concluded that $\alpha^{-1} \cong 137$ is in fact the relevant scaling factor for this selected data set. In Secs. 2.3 and 2.4 we considered some other facets of the spin-0 pseudoscalar meson lifetimes. The PS mesons are the lowest-mass and therefore most fundamental of all the hadronic particle states (and possibly the most misunderstood).

Before extending this lifetime scaling analysis to the full 36-particle data set, we made a detour in Sec. 2.5 to examine a factor-of-2 and factor-of-3 hyperfine (HF) lifetime structure that appears to be essentially independent of the postulated global lifetime α-scaling. This HF structure appears in the s and c flavored quark states, in pretty much the same form, in all three of the unpaired-quark lifetime groupings. When an α-spaced lifetime grid is used that is anchored on the $\pi^\pm$ lifetime (Eq. 2.2.5), these three unpaired-quark lifetime groups occupy the α-levels, $x_i \cong 0$, 1, and 2 (Figs. 2.5.1–2.5.4). We then applied empirical factor-of-2 and factor-of-3 "corrections" to remove the HF structure from the data sets (Figs. 2.5.5 and 2.5.6). The b-flavored unpaired-quark lifetimes do not exhibit HF structure (Fig. 2.5.7). One reason for selecting the PS quintet mesons as a trial basis set in Sec. 2.2 was that they do not exhibit HF structure.

After these preliminary studies, we are now in a position to examine the full 157-particle lifetime data set, using both the original uncorrected data from RPP2006 (Appendix A) and also the data with HF corrections applied (Fig. 2.5.6). The uncorrected lifetimes are displayed in Fig. 2.6.1, where

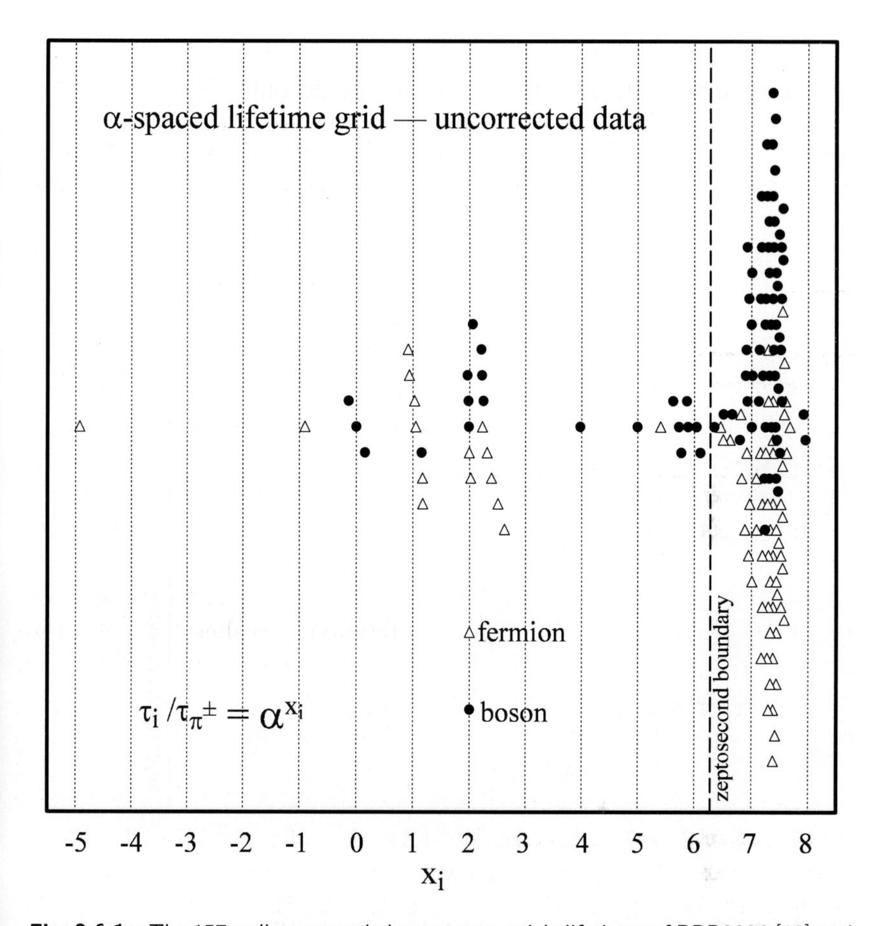

Fig. 2.6.1 The 157 well-measured elementary particle lifetimes of RPP2006 [10] and Appendix A, shown identified as half-integer-spin fermions or integer-spin bosons, and plotted on an α-spaced lifetime grid with $\pi^{\pm}$ as the reference lifetime. The boundary line at $\tau = 10^{-21}$ s (one zs) divides these lifetimes into 36 low-mass threshold states to the left of the boundary, which occur in widely-separated lifetime groups, and 121 higher-mass excited states to the right of the boundary, which occur with essentially a continuum of values.

they are divided into integral-spin bosons and half-integral-spin fermions, and are plotted on an α-spaced lifetime grid. As can be seen, bosons and fermions combine together in one overall lifetime pattern. The 121 lifetimes to the right of the zs boundary line at 10^{-21} s exhibit a continuum of values, at least on this global lifetime plot. By way of contrast, the 36 lifetimes to the left of the zs boundary line occur in discrete groups. Figure 2.6.2 is

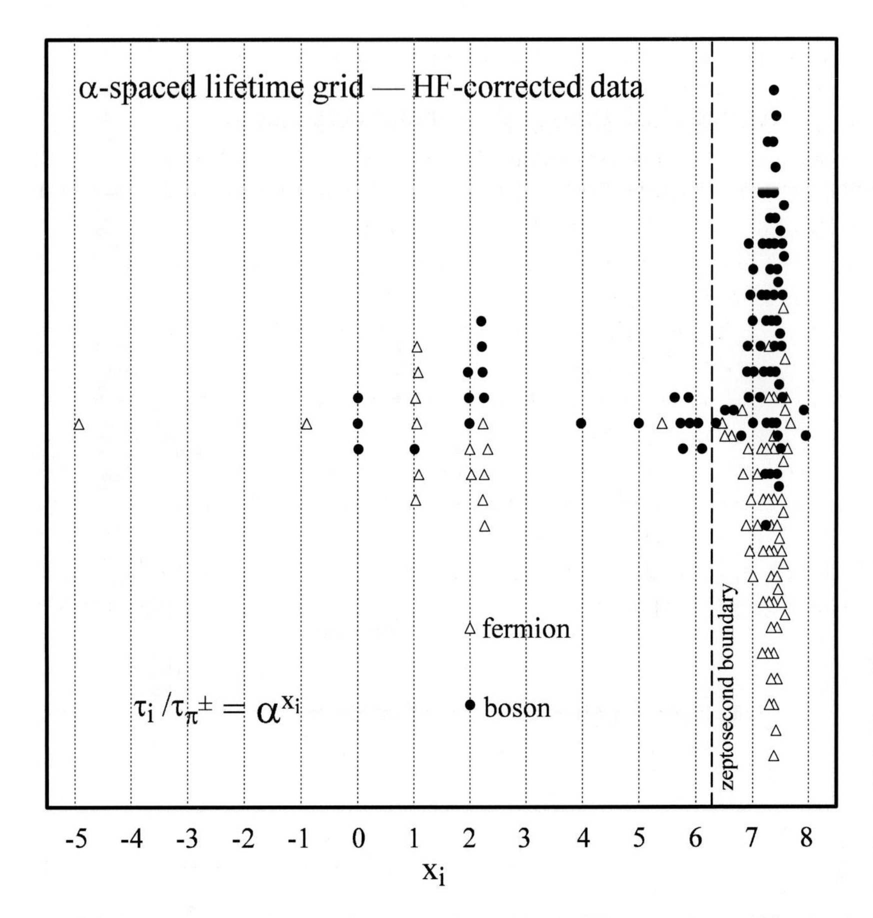

Fig. 2.6.2 The same figure as Fig. 2.6.1, but with the HF corrections of Fig. 2.5.6 applied. The α-quantization of the 36 threshold-state particles to the left of the zeptosecond boundary line is quite clear-cut, and it extends over 11 powers of α, or 23 orders of magnitude, with no "rogue" particle lifetimes in evidence.

the same plot as Fig. 2.6.1, but with HF corrections applied to unpaired-quark states on the $x_i \cong 0$, 1, and 2 grid lines. As can be seen, these HF corrections considerably sharpen the α-quantization.

We now carry out a series of ADI(S) minimization scans, using Eqs. (2.2.2) and (2.2.3) to delineate the lifetime exponents x_i and define the average ADI value for the data set. Calculations are made over a range of S-values. At an ADI minimum, ADI(S_{min}), the exponents x_i have near-integer values, showing that the lifetimes are grouped near the S_{min}-spaced grid lines. The maximum absolute deviation of x_i from an integer value is 0.5, and the minimum is 0, so the average ADI value for a random distribution of lifetimes is 0.25. For S to be a relevant scaling factor, ADI should be well below this value. As an example, the ADI(S) curve for the PS quintet, which is displayed in Fig. 2.2.3, has an average ADI value of 0.0225 at S_{min}. This indicates that the exponents x_i for the PS quintet lifetimes at S_{min} are very close to integers, which is verified in Fig. 2.2.5.

The first ADI(S) scan we carry out is for the full 157-particle lifetime data set, using both the uncorrected data of Fig. 2.6.1 and the HF-corrected data of Fig. 2.6.2. The ADI scan was taken over a range of S values from 10 to 200, and the average ADI values are shown in Fig. 2.6.3. As can be seen, the average ADI value for the oscillating curve is about 0.25, as expected for a random distribution of x_i values. This shows that the ADI sum is dominated by the 121 short-lived particles with lifetimes less than 1 zs. Any grouping effect in the 36 particles with lifetimes greater than 1 zs is swamped by the much greater number of ungrouped short lifetimes. The other thing to note in Fig. 2.6.3 is that the curves for the uncorrected

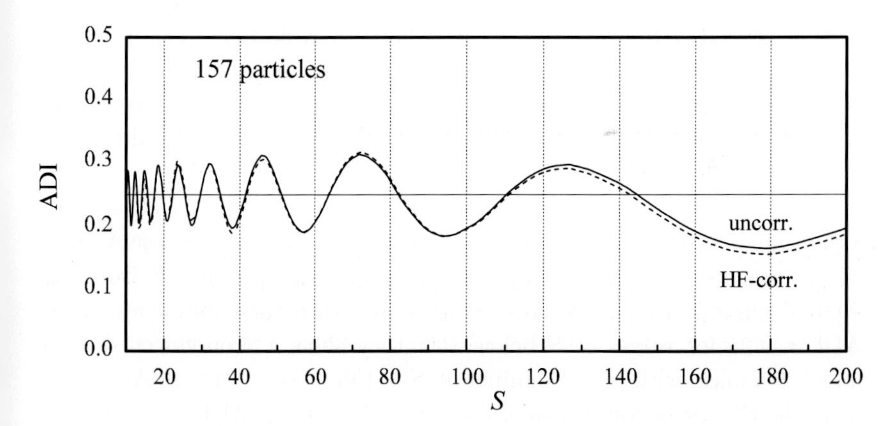

Fig. 2.6.3 An Absolute Deviation from an Integer (ADI) plot for the 157 particle lifetimes of Fig. 2.6.1 (uncorrected) and Fig. 2.6.2 (HF-corrected), calculated as a function of the scaling factor S^{-x_i} (Eq. (2.2.2)), and averaged over the individual ADI values of the exponents x_i (Eq. (2.2.3)). Since the maximum and minimum ADI values for x_i are 0.5 and 0, respectively, a random distribution of exponents will yield ADI $\cong$ 0.25. As can be seen in Fig. 2.6.3, the 157-particle ADI(S) value is in fact centered around ADI $=$ 0.25, for both the uncorrected and corrected data sets. This shows that the ADI average is dominated by the 121 short-lived excited states, whose lifetimes as a function of the scaling factor S are random.

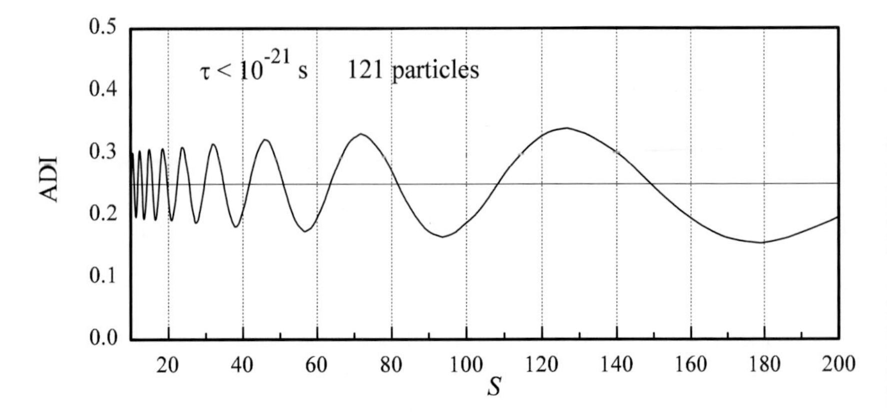

Fig. 2.6.4 The same plot as in Fig. 2.6.3, but just for the 121 short-lived excited-state lifetimes (which require no HF corrections). As can be seen, the average ADI(S) value oscillates around the value 0.25, which demonstrates that these short lifetimes are randomly distributed with respect to the scaling factor S.

data and the HF-corrected data are very similar to one another, so the HF corrections represent a small effect with respect to the entire data set.

The second ADI scan is for the 121 short-lived particles with lifetimes $\tau < 1$ zs. There are no HF corrections for this data set. The ADI(S) curve is displayed in Fig. 2.6.4. As can be seen, the oscillating curve is accurately centered around 0.25, thus indicating that the data are truly random, and also that the ADI sum is being properly calculated.

The third ADI scan is for the 36 long-lived particles with lifetimes $\tau >$ 1 zs, using both the uncorrected and HF-corrected data sets. These ADI(S) curves are displayed in Fig. 2.6.5, and they are quite different from the ADI(S) curve in Fig. 2.6.4. After starting out with variations centered on ADI $= 0.25$ for values of S below 100, they show a pronounced dip for larger S values, with broad minima at $S \sim 135$ in both curves. As can be seen, the HF corrections considerably lower the average ADI value near the dip, but they do not strongly shift the position of the dip.

In order to evaluate the ADI(S) scan for the 36-particle set in more detail, we make fine scans in the region of the minima. Figure 2.6.6 shows the scan for the uncorrected data. This has a minimum at $S = 133.48$, and an average ADI value of 0.1536 at the minimum. The discontinuities in the slope of the ADI curve are not due to coarse zoning for S, but rather to the sparseness of the data set and the fact that ADI values for the individual lifetime data points have discontinuous slopes as integer values

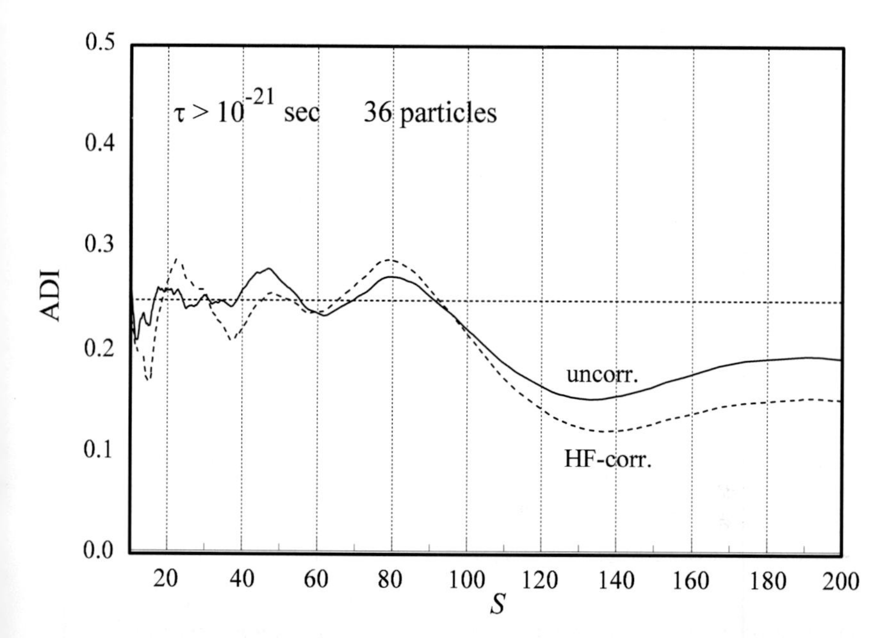

Fig. 2.6.5 The same plot as in Figs. 2.6.3 and 2.6.4, but just for the 36 long-lived thresold-state lifetimes, which occur in widely-separated groups. As can be seen, the ADI(S) value is random for $S < 100$, but then decreases for $S > 100$, with both the uncorrected and corrected lifetime data sets exhibiting flat minima at $S \sim 135$. The HF corrections reduce the ADI value, which reflects the more compact nature of the lifetime groups for the corrected data, but they do not appear to have a major effect on the location of the ADI minimum. The ADI minima are displayed in more detail in Figs. 2.6.6 and 2.6.7.

for x_i are crossed in the scan. Figure 2.6.7 shows the corresponding data plot for the HF-corrected data. The minimum of the ADI curve is now at $S = 136.09$, which is the same minimum as that for the PS quintet mesons of Fig. 2.2.3, and probably reflects the influence of the same data points. The average ADI value at the minimum for this 36-particle data set is 0.1231, whereas the average ADI value at the same minimum for the PS quintet was 0.0225, reflecting the greater accuracy of the scaling for the low-mass PS mesons. If we assume that the "correct" scaling factor for these lifetimes is the inverse of the fine structure constant, $\alpha^{-1} \cong 137.04$, then the HF corrections actually represent a slight improvement in the ADI scaling analysis. The important conclusion we would like to draw from these analyses is not the precise numerical values obtained from the ADI curves, but rather the fact that the fine structure constant α, or its

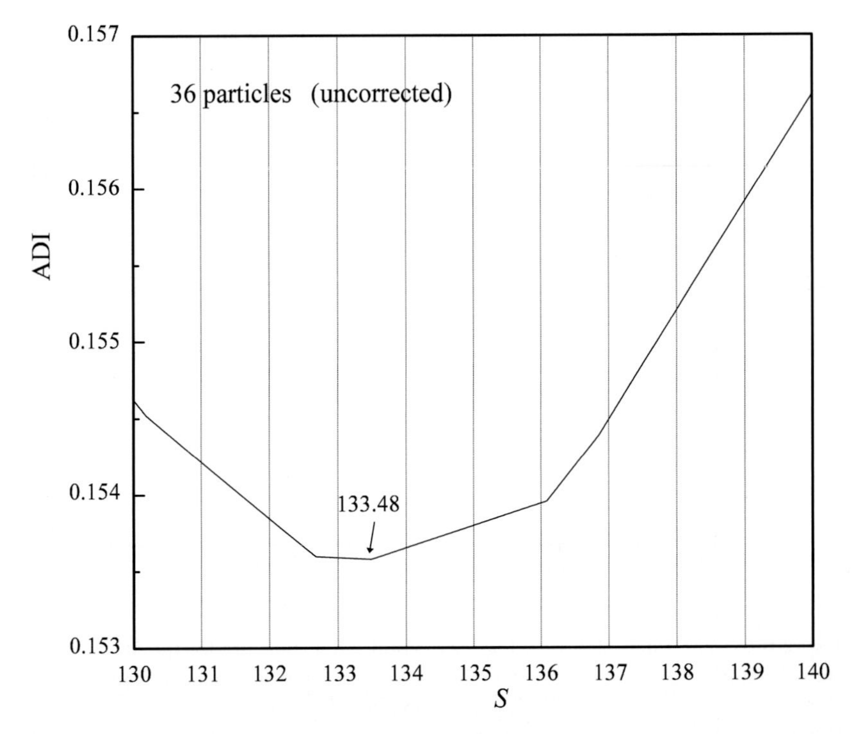

Fig. 2.6.6 A fine-scan plot of the ADI(S) curve for the uncorrected lifetimes of the 36 particles contained in Fig. 2.6.5. The ADI minimum is at $S = 133.48$, which is quite close to the value $\alpha^{-1} = 137.04$ we expect if the fine structure constant α is dictating these lifetime groupings. The discontinuous nature of the ADI curve is not due to insufficient accuracy in the computer calculations, but comes from the small number of data points and the discontinuity in the slope of the absolute value of the lifetime exponent x_i as an integer value is crossed.

inverse α^{-1}, really is the parameter that describes the observed groupings of these threshold-state lifetimes.

In the linear ADI scans, all of the lifetime data points were considered on the same footing. An independent way of carrying out this analysis is to form a quadratic $\chi^2(S)$ fit to the data, where each data point is weighted by the experimental error for that data point (Eq. (2.2.4)). This process only works well for data sets in which all points have roughly the same errors. Otherwise a few very-accurately-measured data points will dominate the χ^2 sum. In Sec. 2.3 we used $\pi^{\pm}$ as the reference lifetime and carried out a $\chi^2(S)$ fit to the π^0, η and η' PS mesons, which require no data corrections. Here we extend this three-particle fit to ten particles by adding in the Σ^-,

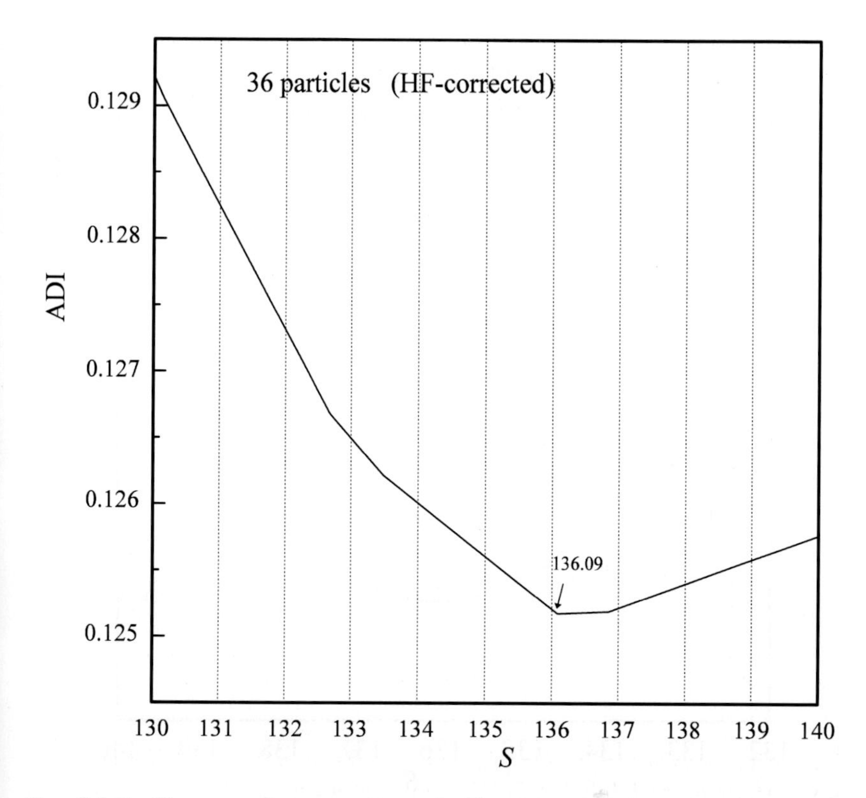

Fig. 2.6.7 The same fine-scan plot as in Fig. 2.6.6, but for the 36 lifetimes of Fig. 2.6.5 with HF corrections applied. In addition to lowering the value of ADI at the minimum, the HF corrections shift the value of the scaling factor S at the minimum from 133.48 to 136.09. If we assume that the "correct" theoretical value for the lifetime scaling factor is $\alpha^{-1} = 137.04$, then the hyperfine corrections, which logically are independent of the α-scaling, represent a phenomenological improvement in the scaling analysis.

Ξ^-, $B^\pm$, B^0, B_s^0, Ξ_b, and Λ_b unpaired-quark resonances, which also require no data corrections. The $\chi^2(S)$ curve for this ten-particle data set is shown in Fig. 2.6.8. As can be seen, the χ^2 minimum is at $S = 136.06$. This is lower than the χ^2 minimum at $S = 138.96$ in Fig. 2.2.4, and is closer to the α^{-1} scaling factor of 137.04.

For the final $\chi^2(S)$ fit, we include the rest of the unpaired-quark hadron that fall in the $x_i \cong 1$ and $x_i \cong 2$ α-grid lifetime groups, all of which require HF corrections. In addition to these HF corrections, we must compensate for the fact that the c-quark excitations are systematically a factor-of-3

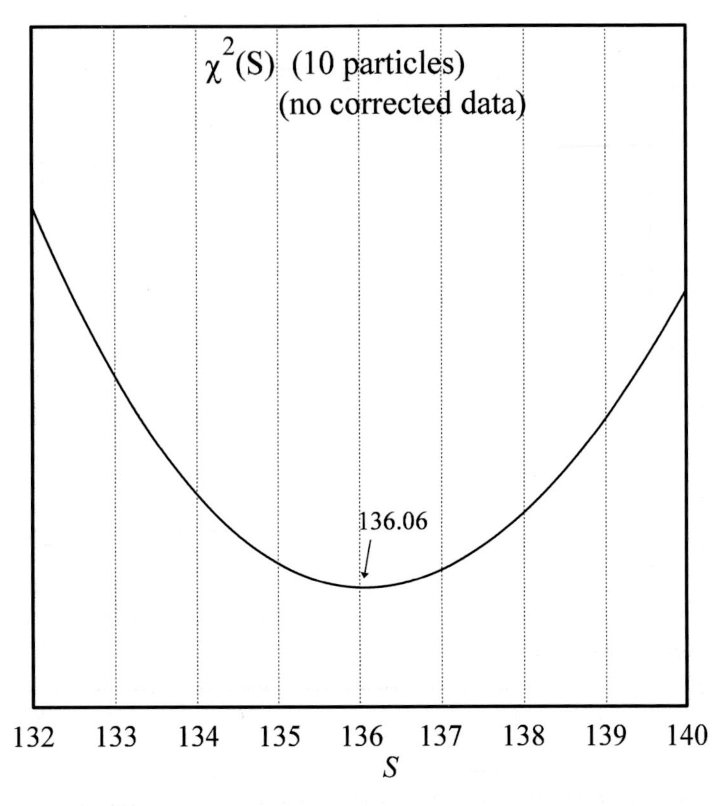

Fig. 2.6.8 A $\chi^2(S)$ analysis of the lifetime scaling, using theoretical fits to the experimental data as weighted by the uncertainties in the data (Eq. (2.2.4)). The ten particles used in the χ^2 fit are the π^0, η and η' uncorrected PS mesons from the $\chi^2(S)$ analysis displayed in Fig. 2.2.4 together with the Σ^-, Ξ^-, $B^\pm$, B^0, B_s^0, Ξ_b, Λ_b unpaired-quark threshold-state particles that require no data corrections (Figs. 2.5.6 and 2.5.7). The $\pi^\pm$ reference lifetime does not enter directly into this χ^2 sum.

shorter in lifetimes than the nearby b-quark excitations (as we demonstrate in detail in Sec. 2.9). The b-quark lifetimes, which require no HF corrections, all lie right in the $x_i \cong 2$ lifetime band (Fig. 2.5.7). Thus the c quarks require a factor-of-3 lifetime "flavor" correction in order to make their χ^2 sums comparable to those of the other data points. Starting with the 10 lifetimes of Fig. 2.6.8, we now add in the K_s^-, Λ^0, Ξ^0, Σ^+ and Ω^- lifetimes (which require HF corrections), and the eight charmed meson and baryon lifetimes near $x_i \cong 2$ (which require both HF and flavor corrections, as shown in Figs. 2.5.6 and 2.5.7). The $\chi^2(S)$ curve for this 23-particle data set is displayed in Fig. 2.6.9. It has a χ^2 minimum at 146.0.

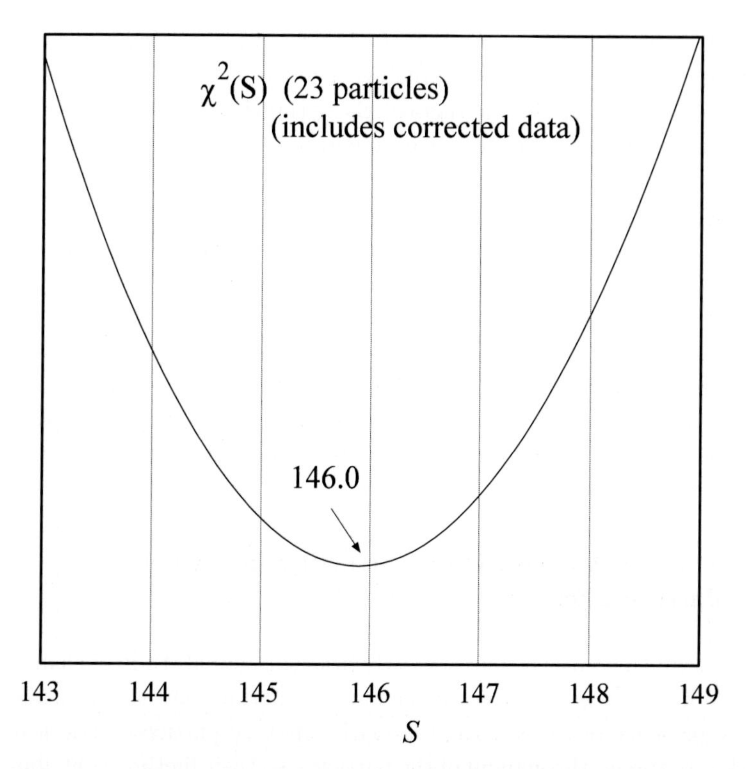

Fig. 2.6.9 The $\chi^2(S)$ analysis of Fig. 2.6.8 extended here to include the thirteen unpaired-quark hadrons on the $x_i \cong 1$ and 2 α-grid levels that require HF and "flavor" corrections. The five particles on the $x_i \cong 1$ α-grid (K_s^0, Λ^0, Ξ^0, Σ^0, Ω^0) require just HF corrections, and the eight charmed mesons and baryons near $x_i \cong 2$ (Fig. 2.5.7) require HF and c–b flavor corrections. The $\chi^2(S)$ minimum shifts from 138.96 to 136.06 to 146.0 as we go from 3 to 10 to 23 particle lifetimes in the χ^2 sum, which mainly reflects the fact that the c–b lifetime flavor correction should actually be 3.21 (Fig. 0.1.12) rather than the value 3 we used here. However, all of these particles remain within the general framework of the global lifetime α-grid.

When we carry out the various chi-squared analyses [34] for these threshold states, the position of the minimum in χ^2 decreases from $S_{\min} = 138.96$ for the three PS meson lifetimes of Fig. 2.2.4 to $S_{\min} = 136.06$ for the ten threshold-state lifetimes of Fig. 2.6.8, and then increases to $S_{\min} = 146.0$ for the twenty three threshold-state lifetimes of Fig. 2.6.9. This increase reflects the facts that (1) the HF-corrected lifetimes near $x_i \cong 1$ have slightly larger x_i values, and (2) the required c–b quark "flavor" correction is actually a little larger that a factor of 3 (Fig. 2.9.1). Thus the phenomenological lifetime "corrections" that we have used in the present work are not exact. But the essential point we are bringing out here is that, as observed on a global lifetime scale, the lifetimes of these threshold-state particles really do fall into lifetime groups which are separated from one another by powers of the fine structure constant $\alpha \cong 1/137$.

The paired-quark J/ψ and Υ particle excitations near $x_i \cong 6$ are slightly less that a factor of α^4 shorter in lifetimes than their unpaired counterparts near $x_i \cong 2$, as is demonstrated in Figs. 0.1.12 and 2.6.1. This α^4 relationship between unpaired-quark and paired-quark lifetimes is a very general result, as we display in detail in Sec. 2.8. But before discussing these α^4 quark relationships, we use Sec. 2.7 to present another remarkable feature of the unpaired-quark excitations near $x_i \cong 0$, 1 and 2: namely, the way in which these long-lived threshold-states are sorted out according to their quark flavor content.

2.7 The *s*, *c*, *b* Quark Group Structure in α-Quantized Particle Lifetimes

One fact which seems to emerge clearly from this global study of elementary particle lifetimes is that the threshold-state lifetimes are strongly affected by the presence of the s, c, and b quarks in these particles. The u and d quarks are spread through all of the particles, so their lifetime contributions are not so readily apparent, and the mass of the t quark is so large that it does not appear in these low-mass particles. The s, c, and b quarks in a sense represent "higher excitations" of the u and d quarks (as observed for example in nucleon excitations), and their inherent instabilities logically tend to dominate the lifetimes of the particles that contain them.

We can turn the argument of the preceding paragraph around and assert that by studying the lifetime systematics of the threshold-state particles, we can learn about their quark content. It was in fact the high mass and long lifetime (narrow width) of the J/ψ meson that signaled the appearance of the c-quark. This scenario was later repeated for the b-quark with the discovery of the Υ meson at three times the mass of the J/ψ. The flavor structure of the unpaired-quark states in the lifetime α-grid levels $x_i \cong 0$, 1 and 2 was already apparent in Fig. 2.5.7. In the present section we bring out this quark structure more clearly.

Figure 2.7.1 is a plot of the HF-corrected lifetimes of the unpaired-quark threshold states, and it shows the dominant quark content of each state. The $\pi^\pm$ meson, the only particle in the figure composed of just u

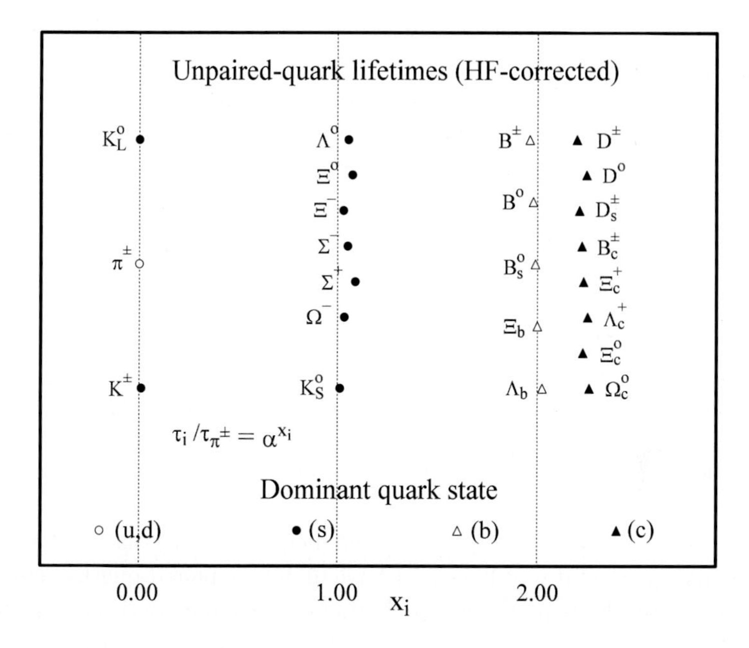

Fig. 2.7.1 A plot of all of the unpaired-quark hadronic elementary particle lifetimes with the HF corrections of Fig. 2.5.6 applied, and with the particles identified as to their dominant quark content with respect to decay rates. (See Fig. 0.1.10.) The universality and accuracy of the data points on this plot is one of the strongest phenomenological pieces of evidence we can offer for the α-dependence of these hadron lifetimes. This plot also demonstrates how a single quark flavor dominates each of the lifetime groups (Sec. 2.10).

and d quarks, anchors the α-spaced lifetime grid at $x_i = 0$. As can be seen, the HF-corrected s-quark excitations are grouped along the $x_i \cong 0$ and 1 vertical α-grid lines. The b-quark excitations, which do not require HF corrections, are compactly grouped along the $x_i \cong 2$ α-grid line. The c-quark excitations also form a compact lifetime group after the HF corrections shown in Fig. 2.5.6 have been applied, but this c-quark lifetime group is displaced toward shorter values by a factor of 3 with respect to the b-quark lifetime group, as displayed in Fig. 2.7.1, and hence does not lie squarely on the α-grid. The s and b quarks have $\frac{1}{3}e$ electric charges, and the particles that contain them are the ones which accurately follow the α-spacing of the lifetime grid. The c quarks have electric charges of $\frac{2}{3}e$, which may be a factor in this shift. Also, the factor-of-3 *lifetime* ratio between b quark and c quark states is echoed by a factor-of-3 ratio in their *mass* values, but the heavy b-quark states are more stable than the lighter c-quark states, which seems contrary to expectation. The striking results which emerge from Fig. 2.7.1 are (a) the comprehensive manner in which the s, c and b quarks dictate the lifetimes of the particles that contain them, and (b) the fact that these flavored-quark lifetimes exhibit an accurate α-spacing on the global lifetime grid.

An interesting feature of the lifetime systematics displayed in Fig. 2.7.1 is the situation that occurs when an excitation contains two different flavors of quarks. Two examples are shown in this data set. The B_s^0 meson contains both a b-flavored quark and an s-flavored quark, and its lifetime is clearly dominated by the b quark. The $B_c^{\pm}$ meson contains a b-flavored quark and a c-flavored quark, and its lifetime is clearly dominated by the c quark. Thus we have a $c > b > s$ priority sequence established with respect to the lifetimes of these states. This subject is discussed in more detail in Sec. 2.10.

A close inspection of the α-quantized lifetimes reveals that the long-lived unpaired-quark states of Fig. 2.7.1 have matching paired-quark excitations and radiative decay modes whose lifetimes are characteristically a factor of α^4 shorter. These relationships are studied in detail in the next section.

2.8 Factor of α^4 Lifetime Ratios between Unpaired and Paired Quark Decays

The particle lifetime groups displayed in Fig. 2.7.1 illustrate the manner in which the lifetimes of these particles are dictated by the flavors of the quarks they contain — s, c or b. Another factor that enters into these lifetimes is the question as to whether the decay involves the annihilation or transformation of a single unpaired quark, or whether a matching pair of particle–antiparticle quarks simultaneously decay. In the latter case the lifetime is shorter by about four powers of α, or eight orders of magnitude. The region between these two types of decay is a *lifetime desert*, in which no particle lifetimes are to be found. In the present section we trace this relationship through the various sets of particles. The universality of the effect is impressive.

Figure 2.8.1 displays the accurate α^4 lifetime ratio between the π^0 and $\pi^{\pm}$ PS meson lifetimes. The mass systematics developed in Chapters 3 and 4 shows that the 140 MeV pions are composed of two spinless 70 MeV mass quanta, which are matching particle and antiparticle substates. The

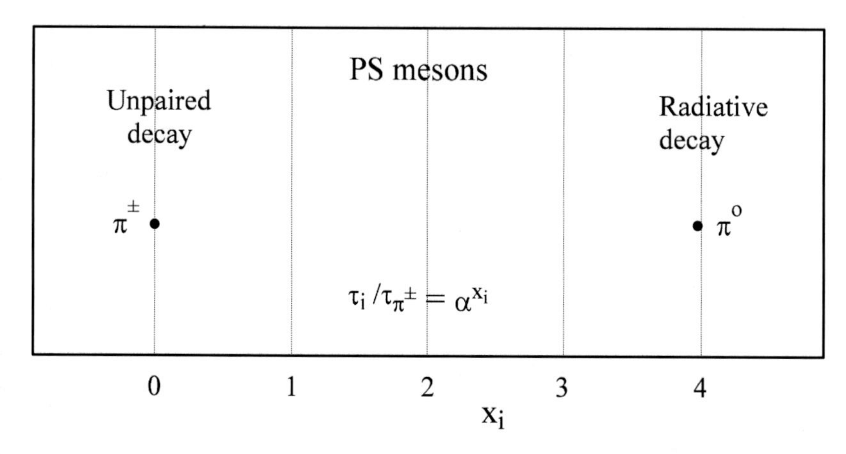

Fig. 2.8.1 A factor of α^4 spacing between the unpaired-quark lifetime of the $\pi^\pm$ and the radiative decay of the π^0. The $\pi^\pm$ meson is composed of matching spinless 70 MeV particle and antiparticle substates. In the decay process, one of these substates is annihilated and the other is set into full relativistic rotation with a calculated spin of $\frac{1}{2}\hbar$ (Sec. 4.3). It also acquires the charge of the $\pi^\pm$, and is converted into a weakly interacting muon. The π^0 radiative decay involves the annihilation of both 70 MeV substates, and it proceeds a factor of α^4 faster than the unmatched quark decay of the $\pi^\pm$.

π^0 neutral meson decay is into two gamma rays, and it clearly involves the annihilation of quark–antiquark mass quanta. The $\pi^\pm$ decay is into a charged muon and a muon neutrino. In the decay process one 70 MeV mass quantum is annihilated, and the remaining one is set into rotation at the relativistic limit, where it has a calculated mass of 105 MeV and a spin angular momentum of 1/2 $\hbar$ (Sec. 4.3). A spin-1/2 neutrino is also produced. The electric charge of the two quarks states in the $\pi^\pm$ pion is transferred to the spin 1/2 final-state 105 MeV mass quantum, which then appears as the charged $\mu^\pm$ muon. From the lifetime ratios of 4:2:1 for the K_L^0, $\pi^\pm$, $K^\pm$ multiplet shown in Fig. 2.5.1, we conclude that the $\pi^\pm$ pion has two decay triggers, which suggests that either of the two mass quanta in the $\pi^\pm$ initial state can trigger the decay. It then follows that the π^0 also has two decay triggers, since its lifetime is accurately scaled by a factor of α^4 with respect to the lifetime of the $\pi^\pm$.

Figure 2.8.2 shows the lifetime α-plot for the Σ^+, Σ^0 and Σ^- mesons. The strange Σ^+ and Σ^- hyperons decay down to the nonstrange p and n nucleons, which is a strangeness-breaking process that leads to long lifetimes for these excitations. The strange Σ^0 hyperon decays down to the strange

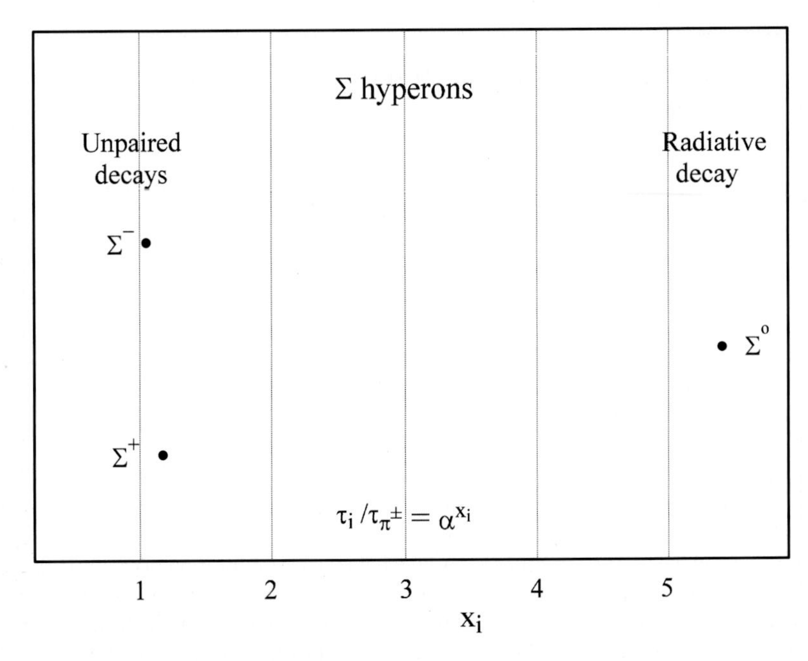

Fig. 2.8.2 A factor of α^4 spacing between the unpaired-quark lifetimes of the Σ^- and Σ^+ and the radiative decay of the Σ^0. The $\Sigma^\pm$ decays are to nucleons, which requires a violation of strangeness conservation and greatly slows the decay rate. The Σ^0 decay is to the Λ^0, which conserves strangeness and proceeds much faster. The lifetime ratio of slightly more than α^4 is similar to the factor of α^4 lifetime ratio observed in Fig. 2.8.1.

Λ hyperon, which is a process that does not break strangeness and requires no charge exchange. Hence its decay is very much faster than the Σ^+ and Σ^- hyperons decays, by a characteristic factor of somewhat more than α^4.

In Fig. 2.8.3 the lifetimes of the unpaired c-quark D mesons are compared with those of the paired-quark $c\bar{c}$ J/ψ_{1S} and J/ψ_{2S} mesons. The unpaired-quark and paired-quark lifetimes are again almost a factor of α^4 different. Of equal interest is the $D^{*\pm}$ lifetime shown in Fig. 2.8.3. This is the longest-lived of the higher-mass D excitations, and it falls right in with the J/ψ_{1S} and J/ψ_{2S} lifetimes. There are no D^* lifetimes in the *lifetime desert* between these two lifetime groups. Thus not only are these lifetime ratios functions of the fine structure constant α, but they are evidently constrained within tight quantization limits. There are no rogue stragglers.

The $B_c^\pm$ meson shown in Fig. 2.8.3 contains both a b-type quark and a c-type quark. Since it has the mass of a B-meson state, it is labeled by the

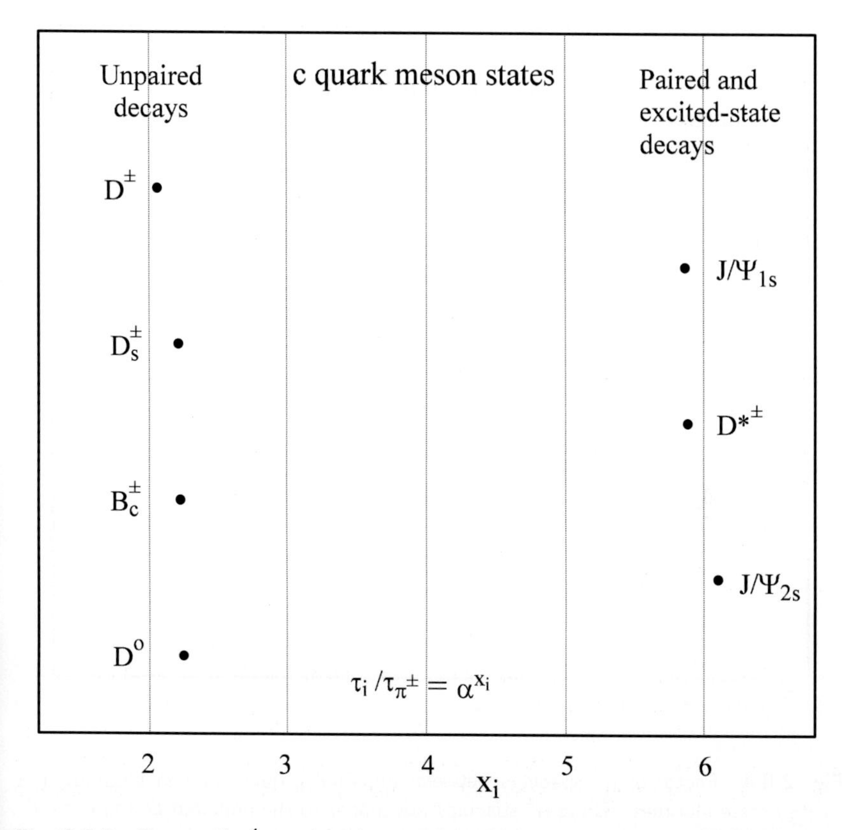

Fig. 2.8.3 Factor of α^4 spacings between the unpaired c-quark D meson lifetimes and the paired-quark $J/\psi \equiv c\bar{c}$ lifetimes. The unpaired $D^{*\pm}$ excited state also falls into this pattern, but is lined up with the paired lifetimes. There are no D-meson lifetimes in the intervening region.

Particle Data Group as a B_c meson. From the standpoint of lifetimes, we would logically label it as a D_b meson, since the c-quark dominates its decay. Again, this example illustrates how sharply these lifetimes are quantized by their quark content. One might suppose that the $B_c^{\pm}$ would have a lifetime that is intermediate between the B lifetimes (which fall essentially on the $x_i = 2$ α-grid line), and the D lifetimes (which fall a factor of 3 to the right of this line, as shown in Fig. 2.8.3). But it does not. It is exactly in line with the $D_s^{\pm}$ and D^0 lifetimes (all of which require no HF corrections). Without an α-quantized lifetime grid as the background for this discussion of global lifetimes, it would be difficult to quantify these phenomenological regularities in the data.

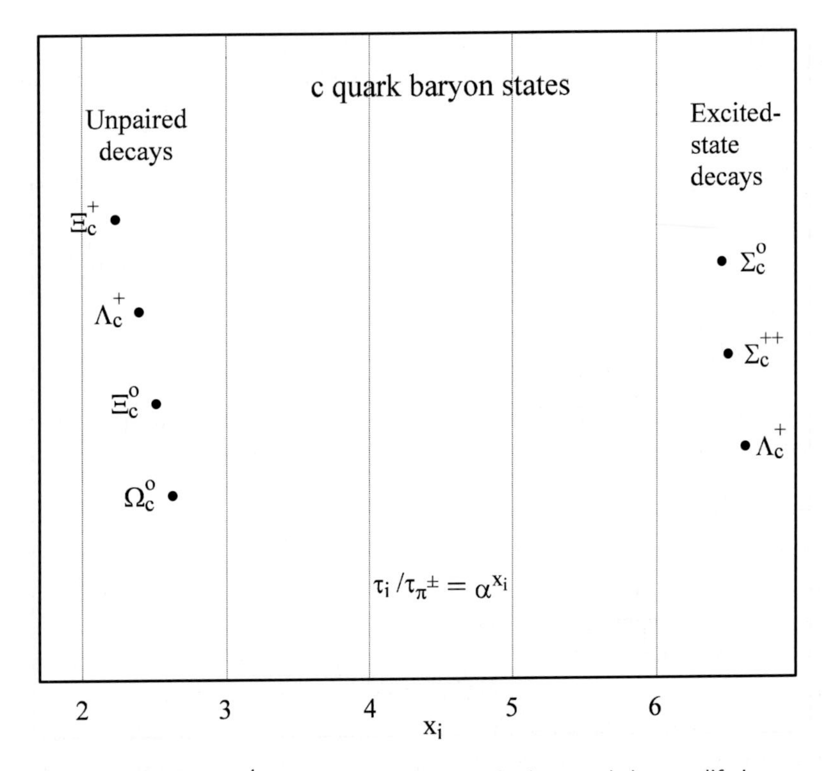

Fig. 2.8.4 Factor of α^4 spacings between unpaired *c*-quark baryon lifetimes and excited-state lifetimes. These α^4 spacings are similar to the unpaired *D* and excited-state D^* lifetime spacing shown in Fig. 2.8.3, and the existence of a *lifetime desert* between the unpaired-quark and excited-state groups is also similar.

In Fig. 2.8.3 we displayed lifetime systematics for unpaired and paired c-quark *meson* states. In Fig. 2.8.4 we extend this c-quark lifetime analysis to the spin-1/2 charmed *baryon* states. Unpaired s-quark hyperons are grouped along the $x_i \cong 1$ α-grid axis, as shown in Figs. 2.6.1, 2.6.2 and 2.7.1. The unpaired c-quark baryon states in Fig. 2.8.4 are grouped at $x_i > 2$, and are created by replacing a strange s-quark with a charmed c-quark, which increases the electric charge e by one unit. It also decreases the lifetime of the state (after HF corrections) by a factor of $\alpha/3$. The c-quark baryon excited states at $x_i > 6$ are the longest-lived of the higher charm excitations, and they are separated from the ground states by a factor of α^4, in agreement with the results shown in the previous figures in this section. In particular, they echo the $D^{*\pm}$ and D excited state and ground state lifetime spacing of α^4.

There is another interesting feature of the baryon excitations shown in Fig. 2.8.4, and it has to do with the basic way that baryons and hyperons are created. The nucleons (the proton and neutron) are each composed of three nonstrange u and d quarks. In the constituent-quark mass systematics displayed in Chapter 3, the u and d quarks have masses of one-third the nucleon mass, or about 315 MeV, and the s quark has a mass that is 210 MeV larger, or 525 MeV. Thus the excitation of a u or d quark into an s quark increases the mass of the particle by about 210 MeV. From this standpoint, the mass of the Λ hyperon indicates that it is a *uds* state with a binding energy of about 40 MeV. The Ξ hyperon, which contains two s-quarks, is expected to appear about 210 MeV above the Λ, which it does. Hence the Λ and Ξ hyperons have masses that characterize them as *basic* hyperon excitations. Extending this result to the triple s-quark Ω hyperon, we expect it to appear about 210 MeV above the Ξ, but it actually appears 140 MeV higher than this value. Similarly, the single s-quark Σ hyperon appears 70 MeV above the single s-quark Λ hyperon, as a sort of higher-mass excitation. The point of this discussion is that when we produce the basic charmed hyperon states shown at $x_i > 2$ in Fig. 2.8.4, we create long-lived Λ_c and Ξ_c states, but no long-lived Σ_c states. These show up only as higher-mass short-lived excitations that decay back down to the Λ_c ground state. This type of excitation hierarchy is not immediately apparent in the Standard Model eightfold-way baryon octet assignments that work so well in reproducing isotopic spins.

Figure 2.8.5 displays the unpaired b-quark and paired $b\bar{b}$ long-lived particle states. These require no HF corrections, and are accurate examples of the lifetime systematics already illustrated for the s and c flavored quarks. The two types of excitation are each tightly bunched, and the lifetime desert between them extends over a factor of almost α^4. Spin-0 mesons and spin-1/2 baryons occur together in the same unpaired-quark grouping at $x_i = 2$ on the lifetime α-grid, which illustrates the dominance of the b-quark itself in dictating the decays of these states. The lifetime factor-of-3 ratio between the unpaired b and c quark states is roughly approximated by a corresponding lifetime ratio between the paired $b\bar{b}$ and $c\bar{c}$ bound states (Fig. 0.1.11).

In Fig. 2.8.6 an unlikely pair of particles are grouped together as our final example of related excitations that exhibit a lifetime ratio of α^4. These two particles are the Compton-sized hadronic neutron and the point-like leptonic muon. The "stable" neutron (as measured in zeptoseconds) appears at $x_i \leq -5$ in the universal lifetime α-grid, and the "metastable"

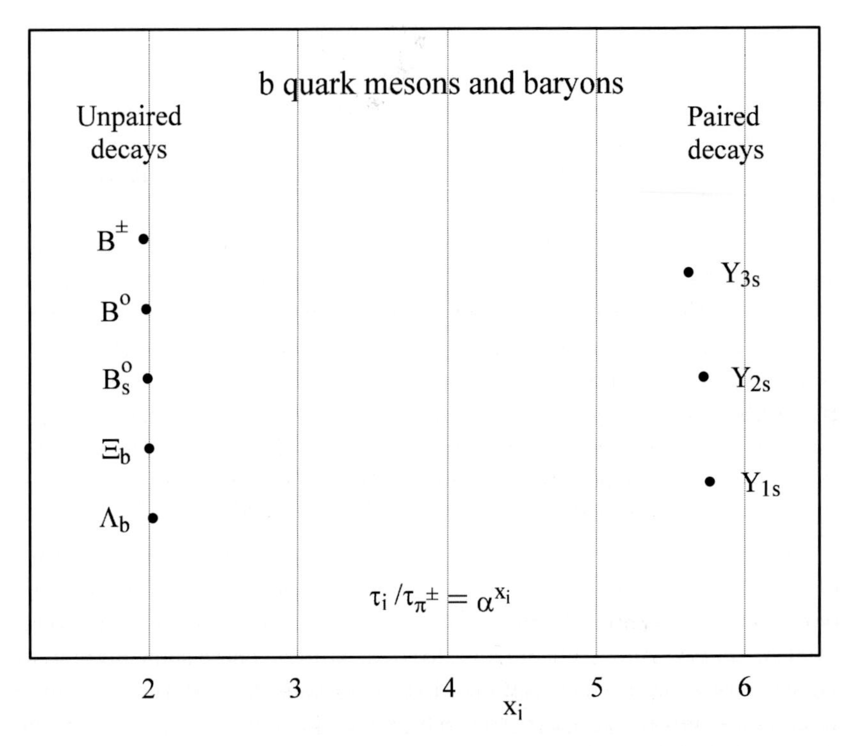

Fig. 2.8.5 Factor of α^4 spacings between unpaired and paired *b*-quark meson and baryon lifetimes. No HF corrections apply here. The *lifetime desert* that is observed in the other flavored-quark groupings is also seen here: no *b*-quark lifetimes occur within a range of lifetimes that spans roughly 7 orders of magnitude, and is bounded at both ends by well-defined lifetime groups.

muon (the lowest-mass and longest-lived massive particle after the electron) appears at $x_i \leq -1$. At first glance these particles seem to have very little in common except their spins of $1/2\ \hbar$. But their masses are in the ratio of 9 to 1, and this innocuous-looking mass ratio is in fact the *raison d'etre* for the present book. In the third week of June 1969, after spending a week poring over the available experimental elementary particle data, the present author concluded that the 105 MeV spin 1/2 muon mass is in fact a fundamental building block for elementary particles, and has been pursuing this point-of-view ever since (see Postscript). The attempt to reconcile the point-like muon with the finite-sized neutron led to the relativistically spinning sphere described in Chapter 4, and its characteristic 3/2 mass ratio between spinning and nonspinning masses led in turn to the 70 MeV

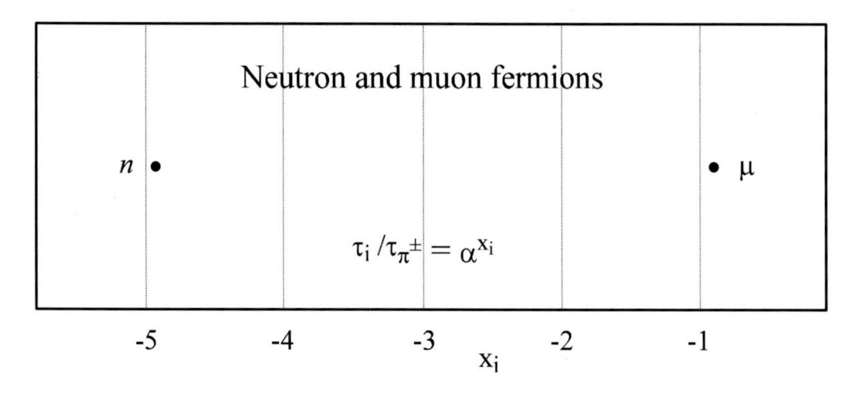

Fig. 2.8.6 A factor of α^4 spacing between the lifetimes of the neutron and muon. *A priori*, these hadronic and leptonic lifetimes would seem to be unrelated, and certainly not related by a ratio that encompasses 9 orders of magnitude. The relationship between these two very different particles comes from their mass values, as is discussed in Sec. 2.8 and developed in detail in Chapter 3.

spinless mass quantum defined in Chapter 3. Our interest in the present chapter is confined to the facts that (1) both the neutron and the muon (quite surprisingly) have lifetimes which fall reasonably accurately on the α-spaced lifetime grid, and (2) these particles have a lifetime ratio of α^4.

After discussing all of these particle sets separately, we combine them in Fig. 2.8.7 on a single α-spaced lifetime grid. This display illustrates the manner in which the α^4 *lifetime desert* between unpaired and paired decays is maintained through all of these lifetime groups, and the way the groups progress from one α-grid to another with changing flavors. The comprehensiveness of these results is as significant as the accuracy of the α-quantization. There seem to be no particles that are wildly outside of the delineated lifetime groupings. The span of lifetimes encompassed in the α-grid is also impressive. These α-dependent lifetimes terminate at the 1 zeptosecond (10^{-21} s) boundary, which is at $x_i = 6.28$ in Fig. 2.8.7. The neutron lifetime is 8.86×10^{23} zs (Table 2.1.1). Thus we have an α-spacing that is maintained over almost 24 orders of magnitude.

Figure 2.8.7 contains 26 of the 36 elementary particles that have measured lifetimes $\tau > 1$ zs. Adding in the other ten particles, and grouping them according to family and flavor type, we have the results displayed in Fig. 2.8.8. The main conclusion we would like to draw in the present section is that these *threshold-state particle lifetimes are universally α-dependent.* The experimental data have been revealing this α-dependence for many

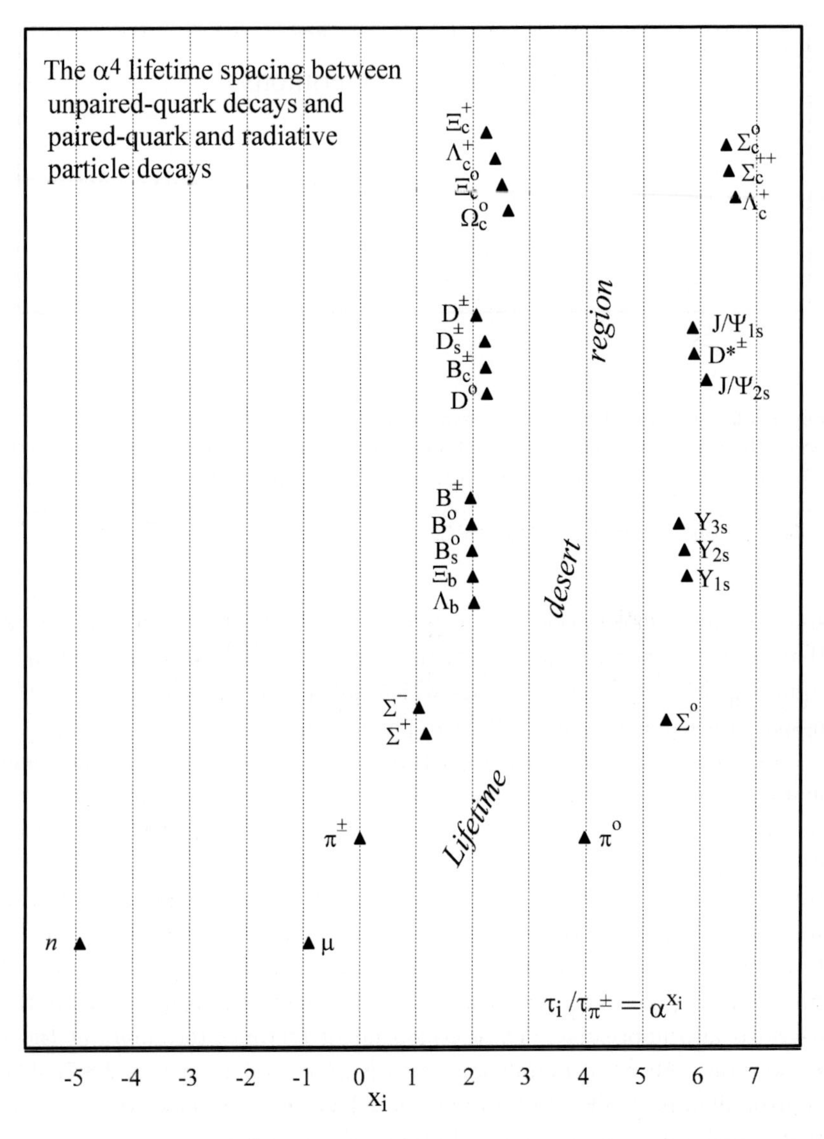

Fig. 2.8.7 Factor of α^4 spacings for all of the particles that were displayed in Figs. 2.8.1–2.8.6, which are shown here arrayed on a common α-spaced lifetime grid with $\pi^{\pm}$ as the reference lifetime. The *lifetime desert region* between the unpaired-quark decay rates and the paired-quark plus radiative decay rates is about eight orders of magnitude, with no "rogue intruders" in this vast span of lifetimes. The comprehensiveness of these results is as important as the span of lifetimes involved, and the non-existence of lifetimes in the lifetime desert region seems as significant as the existence of the lifetime groupings on the desert fringes.

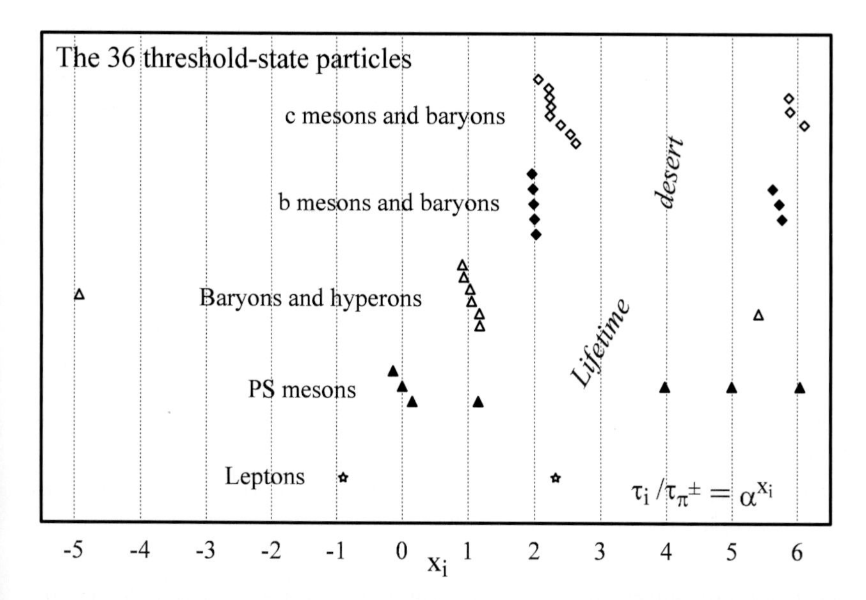

Fig. 2.8.8 The lifetime families of the 36 low-mass threshold-state particles, showing the barrenness of the lifetime desert that separates the unpaired-quark decay rates from the paired-quark and radiative decay rates. The main conclusion we would like to draw from this figure is that all 36 of these fundamental threshold-state particle lifetimes exhibit a dependence on the fine structure constant $\alpha \cong 1/137$, and this dependence cannot be sorted out without the guidance provided by the lifetime α-grid in which they are embedded.

years (see Sec. 2.11), and yet the global α-systematics displayed here has not made its way into the global database [10], and it has not yet come to the attention of SM theorists whose task it is to compare the theoretical SM predictions with the phenomenology of the experimental data.

The α-dependence of the threshold-state elementary particle *lifetimes* is interesting in its own right, but its real significance is in the information it provides with respect to the *mass structures* of these particles.

2.9 The *b*-Quark and *c*-Quark Factor-of-3 Lifetime Flavor Structure

One of the unexpected and interesting results to emerge from this global lifetime study is the clustering of the *b*-quark and *c*-quark unpaired excitations into two narrow and adjacent lifetime bands. This result is shown in

Fig. 2.9.1, where the five observed b-dominant states, which do not require HF corrections, are plotted together with the four observed c-dominant states that also do not require HF corrections. Thus HF corrections play no role in this systematics. The five b-quark states have an average x_i value of 1.99 in this standard α-grid plot, and the four c-quark states have an average x_i value of 2.23. The lifetime ratio of these two average values is 3.21. Thus the b-quark resonances have an average lifetime that is just over a factor of 3 longer than the c-quark resonances. This factor-of-3 difference between the *unpaired* b quark and c quark lifetimes is also observed in the *paired* $b\bar{b} = \Upsilon$ and $c\bar{c} = J/\psi$ lifetimes, as displayed in Figs. 0.1.11 and 2.11.10, but the slopes of the lifetime groups make it difficult to evaluate "average lifetimes" for each of these groups, and thus determine an average lifetime ratio for paired excitations.

In order to ascertain how this b-quark to c-quark lifetime difference carries over to the masses of these states, we use an α-grid *lifetime* plot along the *abscissa* and a linear *mass* plot along the *ordinate*, which gives the results displayed in Fig. 2.9.2. The open circles and triangles denote

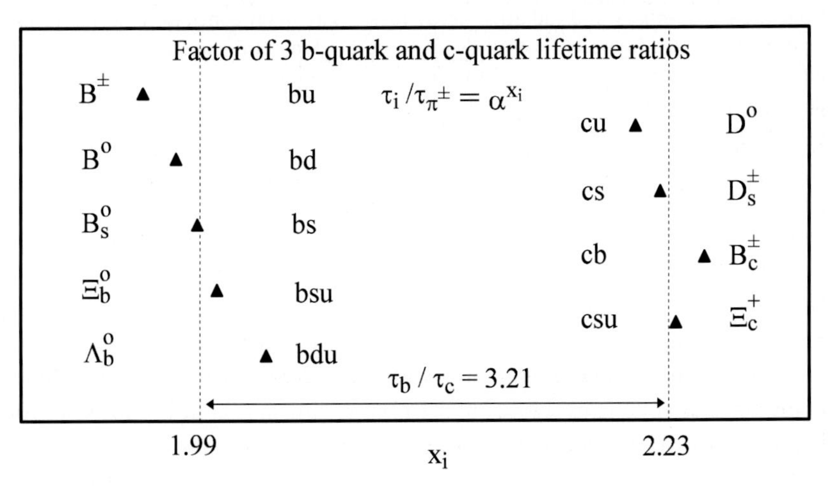

Fig. 2.9.1 A plot of the five measured b-quark meson and baryon lifetimes, shown together with the four c-quark meson and baryon lifetimes that do not require HF corrections. The vertical lines represent the "group-average" lifetimes. The b-quark to c-quark lifetime ratio is 3.21, which is roughly equal to the $\Upsilon_{1S}/(J/\psi)_{1S} \equiv b\bar{b}/c\bar{c}$ mass ratio of 3.05 (Fig. 2.9.2). Interestingly, the higher-mass b-quark states have the longer lifetimes. The $\pi^\pm$ lifetime anchors the global α-grid for this plot and that of Fig. 2.9.2.

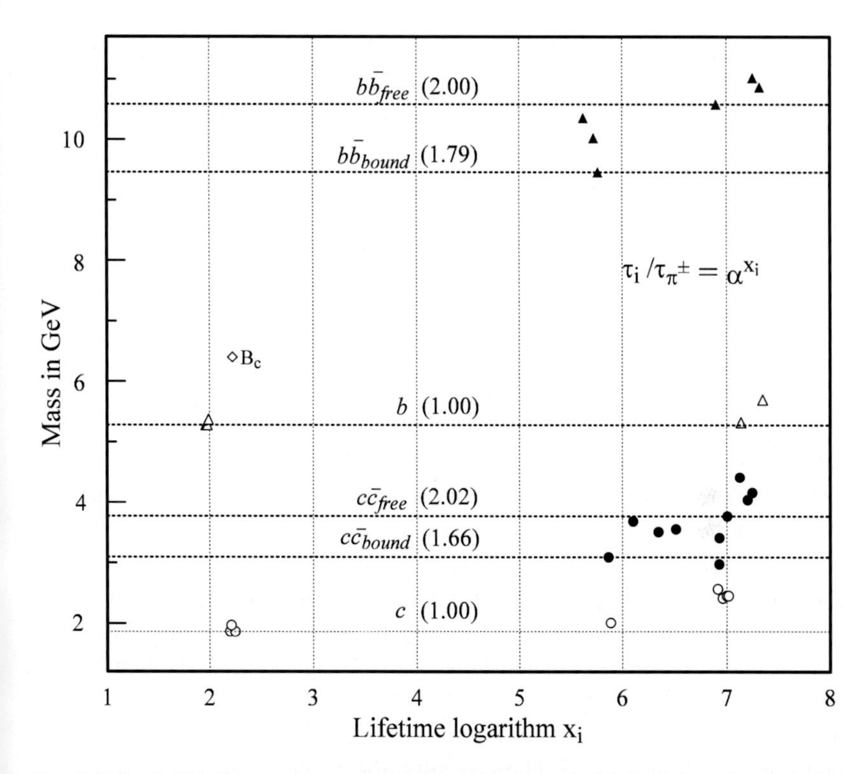

Fig. 2.9.2 A lifetime versus mass plot for the *c*-quark (circles) and *b*-quark (triangles) threshold-state particles (Fig. 2.8.8). The unpaired-quark states (open circles and triangles) have the longer lifetimes. The paired-quark states (filled circles and triangles), which have shorter lifetimes, occur at about twice the masses of the unpaired-quark states, which suggests that these masses are essentially those of constituent quarks. The B_c meson, which contains both *b* and *c* quarks, has the mass of the *b* and lifetime of the *c*.

unpaired *c* quarks and *b* quarks, respectively, and the closed circles and triangles denote the corresponding paired quarks. As can be seen, the *b*-quark masses are generally a factor of 3 higher than the *c*-quark masses. Thus the more massive *b*-quark states are more stable than the lighter *c*-quark states, which is not the result that we usually expect to find.

The characteristic factor-of-3 difference between *b*-quark and *c*-quark lifetimes does not seem to have been noted in the literature except by the present author [38, 39]. It only emerges clearly in the context of a global mapping of elementary particle lifetimes, and a key element in this mapping is the use of an α-quantized lifetime grid to relate these lifetimes to the lifetimes of the other long-lived threshold-state particles. This phenomenological lifetime systematics should serve as a valuable guide in the formulation of theories of hadronic decays. These theories must account for the overall lifetime scaling in powers of α, and also for the hyperfine factor-of-2 and factor-of-3 lifetime perturbations that are superimposed on the overall scaling. The threshold-state particle lifetimes serve as useful benchmarks in the evaluation of competing theories of hadron decay.

2.10 Flavor Substitutions and $c > b > s$ Flavor Dominance in Unpaired-Quark Decays

The experimental data on the various unpaired-quark decays are now complete enough that we can compare one quark flavor against another in order to see how they each affect the lifetime of the state. From the standpoint of quark dominance, there is a clear-cut result: in the case of mixed flavors, one type of favor dominates, and it alone dictates the lifetime (stability) of the particle that contains it. In order to show how this operates in detail, we plot the seven s-quark threshold states at $x_i \cong 1$, the five b-quark threshold states at $x_i \cong 2$, and the four c-quark states that do not require HF corrections at $x_i > 2$, and we indicate the types of quarks that are contained in these states. These results are displayed in Fig. 2.10.1. The u and d quarks appear throughout these particle states, and they do not have any direct effect on the global α-scaling of the lifetime groups. It is the "excited states" of these quarks — the s, c, and b quarks — whose stability dominates the particle decays. These quark excited states operate as follows.

The seven particles grouped near $x_i = 1$ on the α-grid have unpaired s quarks as their common factor. Thus these s quarks dictate the α-spacing of this lifetime group.

The five particles grouped right at $x_i = 2$ on the α-grid each feature an unpaired b quark, which is combined with u, d and s quarks. Since all of these particles have essentially the same lifetime, it is apparent that the b quark is dictating the lifetime. In particular, the B_s^0 and Ξ_b^0 particles, which each contain both a b quark and an s quark, have lifetimes at $x_i = 2$, and not at $x_i = 1$. Hence the flavor dominance here is $b > s$.

The final lifetime group in Fig. 2.10.1 has four HF-uncorrected states at $x_i \approx 2.23$ on the α-grid which each contain an unpaired c quark. Although b-quark and s-quark flavors are also present in this set, the four lifetimes are

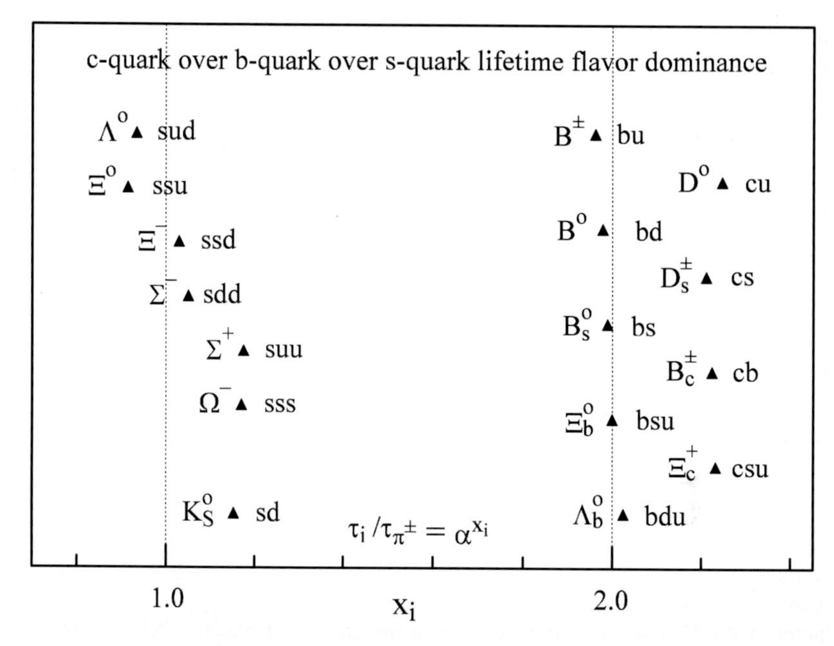

Fig. 2.10.1 A lifetime plot of unpaired-quark mesons and baryons near $x_i = 1$ and 2 on the α-spaced lifetime grid. (The K_S^0 meson is shown here for illustrative purposes with its $s\bar{d}$ Standard Model quark assignment, but this is not the constituent-quark assignment that is given, *e.g.*, in Table 0.10.1.) No lifetime HF corrections have been applied to the $x_i \cong 1$ states, and only the c-quark states near $x_i = 2$ that do not require HF corrections are included. As can be seen, all of the particles that contain c quarks have the c lifetime; the particles that contain a b quark but no c quark have the b lifetime; and the quarks that contain an s quark but no b or c quarks have either the $x_i \cong 1$ or $x_i \cong 0$ (not shown) lifetime. Thus there is a $c > b > s$ lifetime flavor dominance.

all approximately equal, and their x_i values correspond to c-quark decays. Thus we have clear-cut lifetime quark priorities:

The lifetime flavor dominance for unpaired quarks is $c > b > s$.

The $B_c^{\pm}$ meson is denoted as a B meson by the Particle Data Group because its mass value reflects the large mass of the b quark. But if we made lifetimes the primary consideration, we would label it as a $D_b^{\pm}$ meson.

It is interesting to consider the effect on particle lifetimes that is achieved by actually replacing one quark flavor with another. Figure 2.10.2 shows two examples where an s quark is replaced by a b quark: $\Xi^0 \rightarrow \Xi_b^0$ and

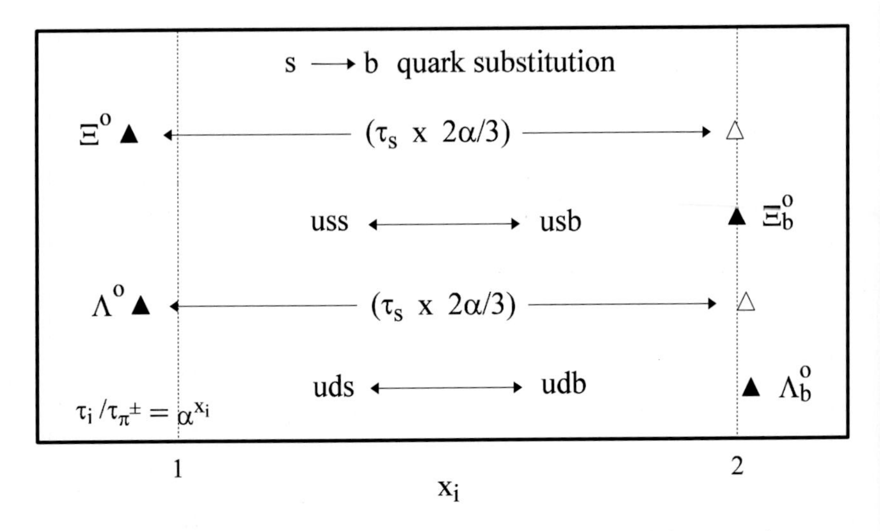

Fig. 2.10.2 A comparison of the lifetimes for two s-flavored baryons when an s quark is replaced by a b quark. Since both the s and b quarks have charges of $-\frac{1}{3}e$, there is no change of charge state, but the lifetime in each case is shortened by a factor of $2\alpha/3$. The open triangles denote the calculated b-quark lifetimes that are obtained when the scaling factor of $2\alpha/3$ is applied to the corresponding s-quark lifetimes.

$\Lambda^0 \rightarrow \Lambda_b^0$. In each case, the lifetime is shortened not only by one power of α, as we expect from Fig. 2.10.1, but also by a factor of 2/3.

Figure 2.10.3 shows four examples where an s quark is replaced by a c quark: $\Xi^0 \rightarrow \Xi_c^+$, $\Lambda^0 \rightarrow \Lambda_c^+$, $\Xi^- \rightarrow \Xi_c^0$ and $\Omega^- \rightarrow \Omega_c^0$. Since the $s \rightarrow b$ quark replacements both led to factors of $2\alpha/3$ shortening in lifetime (Fig. 2.10.2), and since c-quark lifetimes are intrinsically a factor of 3 shorter than b-quark lifetimes (Fig. 2.9.1), we logically expect to find factors of $2\alpha/9$ in the $s \rightarrow c$ quark replacements. The $\Xi^0 \rightarrow \Xi_c^+$ quark substitution shows this factor, but the other three examples show factors of $\alpha/9$. The factor-of-2 difference in the Ξ^0 and Ξ^- lifetimes (Fig. 2.5.2) accounts for the observed factor of $\alpha/9$ in the $\Xi^- \rightarrow \Xi_c^0$ substitution. In order to authoritatively account for the observed $\Lambda^0 \rightarrow \Lambda_c^+$ and $\Omega^- \rightarrow \Omega_c^0$ factors of $\alpha/9$, we should have a comprehensive method for making HF corrections, which has not yet been achieved. The interesting fact that emerges here is that the factor of 1/3 which appeared in the $s \rightarrow b$ quark substitutions of Fig. 2.10.2 also appears in all four of the $s \rightarrow c$ quark substitutions in Fig. 2.10.3, as evidenced by their 1/9 lifetime scaling factors.

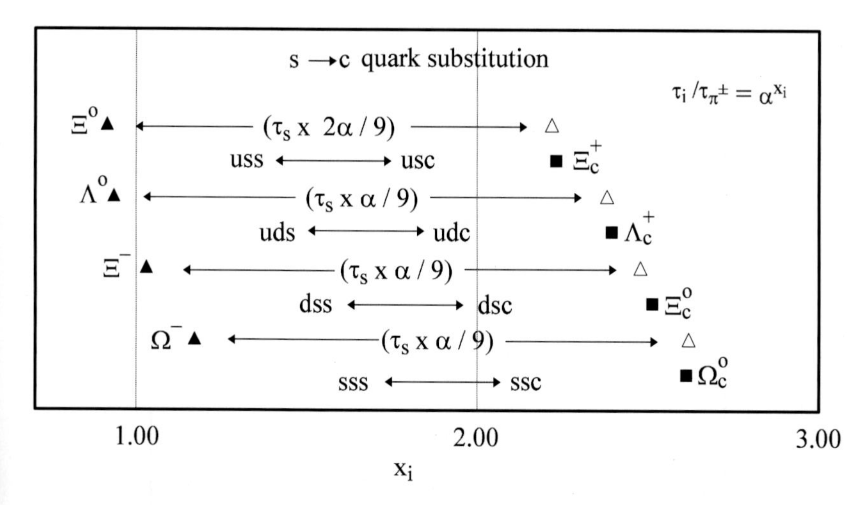

Fig. 2.10.3 A comparison of the lifetimes for four s-flavored baryons when an s-quark is replaced by a c-quark. Since the c-quark has a charge of $+\frac{2}{3}e$, the charge is increased by $+e$ in the $s \rightarrow c$ substitution. The lifetimes are each shortened by a factor of $2\alpha/9$ or $\alpha/9$, as indicated in the figure.

There are two general conclusions to be drawn from these global lifetime studies: (1) the accurate measurements that have been made of the long-lived threshold-state lifetimes contain a wealth of untapped information about the flavor substructures of the particles, and logically also about their mass substructures; (2) these α-spaced lifetime groups are fully consistent with the flavor systematics of the Standard Model, and are in fact a part of that systematics. In the final section of Chapter 2 we summarize the publication history of this α-dependent global lifetime structure, which serves to delineate its experimental evolution in elementary particle physics over the past third of a century.

2.11 The Historical Emergence of α-Quantized Elementary Particle Lifetimes

The α-quantized structure of the low-mass threshold-state elementary particle lifetimes was already evident in the early days of particle physics. The first published paper on this topic appeared in 1970 [26], when only 13 threshold-state particles had measured lifetimes [27]: neutron, muon, $\pi^\pm$, π^0, $K^\pm$, K_L^0, K_S^0, Λ, Σ^+, Σ^-, Ξ^0, Ξ^-, Ω^-. This same basic information on

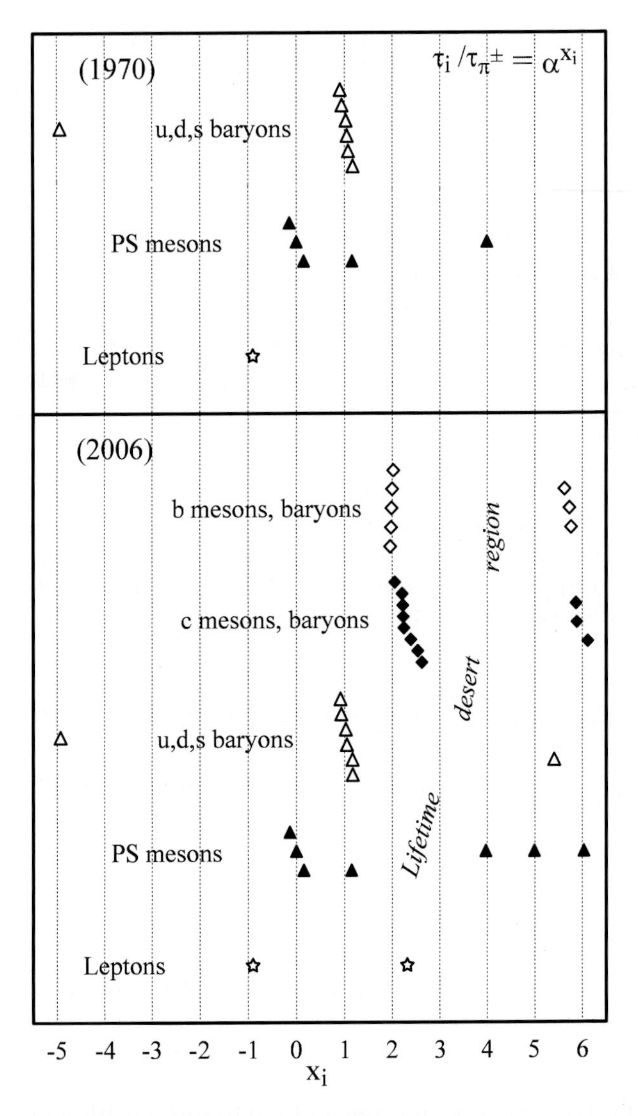

Fig. 2.11.1 A 36-year-apart (1970 and 2006) comparison of α-quantized lifetime plots (Eq. (2.2.5)) for the long-lived ($\tau > 1$ zs) threshold-state elementary particles. The top plot is for the 13 particle lifetimes listed in RPP1970 [27] (see Refs. [26] and [40]), using the lifetime values given there. The bottom plot is for the 36 particle lifetimes listed in RPP2006 [10] (which are summarized in Appendix A of the present book). The same lifetime α-grid fits both data sets. As can be seen, the c-quark states and (especially) the b-quark states have accurately extended this α-quantized lifetime systematics, and have served to delineate the *lifetime desert region* that separates the long-lived unpaired-quark decays from the shorter-lived paired-quark and radiative decays.

α-quantized lifetime structure was also presented as an invited contribution to the January 1971 Coral Gables Conference on Fundamental Interactions at High Energy [40]. The latest data summary, RPP2006 [10], adds 23 more particles to this list: τ lepton, η, η', Σ^0, 11 charm particles, and eight bottom particles. These two data sets are displayed in Fig. 2.11.1, which shows the 13-particle 1970 lifetime data set in the top plot, and the 36-particle 2006 data set in the bottom plot. The 1970 data are displayed with the lifetime values they had at that time [27]. Both data sets are plotted on the same α-quantized lifetime grid, which is anchored on the $\pi^\pm$ lifetime. The fact that the α-grid which was defined in 1970 does not have to be modified in order to accommodate these later particles stands in a sense as a predictive confirmation of the relevance of the fine structure constant α as a scaling factor for these lifetimes.

In addition to the global α-spaced lifetime grid, the 1970 paper also describes the accurate 1:2:4 lifetime ratios of the K_L^0, $\pi^\pm$, $K^\pm$ meson lifetimes and the factor-of-2 lifetime ratios in the Λ, Σ^+, Σ^+, Ξ^0, Ξ^0, Ω^- hyperon states, and it contains a discussion of their significance as "decay triggers" that arise from the substructures within the particles [41]. Thus the basic elements of this threshold-state lifetime phenomenology — a global α-grid with a superimposed factor-of-2 hyperfine (HF) structure — were laid down on the basis of the earliest lifetime measurements [27], and this α-systematics has accommodated the lifetimes of the subsequently discovered c-quark and b-quark mesons and baryons [10]. These higher-mass

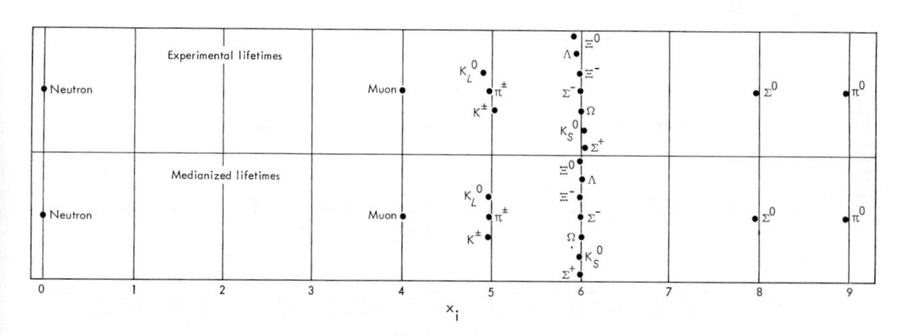

Fig. 2.11.2 Published 1974 α-grid plots [42, 43] of the 13 known threshold-state elementary particles (plus the Σ^0, which is only an upper limit). The top plot shows the original uncorrected data, and the bottom plot shows "medianized" data that have had phenomenological factor-of-2 hyperfine (HF) corrections applied. The neutron serves as the reference lifetime. This publication was just before the discovery of the J/ψ.

charm and *bottom* particles have served to delineate the *lifetime desert region* that separates the unpaired-quark from the paired-quark decays (see Figs. 0.1.13, 0.1.14, 2.8.7 and 2.8.8).

Figure 2.11.2 shows a 1974 particle lifetime plot that was published [42, 43] just prior to the discovery of the charmed c-quark J/ψ resonance peak. The lifetimes are plotted on an α-spaced grid anchored on the neutron lifetime. The upper α-grid in Fig. 2.11.2 displays the lifetimes of the 13 particles that were contained in the 1970 paper [26] together with that of the Σ^0, which was then known only as an upper limit. The lower α-grid shows these same states after they have been "medianized" by the application of factor-of-2 hyperfine HF corrections (as in Sec. 2.5 of the present chapter). Comparing Figs. 2.11.1 and 2.11.2, we see that no new long-lived ($\tau > 1$ zs) particle lifetimes were established between 1970 and mid-1974.

Figure 2.11.3 [44] is an ADI type of plot (see Secs. 2.2 and 2.6 of the present chapter) of the *deviations from integer values* of the exponents x_i for the uncorrected lifetimes of Fig. 2.11.2. The exponents x_i are obtained from the equation $\tau_i/\tau_{\text{neutron}} = S^{-x_i}$, where the scaling factor S is successively assigned the values $(V, W, X, Y, Z) = (14.4, 100, 137, 164, 1000)$,

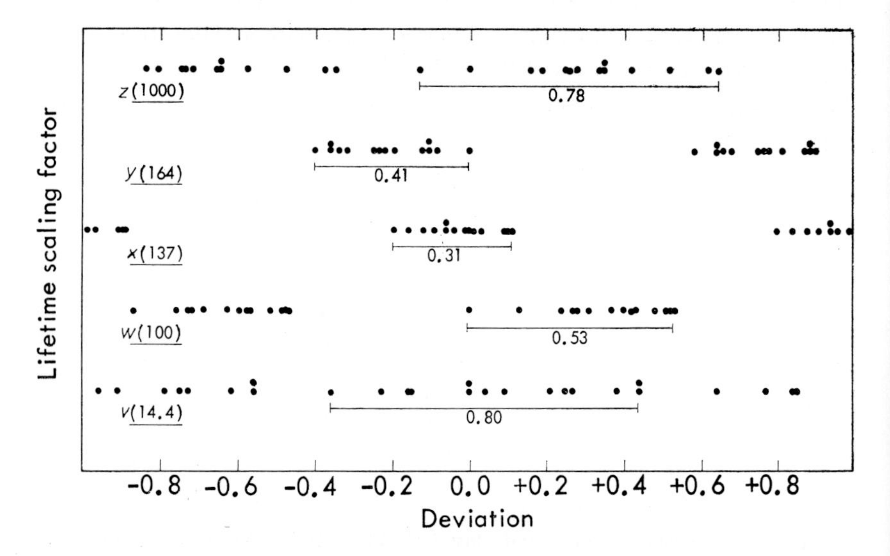

Fig. 2.11.3 A 1974 [44] *deviation from an integer* analysis of the lifetime exponents x_i for various values V, W, X, Y, Z of the scaling factor S, using the uncorrected 13-particle data set at the top of Fig. 2.11.2, with the neutron as the reference lifetime. The spread in the *range of exponent deviations*, $R_{\text{exp dev}}$, is the least ($R_{\text{exp dev}} = 0.31$) for $S = X = 137$.

respectively. The *range of the exponent deviations* $R_{\text{exp dev}}$ is plotted for each S value. As can be seen in Fig. 2.11.3, the range of the x_i deviations from integers is the smallest ($R_{\text{exp dev}} = 0.31$) when the value $S = 137$ is used as the lifetime scaling factor, and they are the most symmetrically balanced in positive and negative deviations. Thus the early (1974) lifetime data were already sufficient to demonstrate the relevance of the fine structure $\alpha \cong 1/137$ in the spacings of these threshold-state lifetime groups.

Figure 2.11.4 [44] repeats the analysis of Fig. 2.11.3, but for the HF-corrected (medianized) lifetimes of Fig. 2.11.2. The range of the exponent deviations $R_{\text{exp dev}}$ is not altered very much by this medianization except for the scaling factor $S = 137$, where it drops from $R_{\text{exp dev}} = 0.31$ to $R_{\text{exp dev}} = 0.12$, and again is more symmetrically balanced than for the other scaling factors. Thus the HF-corrections appreciably affect the grouping of the lifetime exponents x_i only for scaling factors that are near the value $S = \alpha^{-1}$.

In Fig. 2.11.5 [44] the lifetime exponents x_i themselves (not the deviations) are plotted for the HF-corrected data of Fig. 2.11.2 as functions of the scaling factor S, using the same scaling factors as in Figs. 2.11.3 and

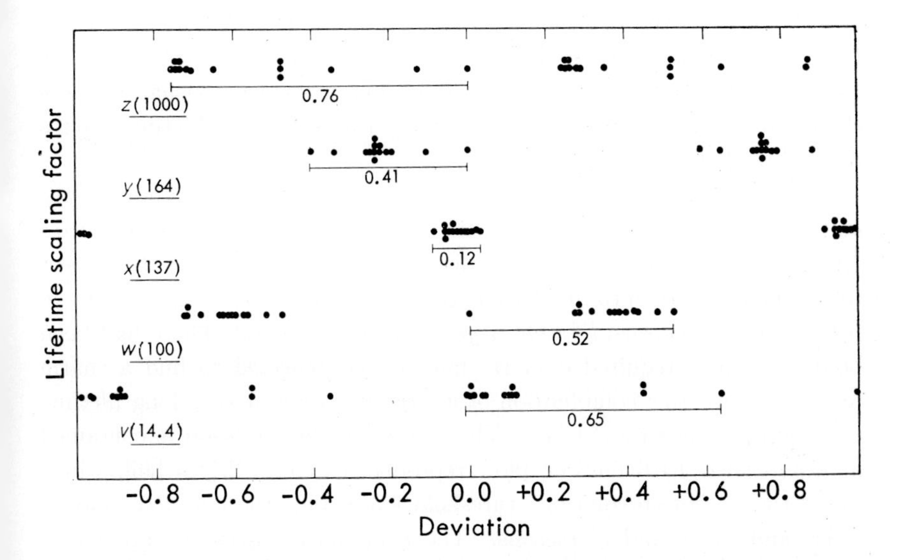

Fig. 2.11.4 The same 1974 plot [44] as in Fig. 2.11.3, but for the HF-corrected data set at the bottom of Fig. 2.11.2. The $R_{\text{exp dev}}$ value is decreased only slightly except for the scaling factor $S = X = 137$, where it drops dramatically from $R_{\text{exp dev}} = 0.31$ to $R_{\text{exp dev}} = 0.12$.

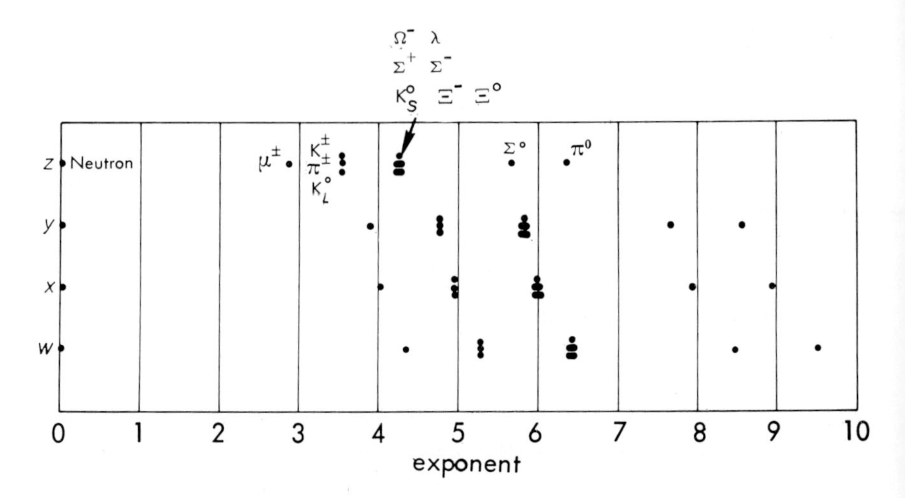

Fig. 2.11.5 A 1974 plot [44] of the HF-corrected lifetime exponents themselves (rather than the deviations of the exponents from integers) for the HF-corrected data of Fig. 2.11.2. As can be seen, the lifetime groups match the integer-exponent vertical lifetime grid only for the scaling factor $S = X = 137$ (see Fig. 2.11.4).

2.11.4, and with the neutron as the reference lifetime. As can be seen in Fig. 2.11.5, the lifetime groups lie on the integer-exponent grid lines when the scaling factor $S = \alpha^{-1}$ is used. This figure provides a visual image of the accuracy of the α-quantization for the ratios of these lifetime groups relative to that of the neutron.

The above figures were published before the J/ψ and ψ' (now J/ψ_{1S} and J/ψ_{2S}) charm peaks made their unexpected appearance in the 1974 "November revolution" [45]. Their discovery raised a question with respect to the known α-quantized lifetime groups: *would these new lifetimes fit into the pre-established lifetime α-grid, and if so, where?* There had been predictions of a required c-quark, but no one expected to find a charm resonance with this combination of a large mass and a very long lifetime (narrow resonance width) [45]. The answer to this question is contained in a published 1976 lifetime plot, reproduced in Fig. 2.11.6 [46], which shows the 13 "pre-revolution" threshold states plus the new J/ψ and ψ' peaks and the η and η' mesons. The η meson occupies the previously empty $X = 10$ lifetime α-grid level, and the J/ψ, ψ' peaks and η' mesons, which all have roughly the same lifetimes, occupy the adjacent empty $X = 11$ level. The logic of the placement of the paired c-quark J/ψ and ψ' lifetimes on the α-quantized lifetime grid would not become apparent until

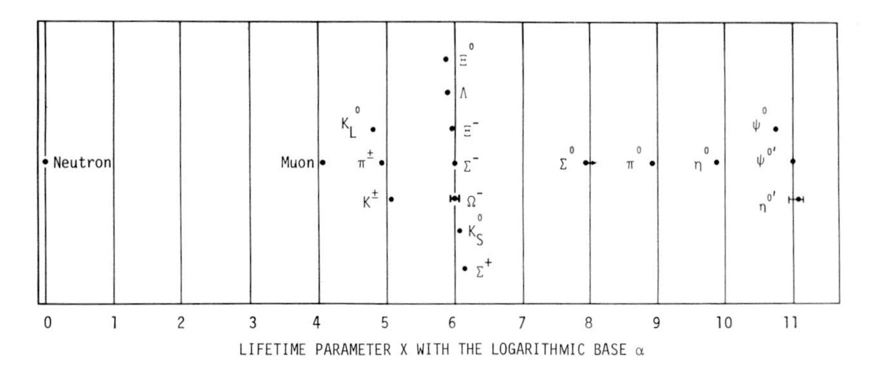

Fig. 2.11.6 A 1976 lifetime α-grid plot of the 13 threshold-state particles of Figs. 2.11.2–2.11.4 plus the J/ψ, ψ', η and η' mesons, with the neutron as the reference lifetime. The new particles occupy the previously empty $X = 10$ and 11 α-grid levels.

after the subsequent discoveries of the unpaired c-quark D mesons and charmed hyperons, and also the closely-related b-quark states, which all follow the α^4 separation rule between unpaired-quark and paired-quark decays (Figs. 0.1.13, 0.1.14, 2.8.7 and 2.8.8).

The effect of adding the J/ψ, ψ', η and η' mesons to the previous collection of threshold states was evaluated by repeating the ADI-type analysis of Fig. 2.11.3, where no HF corrections are applied. The results are shown in Fig. 2.11.7 [47]. The narrowest and best-centered deviation plot $R_{\text{exp dev}}$ was again obtained for the scaling factor $S = 137$, demonstrating that the four new meson states are in good agreement with the threshold-state lifetime α-quantization defined by the original thirteen meson and baryon states.

A reference to these lifetime results is contained in Barrow and Tipler's 1986 book *The Anthropomorphic Cosmological Principle*, which contains the following comment:

> *Mac Gregor's correlation between powers of α and the lifetimes of metastable states is another curious trend.* [48]

The 1990 publication update of α-quantized threshold-state elementary particle lifetimes [49] includes new Υ_1, Υ_2 and Υ_3 Upsilon $b\bar{b}$ resonances, a single (undifferentiated) unpaired b-quark B meson, the unpaired c-quark $D^\pm$, D^0, D_s, Σ_c^+ and Λ_c^+ mesons and baryons, the τ lepton, and also (finally) a lifetime value for the Σ^0. Figure 2.11.8 gives an α-grid plot of all the measured elementary particle lifetimes, with the muon now used as

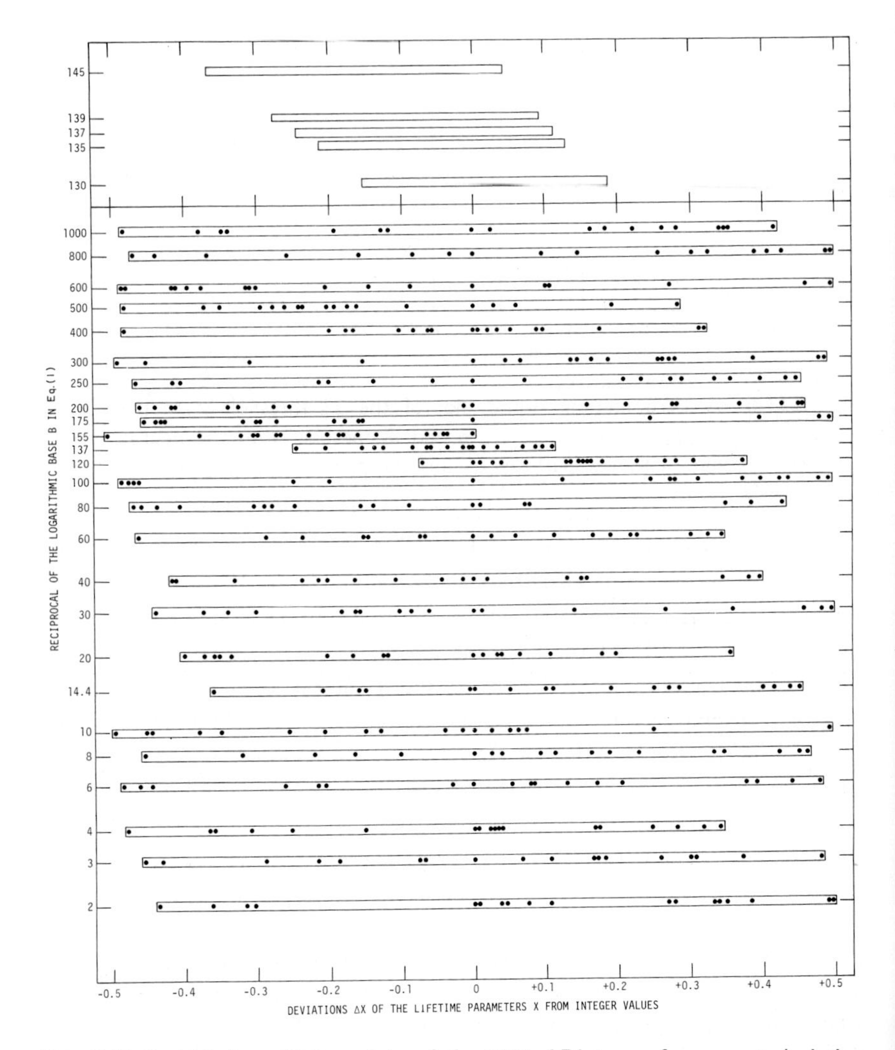

Fig. 2.11.7 This is a 1976 update of the 1974 ADI type of exponent deviation analysis shown in Fig. 2.11.3, with the J/ψ, ψ', η and η' mesons added in. The range of exponent deviations $R_{\text{exp dev}}$ is narrowest and best centered for the lifetime scaling factor $S = 137$.

the reference lifetime [49]. The filled circles denote the original 13 long-lived threshold-state particles of the 1970 compilation (top of Fig. 2.11.1), the open circles denote the threshold-state particles that have subsequently been added, and the dots show the short-lived excited-state particles.

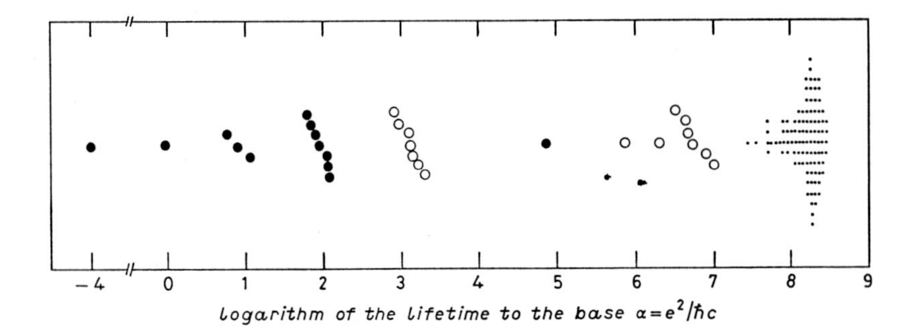

Fig. 2.11.8 This 1990 update [49] of the α-quantized lifetime grid shows the original 1970 long-lived threshold-state particle lifetimes as filled circles, the addition threshold-state lifetimes measured during the next 20 years as open circles, and the continuum of short-lived excited state lifetimes as dots. The muon serves here as the reference lifetime. The manner in which this lifetime α-grid was fleshed out (without counterexamples) over two decades of experiments is impressive.

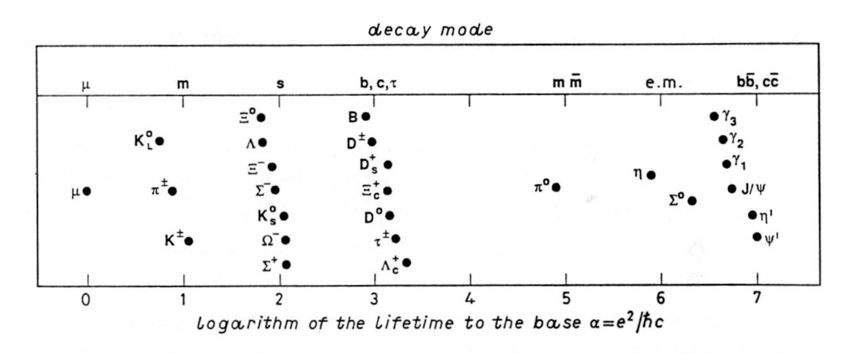

Fig. 2.11.9 This is the 1990 compilation [50] of the 28 threshold-state elementary particles of Fig. 2.11.8, showing how their lifetimes occur in well-localized and well-separated groups. At this time only a single (composite) B-meson lifetime had been measured, but subsequent experimental work at the SLAC B factory led to the identification of five unpaired b-quark excitations whose almost identical lifetimes form one of the most striking examples of accurate lifetime α-quantization (see Figs. 0.1.10 and 2.5.7). The lifetime groups that are dominated by unpaired-quark decays and by paired-quark and radiative (e.m.) decays are indicated at the top of the figure.

The threshold-state particle lifetimes that were known in 1990 are displayed (except the neutron) in Fig. 2.11.9 [50]. If we include the neutron, there were now 28 measured particles with lifetimes $\tau > 1$ zs. As can be seen in Fig. 2.11.9, the addition of these new particles has greatly clarified the

α-quantized nature of these lifetime groups. The three lifetime groups in the $X = 1$, 2 and 3 α-grid levels to the right of the reference ($X = 0$) muon level are each clustered in a tight band (without using HF corrections). As indicated at the top of Fig. 2.11.9, these are unpaired-quark excitations. A matching set of three lifetime groups is observed in the $X = 5$, 6 and 7 α-grid levels. These are paired-quark excitations together with excitations that have radiative (nonquark-changing) decays, which are each displaced by a characteristic factor of α^4 from the matching unpaired-quark excitations (as we discussed in detail in Sec. 2.8).

There are two phenomenological conclusions we can draw from Figs. 2.11.8 and 2.11.9: (a) the spacings of the threshold-state lifetime clusters are in quite accurate powers of α, over a range of 11 powers of α or almost 24 orders of magnitude; (b) there are no counterexamples — all of the 28 particles with lifetimes $\tau > 1$ zs (Fig. 2.11.9 plus the neutron) fall fairly accurately on the α-quantized lifetime grid. In assessing the validity of these threshold-state α-dependent lifetimes, the second conclusion is fully as significant as the first.

Figure 2.11.10, a lifetime α-grid plot published in 2005 [51], brings this historical compilation of α-quantized threshold-state particle lifetimes up to date. This plot (also see Fig. 0.1.11), which uses the $\pi^{\pm}$ lifetime to anchor the α-grid, displays the well-measured elementary particle lifetimes (Appendix A of Ref. [51]), with each particle identified as to its dominant quark flavor content, and with hyperfine (HF) corrections applied (Figs. 0.1.10 and 2.5.6). In this figure, the unpaired-quark states in the α-grid levels $x_i \cong 0$, 1 and 2 appear as an oasis in a "lifetime desert." To their left are the solitary neutron and the muon, which both interestingly fall on the lifetime α-grid. To their right are the electrically neutral π^0 and η PS mesons and Σ^0 hyperon, which are followed by the $b\bar{b}$ bound Υ states, the $c\bar{c}$ bound J/ψ states, a D^* excitation, and the η' PS meson. This lifetime desert ends at the zeptosec boundary, which signals the onset of a continuum of short-lived excited states. These excited states are so transient that flavor lifetime effects are negligible, at least on the global time scales used here.

The τ lepton mass is comparable to the D-meson masses, and its lifetime is comparable to the D-meson lifetimes, as shown in Fig. 2.11.10.

An important feature of Fig. 2.11.10 is that all of the lifetimes to the left of the zeptosecond boundary fit into the α-quantized lifetime grid. As described above, it took a third of a century to complete the lifetime measurements of these long-lived threshold-state particles. There is presently

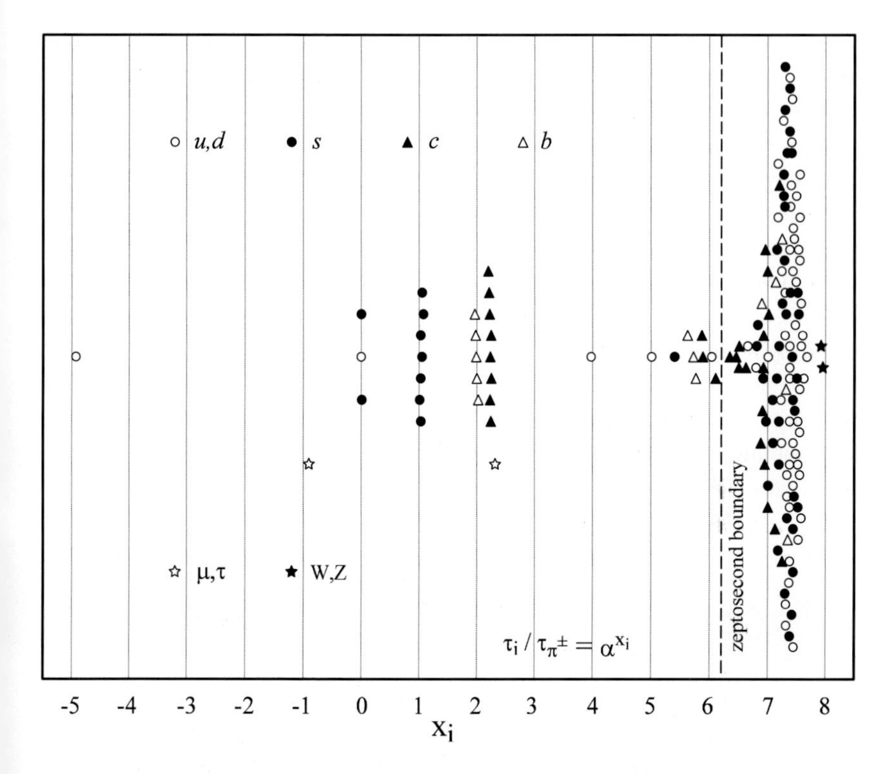

Fig. 2.11.10 This is the 2005 α-quantized lifetime plot [51], which contains the well-determined elementary particle lifetimes that are summarized in Appendix A of Ref. [51]. (See Fig. 0.1.11 for a similar plot.) The quark flavors are identified, and HF corrections (Fig. 0.1.10) have been applied. The number of well-measured threshold-state ($\tau > 10^{-21}$ s) lifetimes increased from 13 in 1970 (Figs. 2.11.1 and 2.11.2) to 28 in 1990 (Fig. 2.11.9), and now to 36 in 2005. All of these states have been accommodated on the 1970 lifetime grid. Mesons, baryons, hyperons and leptons combine together to produce this α-spaced lifetime mapping.

no theory that accounts for the α-dependence of these lifetimes, so that none of the lifetimes could be predicted in advance. However, the lifetime grid itself was already well defined in 1970 on the basis of 13 known threshold-state lifetimes, and the other 23 threshold-state lifetimes were accommodated on this α-grid as they appeared one by one. Since these α-quantized *lifetimes* are conjugate to the *mass widths* of the particles (see Figs. 2.2.1 and 2.2.2 and Tables 2.1.1 and 2.1.2), it follows that the mass widths are also α-quantized. The mass widths are a measure of the stability of the masses themselves, which suggests that the theoretical explanation for the α-dependent lifetimes may reside in α-dependent mass units. This is just a conjecture, but it is a straightforward matter to check this conjecture phenomenologically by examining the masses of the threshold-state particles. This is the task we undertake in Chapter 3.

As we have displayed here, the historical development of α-quantized particle lifetimes has been well documented in the literature, albeit by a single individual. It would be interesting to see these phenomenological results incorporated as a part of the paradigm of modern elementary particle physics (e.g., Ref. [33] by David Akers). The present book was partly motivated by a 2002 e-mail message from an unknown (to me) CERN physicist, Paolo Palazzi, which reads in part: "It has been almost 30 years by now that your amazing lifetime plot in its various incremental incarnations has kept me awake at night ... ."

Chapter 3 The Phenomenology of Reciprocal α^{-1} and α^{-2} Particle Mass Quantization

3.1 What Are the Elementary Particle Lepton and Hadron "Ground States"?

One of the important features of any physical system is its ground state. This is the lowest energy state of the system, and it usually represents the entity from which all the other states are generated, either as excited states (*e.g.*, atomic electron shells) or as compound systems (*e.g.*, atomic nuclei). If we consider the problem of elementary particle masses — the problem that Leon Lederman says "has confounded scientists since antiquity" [5], we naturally look for the lowest massive state, which we expect to be stable. The photon is a zero-mass state, so it is excluded. The neutrino appears to be a massive state, but with a mass so small that we have not as yet been able to accurately measure it. Thus we also exclude neutrinos as candidates for the "elementary particle ground state." This leaves us with roughly 200 massive particle states that have been measured and catalogued in RPP2006 [10]. The mass values for the states with well-measured lifetimes are summarized in Appendix B. When we examine these particles from the standpoint of possible ground states, we encounter an immediate problem: the Standard Model separates them into two distinct groups — leptons and hadrons — with dramatically different properties. *Leptons* have integer point-like electric charges e; they interact only electromagnetically with other particles; and they have no measurable sizes. *Hadrons* also have integer electric charges, but these charges seem to be split into $\frac{1}{3}e$ and $\frac{2}{3}e$ fractions which are distributed on "quark" substates within the particle. Hadrons interact with the long-ranged electromagnetic force, just as leptons do, but in addition they interact with a short-ranged strong "hadronic" force that operates only between hadrons. High-energy collision measurements, which represent scattering from individual quarks, suggest that the $\frac{1}{3}e$ and $\frac{2}{3}e$ charges on the quarks are point-like, just like the lepton charges, but low-energy collision measurements, which reflect the overall size of the particle, indicate Compton sized radii for hadrons. Hadrons also have the frustrating property that the fractional $\frac{1}{3}e$ and $\frac{2}{3}e$ charges carried by the internal quarks are never observed in particle decay products: we can shatter hadrons in high-energy collisions, but the emerging fragments always have integer charges. Thus, although a charge e can be separated into $\frac{1}{3}e$ and $\frac{2}{3}e$ components at very short ranges, these charge components evidently remain firmly bound together over longer distances, and they emerge intact on any collision fragments. Hence it is difficult to measure the mass of a fractionally charged quark. The only case where this has been accomplished is that of the top quark t, as we discuss in Secs. 0.10 and 3.20.

The point in comparing these properties of leptons and hadrons in detail is that these two types of particles appear to be so different from another that they are treated within the Standard Model (SM) as completely different entities, even though leptonic decays can lead to hadron states and hadronic decays can lead to lepton states. The question then arises as to whether there are separate lepton and hadron "ground states," and, if so, what are they? In the case of leptons, the answer seems obvious: the *electron* is the lightest lepton state; it is the *only* stable lepton; and both *muons* and *tauons* — the other two leptonic states — decay back down to electrons. All of these leptons are spin-1/2 particles. Thus the electron logically represents the "*lepton ground state.*" Figure 3.1.1 shows an energy level diagram for the 3-member lepton family — electron, muon and tau. This diagram applies to both leptons and antileptons, which have negative and positive charges, respectively.

The difficulty we have here is that when we now turn to the hadron family and search for a particle to represent the "*hadron ground state,*" there is no obvious candidate. For one thing, hadrons contain both integral-spin bosons and half-integral-spin fermions, and a single ground state must be one or the other. We could of course have both *fermion* and *boson* ground states. The lightest fermionic hadron is the proton, and it is stable, so it can serve as the "*fermionic hadron ground state.*" But this leaves us with the bosonic hadrons, which have no stable members at all. Where is the "*bosonic hadron ground state*"? Figure 3.1.2 displays the energy level diagram for the 34-member hadron threshold states, which are the hadrons with lifetimes $\tau > 1$ zs. (Appendix A lists 34 hadron threshold states, which

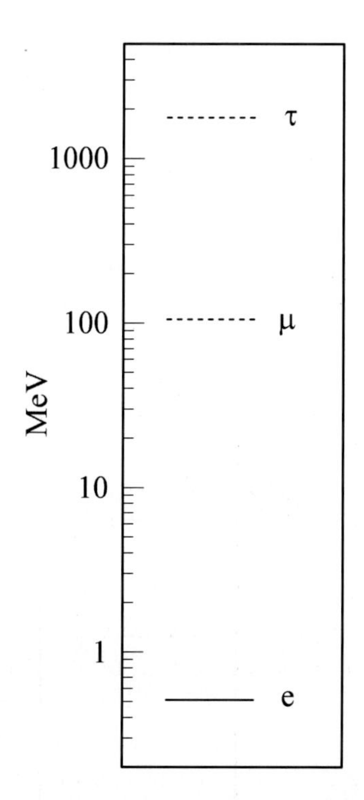

Fig. 3.1.1 The 3-particle lepton mass spectrum, which has both particle (e^-, μ^-, τ^-) and antiparticle (e^+, μ^+, τ^+) components. Their mass values are: m_e = 0.51099892 MeV; $m_\mu = 105.658$ MeV: $m_\tau = 1776.99$ MeV. There is no explanation for these masses or their mass ratios within the context of the Standard Model, which places leptons and hadrons in separate mass-formalism categories. The decays of the unstable $\mu^\pm$ and $\tau^\pm$ are back down to the stable $e^\pm$ "ground state," so they are clearly-related particles. In this and the following three figures, solid lines indicate stable mass levels and dashed lines indicate unstable levels.

are displayed in Fig. 2.8.8, and to this we add the stable proton, and we remove the Ξ_b, whose mass is not measured.) As can be seen in Fig. 3.1.2, the boson states π, K and η are lighter than the proton, which is the only stable level. Figure 3.1.3 contains the *fermionic* hadrons of Fig. 3.1.2, which form a conventional-looking hadronic energy level diagram that is comparable to Fig. 3.1.1 for leptons. Figure 3.1.4 contains the *bosonic* hadrons of Fig. 3.1.2, which are all unstable. Thus we have a class of particles — hadronic bosons — that are just "floating in air": they have no apparent ground state — no particle generator. Faced with this situation, SM theorists have turned to a hypothetical particle — the Higgs gauge

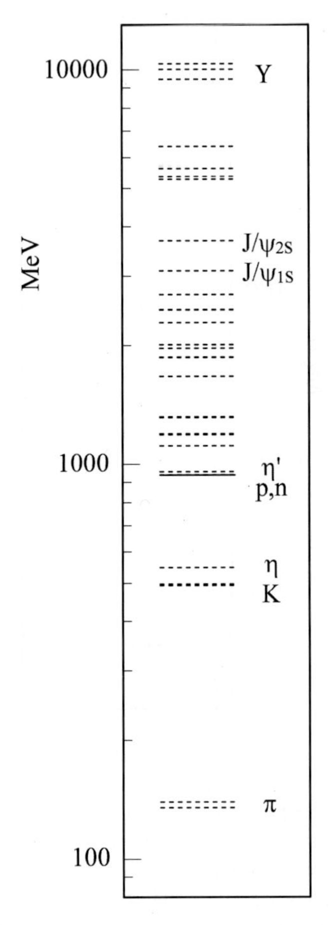

Fig. 3.1.2 The 34-particle mass spectrum of the long-lived ($\tau > 10^{-21}$ s) threshold-state hadrons. This spectrum includes both half-integral-spin baryons and hyperons and integral-spin mesons. There are also matching antiparticle states. The baryonic proton is the only stable hadronic particle. Its mass is greater than the masses of some of the mesons.

boson — which is required for QCD renormalization purposes, and they have speculated that it may also serve as the hadron mass generator. The Higgs mass is unknown, but since it has not been observed at energies below 100 GeV, it must be higher than that value, but not too much higher [9]. We thus arrive at a situation in physics which seems unprecedented: a class of particles is assumed to be generated by a very massive particle. We have physics from the top down, not from the bottom up.

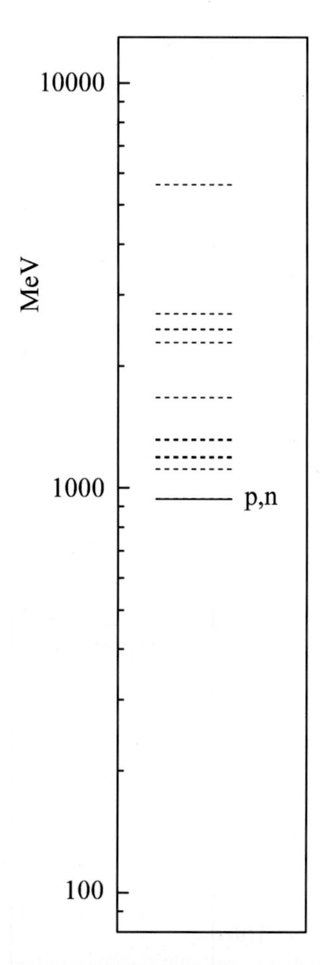

Fig. 3.1.3 This plot displays the half-integer-spin threshold-state baryons and hyperons of Fig. 3.1.2. The decays of the higher-mass baryons and hyperons are back down to the stable proton "ground state." This is a conventional-looking energy level diagram.

If we concentrate on the two elementary particle systems that do seem to have conventional ground states — leptons and baryonic hadrons, as shown in Figs. 3.1.1 and 3.1.3 — there is one more question that should be raised. Do the electron and proton, the putative ground states for these systems, bear any relationship to one another? These two particles are displayed in Fig. 3.1.5. Since one is a lepton and the other is a hadron, the Standard Model does not establish a relationship between them. This is

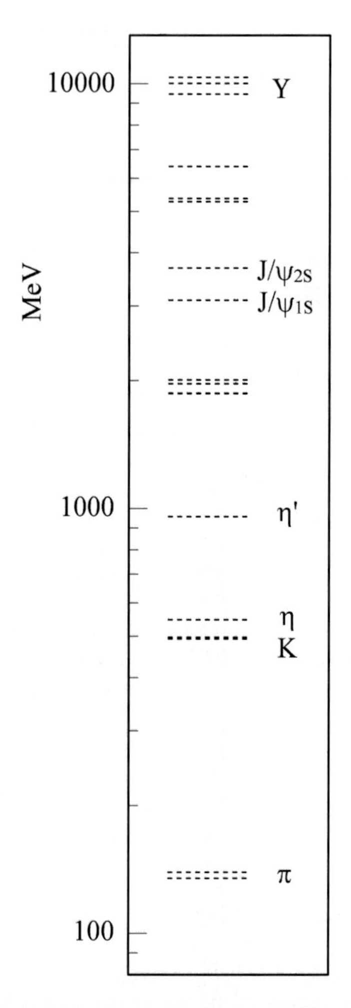

Fig. 3.1.4 This plot displays the integer-spin threshold-state mesons of Fig. 3.1.2. All of these mesons are unstable, so none of them can serve as the "ground state" excitation. The non-radiative and non-proton-producing decays of these levels are all back down to electrons and/or positrons. These hadronic mesons lack a "ground state" — a "generator."

the topic that we discussed in Sec. 1.8. We will demonstrate in Chapter 3 that mass α-quantization does in fact combine these two particles together.

In view of this paradoxical situation with regard to elementary particle ground states, it behooves us to step back a little and ask if there is some information — some part of the phenomenology — that has been overlooked.

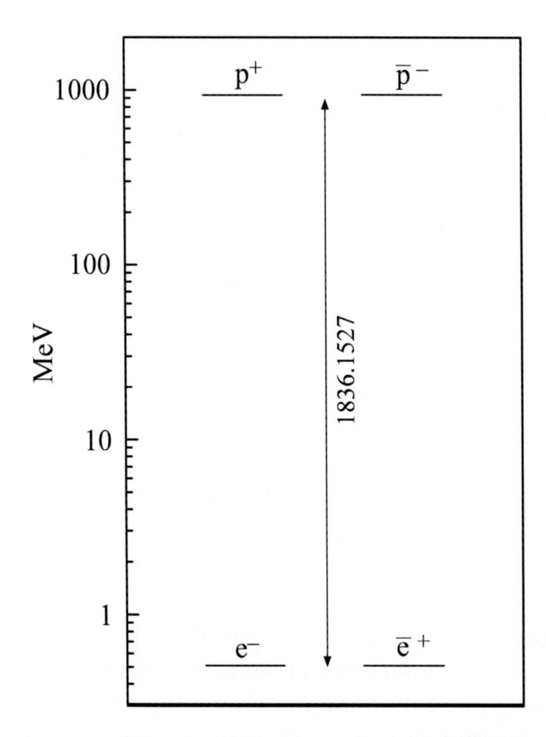

Fig. 3.1.5 The stable *electron* and stable *proton*. These are the only two stable massive hadrons (together with the *positron* and *antiproton*), and they logically serve as ground states for *leptons* (Fig. 3.1.1) and *baryons* (Fig. 3.1.3), respectively. Are they related to one another? An affirmative answer requires us to complete two tasks: (1) account for the observed mass ratio of 1836.1527; (2) explain the difference in the signs of the electric charges for their respective particle (and antiparticle) states. The first task is the numerical challenge that was discussed in Sec. 1.8, and its solution has eluded physicists for the past century. The second task has apparently not been considered to any extent in the literature. By applying the α-quantization of threshold-state particle lifetimes also to particle masses, we can in fact provide a rationale for dealing with both of these tasks, as we demonstrate with the systematics developed in Chapter 3.

We have what Niels Bohr might characterize as a "contradiction" with respect to particle ground states, and this contradiction is possibly telling us to broaden the scope of our investigation. Our focus here has been on the particle *masses* themselves, but there is another related area that may be helpful — the *stability* of particle masses. Mass stability, which experimentally means the length of time that a certain mass configuration persists, is determined by measuring the mass width $\Delta m \equiv \Gamma$ of the resonance cross section. This in turn is related to the particle lifetime or mean life $\Delta t \equiv \tau$ by the limiting Heisenberg equation $\Gamma \cdot \tau = \hbar$, where Γ is the *full width* [10] of the resonance at half maximum. Having just completed a study of particle lifetimes in Chapter 2, we should see what information can be extracted from these lifetimes and carried over to the mass widths, and hopeful also to the masses themselves. The unique feature which emerges from the lifetimes and mass widths of the long-lived threshold-state particles is their quantization in powers of the fine structure constant $\alpha = e^2/\hbar c$. Does the experimental α-dependence of particle mass widths Γ have application to particle masses? And, specifically, what does it tell us about particle ground states? This is the general topic we treat in Chapter 3.

3.2 The Correlation between Particle Mass-Widths and Particle Masses

In Chapter 2 we presented experimental evidence for the α-quantization of the elementary particle lifetimes τ that are longer than 1 zs (10^{-21} s). This lifetime α-quantization is comprehensive, and it is quite accurate when viewed in a global context, especially when a superimposed factor-of-2 hyperfine (HF) structure is taken into account. The lifetime α-grid, with its 1/137 lifetime scaling steps, was in fact in evidence as early as 1970, when only the u, d, s quarks were known [26], and it has been filled in by the subsequent appearances of the c [46] and b [49] quarks in their various combinations. Although this particle lifetime systematics is of interest *per se*, it has a greater significance in the information it conveys with respect to the mass structures of these particles. This follows from the facts that (a) the particle *lifetimes* are reciprocally linked to the particle *mass widths* by the Heisenberg uncertainty principle (Eq. (1.7.1)); and (b) the *mass widths*, which represent the stability of the mass structures, logically bear a close relationship to the properties of the *masses* themselves.

Figure 3.2.1 shows a plot of the mass widths Γ_i for the 157 well-measured particles listed in Appendix A, where the data are taken from RPP2006 [10]. The widths Γ_i are expressed as ratios to the $\pi^{\pm}$ width $\Gamma_{\pi^{\pm}}$, where these ratios are in powers x_i of $\alpha^{-1} \cong 137$. This mass-width plot is the counterpart of the lifetime plot of Fig. 2.11.10, and it similarly uses HF-corrected data (Fig. 2.5.6) and shows the particles identified by their dominant quark content. The vertical dashed line denotes the width $\Gamma_{\text{zs}} = 0.6582$ MeV, which corresponds to the 1 zs boundary that separates the short-lived broad-

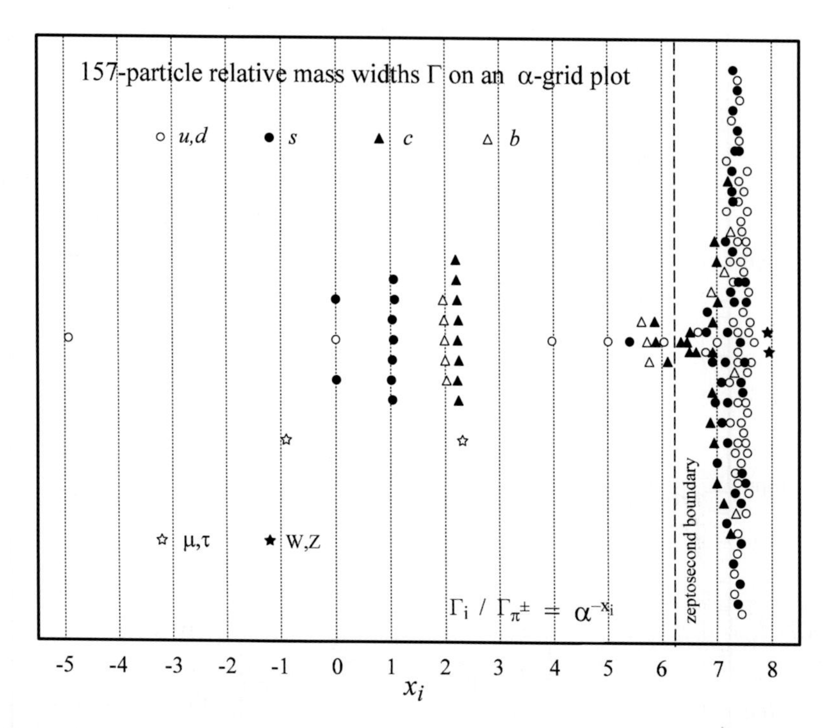

Fig. 3.2.1 This is the comprehensive α-quantized lifetime plot of Fig. 2.11.10 converted into mass widths Γ, using Eq. (1.7.1). The layout of the data points is the same in the two figures, but the lifetimes τ decrease in powers of $1/137$ as we move to the right on the α-grid, whereas the mass widths Λ broaden in powers of 137. Hyperfine corrections have been applied to the data (Fig. 2.5.6). The mass widths are dictated by the flavor content of the dominant quarks in the particles.

width *excited-state* particles at the right from the long-lived narrow-width *threshold-state* particles at the left. This figure demonstrates the manner in which the mass widths Γ of the threshold states are quantized in powers of α^{-1}. It also shows how the various quark flavors affect this α-quantization.

The task which now arises is to ascertain if there is a correlation between the mass widths $\Gamma \equiv \Delta m$ and the masses m themselves. We can address this issue graphically. The plot in Fig. 3.2.1 has the mass widths Γ displayed along the abscissa, with the particles themselves arbitrarily spread out along the ordinate. We now modify that display by using the particle masses to represent the ordinates, which yields a plot of mass widths $\Gamma \equiv \Delta m$ versus masses m. This plot is shown in Fig. 3.2.2. As can be seen, the mass-width points form two diagonal bands which are separated along the abscissa by

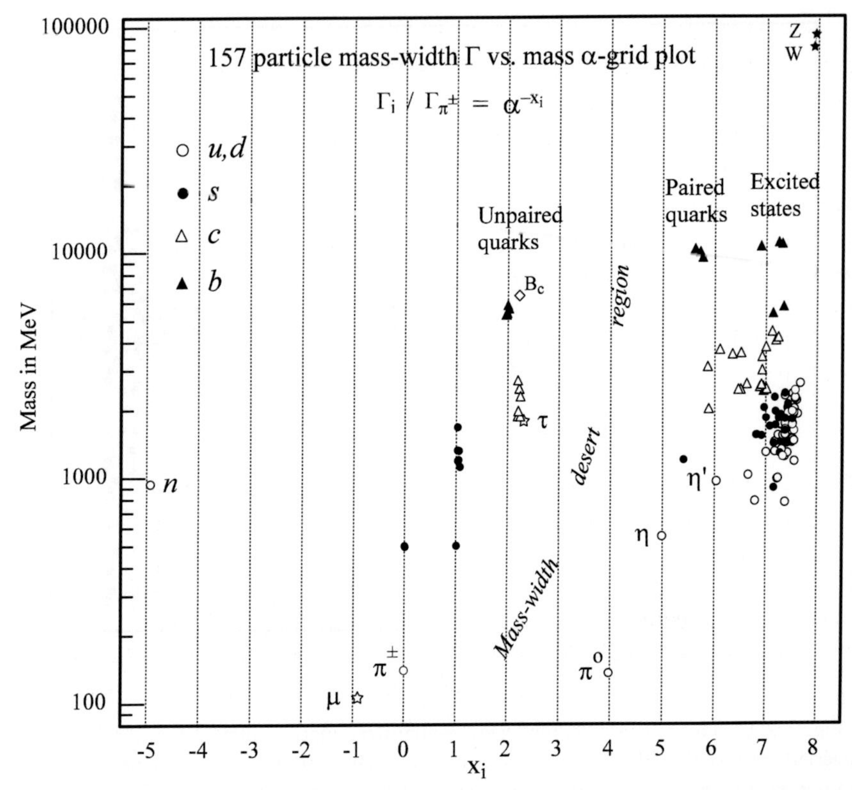

Fig. 3.2.2 This figure uses the mass width Γ abscissa of Fig. 3.2.1, but the ordinate is now the mass of the particle in MeV, so that we have a mass vs. mass-width plot of the threshold-state resonances. There is a clear-cut relationship between masses and widths, with the unpaired-quark states forming a diagonal band of narrow-width states, the paired-quark, radiative decays and excited states forming a second diagonal band of much-broader-width states (broader by a factor of $\alpha^{-4} \sim 10^8$), and the region between them forming a *mass-width desert* (Fig. 0.1.14). These results indicate a correlation between masses and mass-widths (lifetimes), but they do not reveal much about how an α-quantization of the widths is reflected in the masses themselves. For that information, we must go directly to the masses, whose α-quantization awaits our inspection.

factors of about α^{-4} ($\sim 10^8$) in relative width, with a mass-width (lifetime) desert region between them (see Figs. 0.1.13, 0.1.14, 2.8.7 and 2.8.8). This plot displays the manner in which the groups of quantized mass widths Γ are systematically shifted to adjacent α-grid positions and broader widths with increasing mass.

Figure 3.2.2 can be used to infer general relationships between masses and mass widths: larger masses are less stable, and the mass widths increase in quantized powers of α as the mass increases. However, this figure supplies little direct information about what kind of α-dependence is actually contained in the masses. But, encouraged by these results, we can answer this question by studying the masses themselves. Specifically, we can look for mass ratios that contain factors of 137. Since the lifetime scaling in α is spread throughout the threshold-state region of lifetimes $\tau > 1$ zs, we expect factor-of-137-generated mass units to be scattered throughout the masses of these threshold-state particles. This search for α^{-1}-scaled masses is the subject of the next section.

3.3 Electrons, Muons and Pions: The "Rosetta Stones" of α-Quantized Masses

In the present section we lay the foundation for the phenomenology of *first-order* α-generated elementary particle mass quanta. These results are guided by the lifetime and mass-width systematics studied above, but they emerge here directly from the systematics of the masses themselves, including both hadrons and leptons. When viewed properly, the elementary particle mass data base could hardly be more revealing in what it has to tell us. *Second-order* α-generated particle mass quanta, which are required in order to account for particle states above 12 GeV, are discussed in Sec. 3.20.

The lifetimes of the 36 long-lived ($\tau > 10^{-21}$ s) threshold-state elementary particles are grouped in clusters which are globally separated from one another by scaling factors of $\alpha \cong 1/137$, as we demonstrated in detail in Chapter 2. Thus the mass widths of these same particles are similarly grouped, and are reciprocally separated from one another by factors of $\alpha^{-1} \cong 137$ (Fig. 3.2.1). Since there appears to be a correlation between mass widths and the masses themselves (Fig. 3.2.2), we logically expect to find factor-of-137 mass ratios among the threshold-state masses. Figure 3.3.1 displays a plot of the mass values of these 36 unstable threshold-state particles. As can be seen, their mass values extend from 105 MeV for the muon to 10.355 GeV for the Υ_{3S} upsilon resonance. At the right in Fig. 3.3.1 we have plotted the 105 MeV muon level together with a calculated hypothetical mass level at 105 MeV $\times$ 137 = 14.385 GeV. These two levels represent the $\alpha^{-1} \cong 137$ mass ratio we are looking for. However, the higher-mass level is well above any of the 36 threshold-state masses, which

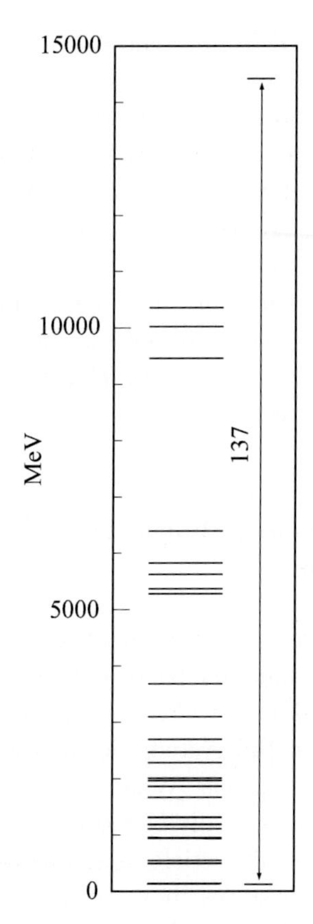

Fig. 3.3.1 An energy level diagram for the 36 threshold-state particles of Fig. 2.8.8, which are the particles that exhibit α-quantized lifetimes. These 36 particles are listed at the beginning of Appendix A, and their mass values are given in Appendix B. The two levels shown at the right in Fig. 3.3.1 represent the 105 MeV muon level (the lowest one in the figure) together with a calculated mass level at 14.385 GeV that is a factor of 137 larger in mass. These two levels give the α^{-1} mass ratio we are looking for. But since this α-enhanced 14 GeV mass level lies well above all of the experimental levels displayed in Fig. 3.3.1, it is apparent that none of the 36 unstable threshold-state masses exhibit 137 MeV mass ratios. Hence these 36 particles do *not* contain the α^{-1}-quantized masses that are needed to explain their α-quantized lifetimes.

illustrates the fact that there are *no* factor-of-137 mass ratios to be found among any of these 36 threshold-state particles. Hence, if we want to find α^{-1}-quantized masses, we must look elsewhere.

The two long-lived massive elementary particles that we have not included in Fig. 3.3.1 are the electron and proton. Since these two particles are stable, they did not enter into the α-quantized logarithmic lifetime systematics of Chapter 2. The proton mass falls well inside the mass domain of the 36 unstable threshold-state particles of Fig. 3.3.1, as shown in Fig. 3.1.2. Thus adding it to the data compilation does not create any factor-of-137 mass ratios. But the electron is a different story. Its mass is so small, 0.511 MeV, that it is a ready candidate for α-quantization. What is remarkable is that not only does the electron create α-quantized masses, as we will demonstrate, but it does so in two different forms, m_f (fermion) and m_b (boson), and with the two next-lowest-mass particle states — the muon and pion, respectively. Furthermore, as we describe in detail in the present chapter, the muon and pion in turn serve as "platform masses" upon which the other threshold-state particles are generated via the "supersymmetric" platform excitation quantum $X = 420$ MeV. And, perhaps even more remarkably, the basis-state masses $m_b = 70$ MeV (spin 0) and $m_f = 105$ MeV (spin 1/2) are themselves mathematically linked together by the systematics of the relativistically spinning sphere, as we calculate in Chapter 4. Hence the electron, muon and pion function together as "Rosetta stones" that furnish vital keys for unlocking the secrets of elementary particle mass α-quantization.

As we have just described, any possible $\alpha^{-1} \cong 137$ mass ratios among the long-lived threshold-state particles must involve the electron mass of 0.511 MeV. But when we consider the creation of elementary particles directly out of "pure energy," which is what we are implicitly considering here, electrons and positrons (anti-electrons) must be created in matching pairs, so electrons first appear at an excitation energy of 1.022 MeV. The first mass level above the electron is that of the muon at 105.66 MeV. Muons, like electrons, are fermions, and they must also be produced in matching particle–antiparticle pairs. Hence the lowest energy at which muons appear is 211.32 MeV. Thus the first observable particles above the electron level are the π^0 and $\pi^\pm$ pions at 134.98 and 139.57 MeV, respectively. The pion has balanced particle–antiparticle symmetry, which means that it is its own antiparticle. Hence it must contain at least two matching subquanta — one particle and one antiparticle. The quantum numbers of the electron do not match those of the pion, and mass ratios only make sense if they are between particle states of the same type. The electron is a spin-1/2 particle, whereas the pion is a zero-spin particle. In order to match the pion quantum numbers by using electrons, we must start with an electron–positron "ground state," which has balanced particle-antiparticle symmetry, and we must combine the spin-1/2 electron and spin-1/2 positron in a spin-0 configuration in order to match the angular momentum of the spin-0 pion. Since the mass of an electron–positron pair is 1.022 MeV, the mass ratio between a charged electron pair and a charged 139.57 MeV $\pi^{\pm}$ meson is 136.57, which is within 0.4% of the mass ratio $\alpha^{-1} = 137.04$ that we are looking for! This is encouraging news in our search for α^{-1} mass ratios. However, the 134.98 MeV π^0 meson has a slightly different mass, and the pion itself, unlike the electron, is a compound structure with an extended spatial distribution [21]. Hence we are not looking for precise numerical values when we combine "simple" electron masses with "complex" pion masses, but rather just for a general indication that these particle states constitute an example of mass α-quantization.

As a note to the reader, we often label the e^-, e^+ "electron–positron pair" that occurs in the creation of particles out of "pure energy" (which must be done in a particle–antiparticle-symmetric manner) simply as an "electron pair." The positron is in fact a "positive electron." In superconductivity, where "Cooper pairs" are coherent spin-up, spin-down pairs of negatively charged electrons, the term "electron pair" has a different meaning, but superconductivity occurs in a realm of physics that contains only "particle" states. In particle physics, a "pair" more naturally means a "particle–antiparticle pair."

Let us now proceed from the *ansatz* that the $\alpha^{-1} \cong 137$ mass quantization does in fact exist, and see where this leads us in our examination of the experimental data. The most straightforward α^{-1} particle mass we can generate from the electron mass m_e is the mass quantum $m_e/\alpha = 70.025$ MeV. (This is the mass unit that is related via the Compton radius $R_C = \hbar/mc$ to Werner Heisenberg's postulated "universelle Lange" [52].) We can denote this as the basic "α-mass" $m_{\text{boson}} \equiv m_b$, since we are dealing here with integral-spin bosonic pions. If we generate a matching antiparticle mass $\bar{m}_b$ from the positron mass $\bar{m}_e$ and then combine them together, we obtain a balanced particle–antiparticle $m_b\bar{m}_b \equiv m_b + \bar{m}_b$ mass of 140.050 MeV, which is quite close to the measured pion masses. If we further add in the initial 1.022 MeV mass of the electron ground state, we obtain a total mass of 141.072 MeV. (It is not immediately obvious why we should do this, but it will become clear when we study the muon.) This result is shown graphically in Fig. 3.3.2, where it gives a pictorial view of the accuracy of the calculation. This is the first step in our "Rosetta stone" search for α-quantized masses.

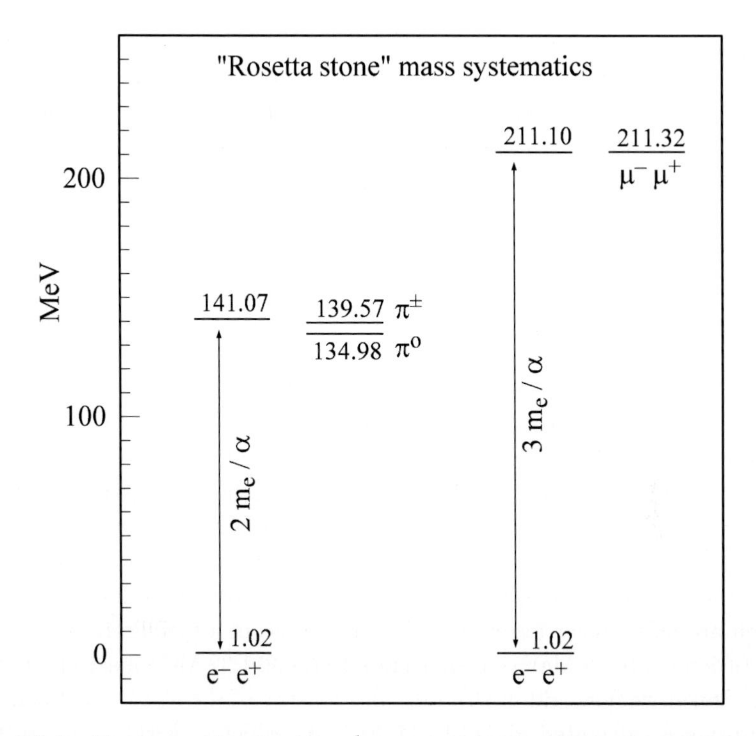

Fig. 3.3.2 The mass values for the α^{-1}-quantized generation of pion and muon-pair threshold states from an electron-pair "ground state." We start with the pion. The fundamental α^{-1}-quantized mass unit, $m_e/\alpha = 70.025$ MeV, is about half the mass of a pion. We denote this as the "α-mass" m_b, where the subscript b stands for "boson," and refers to the fact that the pion is an integral-spin boson. The pion, which is its own antiparticle, must be an $m_b\bar{m}_b$ state. If we use an α^{-1}-quantized mass interval of $2m_e/\alpha = 140.05$ MeV for the $m_b\bar{m}_b$ "α-leap" from a spin-0 $m_e\bar{m}_e$ initial state, and then add in the 1.022 MeV mass of the e^-e^+ state, we obtain a calculated mass of 141.07 MeV, which is a few percent higher than the observed masses of the two pion levels. This mass difference can be attributed to a "hadronic binding energy" (HBE) that operates between the matching hadronically bound m_b and $\bar{m}_b$ particle and antiparticle pion substates (Fig. 0.2.16).

Now consider the muon pair. This excitation level is at half again the mass of the pion, so we define a second α-mass, $m_f = 3/2\ m_b = 105.038$ MeV, where the subscript f stands for "fermion", since the muon is a half-integral-spin fermion. An $m_f\bar{m}_f$ "α-leap" from a spin-0 or spin-1 $m_e\bar{m}_e$ initial state takes us to 210.08 MeV. Adding in the 1.02 MeV mass of the e^-e^+ state gives a calculated mass of 211.10 MeV, which is 0.1% below the observed 211.32 MeV mass of a muon pair. Hence the leptonic muon-antimuon state has essentially zero binding energy within this level of accuracy. The factor of 3/2 mass difference between the m_b and m_f α-masses can be attributed to their different spin values (Chapter 4).

The second step in the Rosetta stone search occurs when we move on to the muons, which are the second type of particle to appear above the electron pair level. The $\mu^{\pm}$ muons are spin-1/2 leptons, just like the spin-1/2 $e^{\pm}$ electrons, so we can form the mass ratio of a single electron with a single muon instead of working with e^-e^+ pairs as we did in the case of the pion. However, it is instructive to work at first with an electron–positron pair and a muon–antimuon pair, so that we can see the relationship between the pion and muon Rosetta-stone α-excitations more clearly. The mass of a muon pair is 211.32 MeV, which is about half again the mass of the pion, as shown in Fig. 3.3.2. Thus we could reach this level by using three 70 MeV m_b and $\bar{m}_b$ masses instead of two as we did for the pion (see Fig. 3.3.2). However a combination of three m_b masses would not have particle–antiparticle symmetry, and thus could not logically produce a symmetric muon–antimuon pair. Hence we postulate the generation of a second type of α-mass from the electron — the mass quantum $3m_e/2\alpha =$ 105.038 MeV. To distinguish this α-mass from the pion α-mass, we denote it as the mass quantum $m_{\text{fermion}} \equiv m_f$, because the spin-1/2 electron and muon are half-integer fermion states. The mass of a symmetric $m_f\bar{m}_f$ α-excitation is 210.08 MeV, which is close to the 211.32 MeV mass of a muon pair. But if we now add in the 1.02 MeV energy of the e^-e^+ ground state, we obtain a calculated mass of 211.10 MeV, which is within 0.1% of the experimental mass of the muon pair (Fig. 3.3.2). This mass accuracy does not appear to be accidental, and it suggests that the α-mass m_f represents an excitation quantum which is to be added to the m_e ground state mass in order to obtain the mass of the muon excitation level. When we are dealing with electrons and muons, we have only a single charge state in each to consider, and we do not have a hadronic binding energy involved. Also, the electron and muon are both particles that have no measurable substructures. Thus they represent a particularly clear-cut situation with respect to the particle excitation process. Hence the accurate agreement (0.1%) between their calculated and experimental mass ratios is significant.

It may seem as if we have simply pulled the $m_f/m_b = 3/2$ mass ratio out of a hat in order to reproduce the mass of the muon, but this is not the case. When we come to a study of the mathematical phenomenology (mathology) of spinning masses, which we present in Chapter 4, we discover that a relativistically spinning spherical mass which is rotating at the full relativistic limit (where the equator is moving at, or infinitesimally below, the velocity c) is half again as massive as it was at rest: $m_s = 3/2\ m_0$. Furthermore, if the radius of the spinning sphere is the Compton radius $R_c = \hbar/m_s c$, then its calculated spin angular momentum is $J = 1/2\ \hbar$. These are straightforward (but unfamiliar) results of special relativity, and they emerge here from the mass distributions in spinning muons as compared to those in nonspinning pions (Sec. 4.3). Nature is telling us something here about the workings of special relativity.

The above discussion has centered on just the three "Rosetta stone" particles — the electron, muon and pion. In order to demonstrate how the other threshold-state particles fit into this general excitation scheme, we first divide these states into $J = 0$ and $J = 1$ electron-pair excitations, which are placed in "excitation columns." Then we select the fermion states from $J = 1$ excitation column and portray them in matching paired $J = 1/2$ excitation columns (only one of which has to be displayed). The experimental mass levels for the primary $J = 0$ and $J = 1$ excitation columns, together with a deduced $J = 1/2$ fermion excitation column, are plotted in Fig. 3.3.3. As can be seen, the primary $J = 0$ or 1 excitation in each case is an $m_b\bar{m}_b$ or $m_f\bar{m}_f$ α-mass pair from an e^-e^+ ground state, with deduced $J = 1/2$ m_f and $\bar{m}_f$ excitations for μ^- and μ^+ fermions. These initial excitations are denoted here as "α-leaps" to "platform states" M. The higher excitations are also composed of m_b and m_f mass quanta, but they feature an "M^X" platform excitation formalism based on an α-quantized excitation quantum $X = 6m_b = 4m_f$, as we demonstrate in later sections of this chapter.

3.4 The First-Order $m_b = 70$ MeV Boson and $m_f = 105$ MeV Fermion "α-Leap" Masses

In Sec. 3.3 we introduced the α-masses m_b and m_f as the basic mass quanta that are required in order to reproduce the pion and muon from the electron by the operation of the coupling constant α on the electron mass. This is analogous to the operation of the coupling constant α on the electron mass to produce the photon in QED. We formally define these fundamental m_b and m_f α-mass states as follows:

$$(m_b, \bar{m}_b) \equiv (m_e, \bar{m}_e)/\alpha = 70.025 \text{ MeV} \quad (\text{"Boson } \alpha\text{-mass," spin-0 channel}), \tag{3.4.1}$$

$$(m_f, \bar{m}_f) \equiv (3/2)(m_b, \bar{m}_b) = 105.038 \text{ MeV} \quad (\text{"Fermion } \alpha\text{-mass," spin-1/2 channel}), \tag{3.4.2}$$

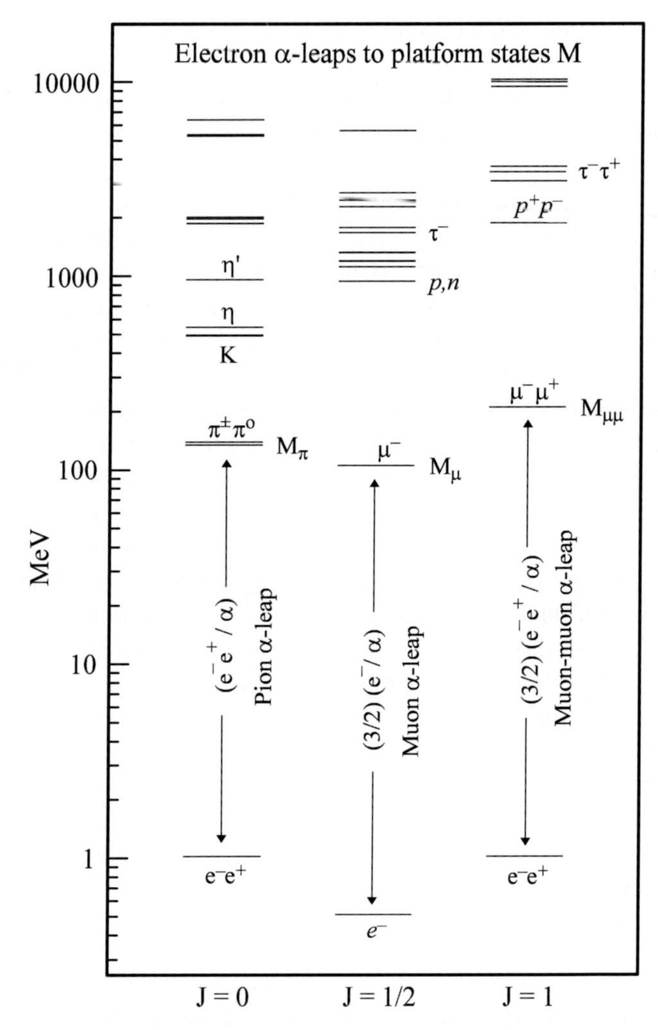

Fig. 3.3.3 The mass values of all massive elementary particles with half-lives longer than 10^{-21} s, shown arranged in spin-channel excitation towers. Also indicated are the quantized "α-leap" mass intervals above the electron "ground states." The $J = 0$ and $J = 1$ e^-e^+ spin channels contain the particle–antiparticle primary excitations, and the $J = 1/2e^-$ channel represents one branch of the $J = 1$ channel after it has been conceptually separated into particle and antiparticle branches. The particle–antiparticle pairs that are contained in the $J = 1$ channel are very lightly bound (Fig. 0.2.16), so that each branch can be treated separately from the standpoint of calculating particle masses. The particle levels at the tops of these α-leaps represent "platform" masses M, which are defined in detail in Sec. 3.5. The particle states that are built upon these platforms follow a quite different "M^X" excitation pattern, as described in Secs. 3.6–3.13.

where m_e is the electron mass and $\alpha = e^2/\hbar c$. As we demonstrate in succeeding sections, these two α-mass states occur in various combinations to create the basic threshold-state particles that dominate the formation of the higher-mass elementary particle excitations.

We have not assigned any charge dependence to the m_b and m_f α-masses in Eqs. (3.4.1) and (3.4.2). Thus we are unable at this level of mass phenomenology to reproduce the charge splittings of the masses in the isotopic spin multiplets. The pion isotopic spin doublet consists of the pion states π^0 and $\pi^{\pm}$, which have masses that differ by 3.3%. Hence the α-mass substates m_b and $\bar{m}_b$ that compose the pion should logically have a charge dependence of some kind. When we examine the charge splitting of the pion masses, we discover an important fact: this charge splitting is almost precisely equal to nine electron masses, as we illustrate graphically in Fig. 3.4.1. The accuracy of this result, 0.1%, combined with the various factors of 3 and 9 that can be observed in both the lifetime and mass ratios, as well as in the quark charges, suggests that this is more than just a coincidence. But knowing how to incorporate this information into a model which reproduces the experimental charge splitting is a different matter. For example, the $K^0 - K^{\pm}$ charge splitting is almost as large as the $\pi^{\pm} - \pi^0$ charge splitting, but is of opposite sign. And other charge splittings show a variety of values. In lieu of a theoretical formulation for charge splittings of the masses, we simply take the average mass of each isotopic spin charge multiplet and denote it as the "charge-independent" mass (CI mass) of the state. The CI mass of the π^0 and $\pi^{\pm}$ pion states is 137.27 MeV, and the CI mass of the K^0 and $K^{\pm}$ kaon states is 495.66 MeV.

One more result we can extract from the "Rosetta stone" systematics of Sec. 3.3 is some information about hadronic binding energies. The muon mass calculation displayed in Fig. 3.3.2 does not seem to involve any binding energy effects, at least at the level of accuracy we are considering here. However, the calculated pion mass is higher than the experimental π^0 and $\pi^{\pm}$ masses by a few percent. Since the pion contains both particle and antiparticle substates in a hadronically bound configuration, it is plausible that there is an appreciable *hadronic binding energy* (HBE) which must be taken into consideration. One experimental result we can point to which gives a direct measurement of hadronic binding energies is the annihilation of an antiproton with a neutron: detection of the total energy of the decay products reveals an HBE of 4.4% [53], which is the largest binding energy ever recorded. The threshold states that we consider in the present studies have HBE's of roughly 3%, as we summarize in Sec. 3.15. In order to eval-

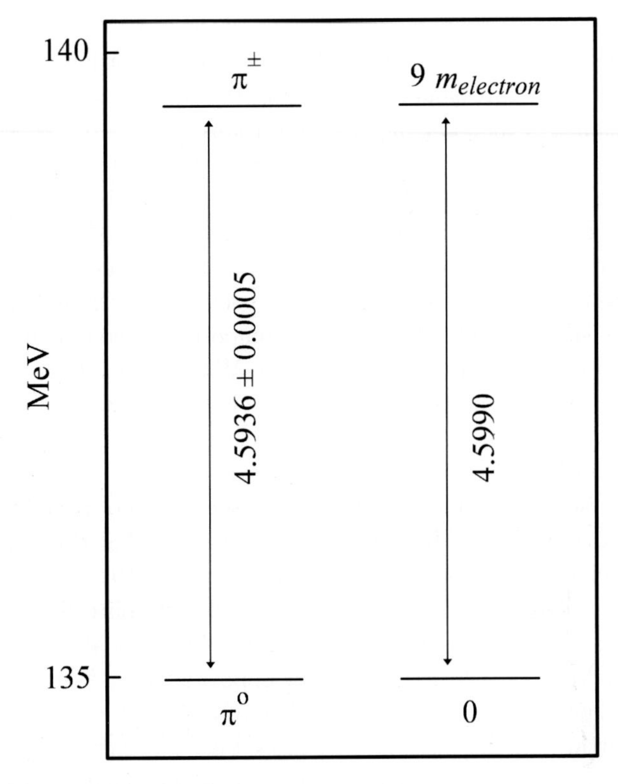

Fig. 3.4.1 A graphical comparison of the $\pi^{\pm}$ and π^{0} mass difference with the mass of nine electrons. The agreement to 0.1% between these two mass intervals does not seem accidental. Lacking a comprehensive formalism for reproducing isotopic spin mass splittings, we use "charge-independent" (CI) average mass values for isotopic spin multiplets when calculating particle masses (Appendix B).

uate the pion HBE, we must compare the (calculated) unbound mass with the experimental mass. The calculated pion mass displayed in Fig. 3.3.2 is 141.07 MeV. For the experimental pion mass we use the CI mass of 137.27 MeV [10]. This gives a calculated HBE of 2.7%. The binding energy systematics of the pion and muon "Rosetta stone" excitations is portrayed graphically in Fig. 3.4.2.

In order to handle the α-quantization of the higher-mass threshold-state particles, we need to formalize the concept of "platform excitations," which we do in the next section. These platform states emerge directly from the Rosetta stone α-leap excitations described above, as we indicated in Fig. 3.3.3.

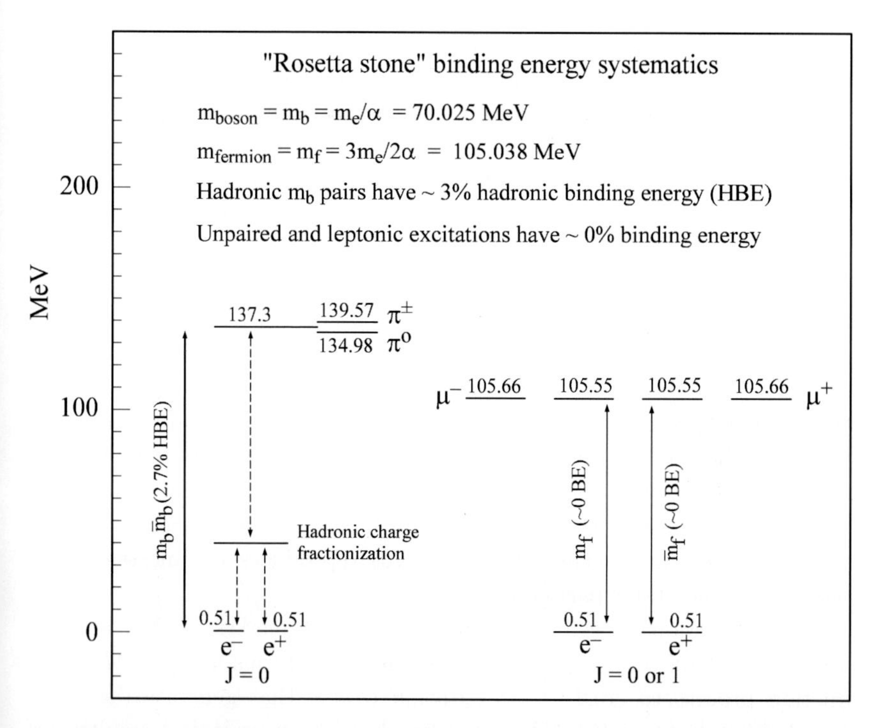

Fig. 3.4.2 The binding energy systematics of pion and muon α-generation from electron ground states. The m_b and $\bar{m}_b$ substates in the pion presumably carry fractional charges which are created when the electron ground states merge in a pairwise manner with the spinless m_b excitation quanta. The fractional charges lead to the correct isotopic spin rules for these states, and also to a few percent hadronic binding energy (HBE) between particle and antiparticle substates. Matching to the average CI mass of the pion isotopic spin doublet requires an HBE of 2.7%. The muon mass is accurately reproduced by assuming HBE = 0.

3.5 Symmetric (M_π, M_ϕ, $M_{\mu\mu}$) and Asymmetric (M_K, $\bar{M}_K$, M_μ, $\bar{M}_\mu$) "Platform" States

The search for factor-of-137 mass ratios in the threshold-state elementary particles has turned up two candidates — the 137 MeV α-leap from an electron pair to the pion, and the 211 MeV α-leap from an electron pair to a muon pair. In terms of *single-electron* excitations, these electron-pair α-leaps correspond to the production of excitation quanta $m_b \cong 70$ MeV or $m_f \cong 105$ Mev (Eqs. (3.4.1) and (3.4.2)), which are combined with matching $\bar{m}_b$ or $\bar{m}_f$ states to form $m_b\bar{m}_b \cong 140$ MeV or $m_f\bar{m}_f \cong 210$ MeV excitations. We might think that higher-mass electron-pair α-quantization excitations would consist of additional 140 or 210 MeV α-leaps, but this is *not* what is observed experimentally. Additional excitations typically occur in mass units of $Q = 210$ MeV (Figs. 0.2.4 and 0.2.12) or $X = 420$ MeV (Figs. 0.2.19 and 0.2.20), where Q and X are α-quantized excitation quanta. Conceptually, the combination of the initial electron-state masses plus the m_b or m_f excitations that correspond to the α-leap create what we denote here as "platform masses" M, which establish the basic quantum numbers for the production channel (Figs. 0.2.15–0.2.17). These platform masses serve as "ground states" for the construction of "excitation towers" formed from mass quanta Q or X. The supersymmetric platform excitation quantum X is displayed graphically in Fig. 0.2.18, and is formally defined in Sec. 3.9. The combination of an α-leap from an electron state to a platform M plus the creation of additional X-quanta to reach higher-mass states constitutes the two-step "M^X" production process described in Secs. 3.7–3.13, which uniquely and accurately (Sec. 3.16) reproduces the masses of the basic threshold-states particles.

In this section we define the various platform masses M that are created by electron α-leaps. We later use these platforms and the M^X production process to create the excitation towers that generate the other threshold-state particles. These are purely phenomenological results: we first generate the platforms M and then select the required number of X quanta to reproduce the particle masses. However, the results we obtain are in the form of an M^X excitation pattern that clearly is not random (Figs. 0.2.19 and 0.2.20), and the accuracy of the results attests to their relevance. The most remarkable aspect of these results is that they *combine leptons and hadrons together in one unified mass formalism.* In fact, without combining leptons with hadrons we could not ascertain the details of this formalism.

We start by conceptually defining generic *boson* and *fermion* platform states M that are generated by α-leaps from electron ground states. Then we relabel these platforms in terms of the lowest-mass particles that they produce. The direct conversion of energy into particle mass is a particle–antiparticle-symmetric process. Thus an electron-based α-leap excitation process must start with a symmetric electron–positron pair, and the platform state it produces must be particle–antiparticle symmetric. These *symmetric* platforms can be separated into *asymmetric* particle and antiparticle components, which we now define. The generic asymmetric *boson* platform components are

$$M_b \equiv m_e + m_b\,, \quad \bar{M}_b \equiv \bar{m}_e + \bar{m}_b\,, \qquad (3.5.1)$$
70.54 MeV ($J = 0$) "asymmetric boson platforms."

The generic asymmetric *fermion* platform components are

$$M_f \equiv m_e + m_f\,, \quad \bar{M}_f \equiv \bar{m}_e + \bar{m}_f\,, \qquad (3.5.2)$$
105.55 MeV ($J = 1/2$) "asymmetric fermion platforms."

These generic platform states are illustrated graphically in Fig. 0.2.15. The corresponding generic *symmetric* platform masses are

$$M_b + \bar{M}_b \equiv (m_e + \bar{m}_e)(1 + 1/\alpha) = 141.07\ \text{MeV}\,, \qquad (3.5.3)$$
"symmetric boson platform,"

and

$$M_f + \bar{M}_f \equiv (m_e + \bar{m}_e)(1 + 3/2\alpha) = 211.10\ \text{MeV}\,, \qquad (3.5.4)$$
"symmetric fermion platform."

We denote these symmetric platform masses notationally as the generic α-quantized platform states $M_b\bar{M}_b \equiv M_b + \bar{M}_b$ and $M_f\bar{M}_f \equiv M_f + \bar{M}_f$.

We now relabel these generic symmetric and asymmetric platform states in terms of the lowest-mass particles that occur in their excitation towers. The symmetric boson platform $M_\pi = M_b\bar{M}_b$ is produced by matching m_b and $\bar{m}_b$ α-leaps from an electron–positron pair, and it contains the π^0 and $\pi^\pm$ pions. Its M_b substates are hadronically bound together. Since the pion is a spin 0 particle, the initial electron–positron pair must be in a $J = 0$ spin state. The symmetric pion platform M_π is formally defined as follows:

$$M_\pi \equiv (m_e + \bar{m}_e)(1 + 1/\alpha) = 141.07 \times (1 - \text{HBE})\ \text{MeV}\,, \qquad (3.5.5)$$
$J = 0$ "pion platform,"

where HBE is the hadronic binding energy of 2.7% (Fig. 3.4.2).

Symmetric fermion platforms $M_f\bar{M}_f$ come in two forms, M_ϕ and $M_{\mu\mu}$. The M_ϕ platform is hadronically bound, like the M_π platform, but the $M_{\mu\mu}$ platform is unbound. The symmetric $M_\phi = M_f\bar{M}_f$ hadronic platform is created by matching m_f and $\bar{m}_f$ α-leaps from an electron–positron pair. It does not represent an observed particle, but when an X quantum is added to each of its M_f substates, the hadronic $\phi = s\bar{s}$ vector meson is formed, where s is the strange flavored quark. Since the ϕ is a spin-1 vector meson, the initial electron–positron pair in the M_ϕ production mode must be in a $J = 1$ spin state. The symmetric and hadronically bound phi meson platform M_ϕ is

$$M_\phi \equiv (m_e + \bar{m}_e)(1 + 3/2\alpha) = 211.10 \times (1 - \text{HBE})\ \text{MeV}\,, \qquad J = 1 \text{ ``phi platform.''} \tag{3.5.6}$$

The symmetric $M_{\mu\mu} = M_f \bar{M}_f$ platform is not hadronically bound. It is created by matching m_f and $\bar{m}_f$ α-leaps from an electron–positron pair, just like the M_ϕ platform, but is occupied by a leptonic $\mu^-\mu^+$ muon pair. It can be in either a $J = 0$ or $J = 1$ spin state. The formal definition of the symmetric $M_{\mu\mu}$ platform state is

$$M_{\mu\mu} \equiv (m_e + \bar{m}_e)(1 + 3/2\alpha) = 211.10\ \text{MeV}\,, \qquad J = 0 \text{ or } J = 1 \text{ ``mu} - \text{mu platform.''} \tag{3.5.7}$$

The generation processes for these three symmetric platform states are displayed graphically in Fig. 0.2.16. We sometimes refer to the leptonic platform state $M_{\mu\mu}$ as being in the "$J = 1$ electron pair channel" to distinguish it from the "$J = 0$ electron pair channel" platform state M_π that corresponds to pion production. These two platforms contain the two lowest particle production levels above the electron–positron ground-state level.

The muon and antimuon leptons that occupy the $M_{\mu\mu}$ platform are essentially unbound. This means that the $M_f \equiv \mu$ and $\bar{M}_f \equiv \bar{\mu}$ spin-1/2 particles of the $M_{\mu\mu}$ platform act more-or-less independently after their joint production process, and we can study them as separate excitation channels instead of using the combined form shown in Eq. (3.5.7). This is particularly useful when calculating mass values, since the excitations in each channel occur right on the mass shell. We denote these separated fermion channels as the asymmetric platform states $M_\mu = M_f$ and $\bar{M}_\mu = \bar{M}_f$. Their formal definition is

$$M_\mu \equiv m_e(1 + 3/2\alpha)\,, \quad \bar{M}_\mu \equiv \bar{m}_e(1 + 3/2\alpha)\,, \qquad 105.55\ \text{MeV}\ J = 1/2 \text{ ``muon platforms.''} \tag{3.5.8}$$

As we will see, these asymmetric M_μ platforms enable us to accurately reproduce the masses of the proton and the τ lepton. This is the prime example of how lepton and hadron masses interleave together to form a coherent elementary particle mass grid.

The M_b and $\bar{M}_b$ platform states are conceptually created by splitting apart the hadronically bound M_π platform. The M_b and $\bar{M}_b$ platform masses, like the M_ϕ platform mass, do not correspond to observed particles, but when an X quantum is added to each, they become the strange K mesons, or kaons. Thus we relabel M_b and $\bar{M}_b$ as the M_K and $\bar{M}_K$ platform states, which are defined as follows:

$$M_K \equiv m_e(1 + 1/\alpha)\,, \quad \bar{M}_K \equiv \bar{m}_e(1 + 1/\alpha)\,, \qquad (3.5.9)$$
$$70.54 \text{ MeV } J = 0 \text{ “kaon platforms.”}$$

These platforms lead to a crucial phenomenological result. The $M_K + X \equiv K$ meson is spinless and the X quantum is spinless. Thus the M_K platform mass must also be spinless, including its m_b α-mass, even though it is generated from the spin-1/2 electron. The reason this can happen is that the M_K or M_b mass elements are always produced in pairs when created from an e^-e^+ ground state, and if the electron pair is in a $J = 0$ spin configuration, then overall angular momentum is conserved in the α-induced M_π platform production process that leads to the separated M_K platforms. The production channel for the creation of the M_π platform state is depicted in Fig. 0.2.16. This may at first glance appear to be a very convoluted way of producing pions, but it is what the experiments require, and it has ramifications that extend through all the threshold-state production processes. The M_π separation into M_K and $\bar{M}_K$ substates is displayed graphically in Fig. 0.2.17.

Having defined the various α-leap platform states, we go on in the following sections to illustrate the excitation towers that are erected on these platforms. We start by sorting out the small number of particle states that are available to populate these excitation towers.

3.6 The Spin and Flavor Hierarchy of the M^X Platform Excitations

The α-quantization of elementary particle states first became clearly evident in the lifetimes or mass widths (Eq. (1.7.1)) of these states, as we discussed in detail in Chapter 2. We demonstrated that the lifetimes or mean lives of these states divide into two groups: (1) the short ($\tau < 10^{-21}$ s) lifetimes, which occur as essentially a continuum of closely-spaced values, and which correspond to excited-state resonances; (2) the long ($\tau < 10^{-21}$ s) lifetimes, which occur in widely-spaced lifetime groups that are separated by powers of α, and which correspond to threshold-state particles. These two groups of lifetimes are displayed in Fig. 2.11.10. Since the *lifetime* scaling in powers of $\alpha \cong 1/137$ occurs only among the long-lived threshold-state particles, we should expect a reciprocal *mass* scaling in powers of $\alpha^{-1} \cong 137$ to occur among these same threshold-state particles. The first 36 particles listed in the lifetime compilation of Appendix A are the threshold-state particles displayed in Fig. 2.8.8. For mass studies we should also add in the stable electron and proton, which have infinite lifetimes. We also include the $\phi(1020)$ meson. Its lifetime is slightly shorter than the 1 zs threshold-state boundary, but, as we will demonstrate, the $\phi = s\bar{s}$ meson is a true threshold-state particle which is composed of a "strange" $s\bar{s}$ quark–antiquark pair. Since a $K\bar{K}$ kaon pair is also strange, and is less massive than the ϕ (the $K\bar{K}$ bound state is the $\eta'(958)$ meson), the ϕ can decay rapidly into a pair of kaons (its main decay mode) without breaking strangeness, which gives it an anomalously short lifetime. Finally, we remove the composite Ξ_b excitation. We thus have a total of 38 particles that logically exhibit α^{-1} mass ratios. These particles are listed in Table 3.6.1, where they are broken down into boson (integer) and fermion (half-integer) spin states, and are assorted into variously flavored (s, c, b) quark states. Corresponding equal-mass antiparticle states also occur, but for simplicity are not included here.

Our purpose in listing these 38 long-lived particle states is to evaluate them with respect to their degree of importance in the hierarchy of the elementary particle mass generation process. The ones that are generated first are the most fundamental in our search for an α-quantization of the

Table 3.6.1 The 36 threshold-state particles with α-quantized lifetimes $\tau > 1$ zs, plus the $\phi(1020)$ meson and the stable electron and proton, and minus the Ξ_b. These 38 particles logically exhibit α-quantized masses. They are shown here separated into boson and fermion spin states and sorted into flavor types.

(1a) Nonstrange PS bosons: $(\pi^{\pm}, \pi^0, \eta, \eta')$
(1b) Leptons and nonstrange fermions: $(e^{\pm}, \mu^{\pm}, p^{\pm}, n^0, \tau^{\pm})$

(2a) Strange PS bosons: $(K^{\pm}, K_L^0, K_S^0)$
(2b) Strange vector bosons: $(\phi(1020))$
(2c) Strange fermions: $(\Lambda^0, \Sigma^+, \Sigma^0, \Sigma^-, \Xi^0, \Xi^-, \Omega^-)$

(3a) Charm bosons: $(D^{\pm}, D^0, D_s^{\pm}, D^{*\pm}(2010), J/\psi_{1S}, J/\psi_{2S})$
(3b) Charm fermions: $(\Lambda_c^+, \Xi_c^+, \Xi_c^0, \Omega_c^0)$

(4a) Bottom bosons: $(B^{\pm}, B^0, B_s^0, B_c^{\pm}, \Upsilon_{1S}, \Upsilon_{2S}, \Upsilon_{3S})$
(4b) Bottom fermions: (Λ_b^0, Ξ_b) (Ξ_b has no measured mass)

masses. Ones that are generated later as additions to or modifications of the "first generation" are not likely to be as informative in revealing the essence of particle mass generation. By "particle mass generation" we mean the straight transformation of energy (E) into matter (mc^2), as defined by Einstein. This process is, as far as we know, particle-antiparticle symmetric. The *first generation* particles that fall in this category are the nonstrange pseudoscalar bosons of group (1a) in Table 3.6.1, which are particle–antiparticle self-symmetric, and the leptons and nonstrange fermions of group (1b), which must be produced in matching particle–antiparticle pairs. Group (1a) is analyzed in Sec. 3.7, and group (1b) is analyzed in Sec. 3.8.

The *second generation* particles are the "strange" bosons and fermions of groups (2a)–(2c). These occur in two tiers. The low-mass tier consists of groups (2a) and (2b), which can be thought of as composed of matching particle and antiparticle "strange" components of "non-strange" excitations: $K\bar{K} = \eta'$ (2a) and $s\bar{s} = \phi$ (2b). The high-mass tier is the (2c) group, which features hyperon (Λ, Σ, Ξ, Ω) excitations of baryon (p, n) target states. These are asymmetric interactions, and the structures of these particles, although falling within the general framework of α-quantized states (as we will briefly discuss in Sec. 3.23), are more complex with regard to binding energies and the mixing of bosonic and fermionic mass units. Group (2a) is analyzed in Sec. 3.10, and group (2b) is analyzed in Secs. 3.11 and 3.12.

The *third generation* particles are the "charm" bosons and fermions of groups (3a) and (3b), and the *fourth-generation* particles are the "bottom" particles of groups (4a) and (4b). From a phenomenological mass-generation viewpoint, the J/ψ_{1S} and Υ_{1S} occur as the result of successive mass triplings of the ϕ meson, as is discussed in Sec. 3.17. This is a different type of mass generation from the M^X excitation towers that are displayed in Secs. 3.7, 3.8, 3.10–3.12.

The point we are leading to in this discussion is that when we go to fill in the excitation towers which are erected on the platform states defined in Sec. 3.5, there are not many particles available to occupy each tower. The M_π platform tower contains the group (1a) particles. The $M_{\mu\mu}$ platform tower has the group (1b) particles plus their corresponding antiparticles, which are generated simultaneously in matching M_μ and $\bar{M}_\mu$ tower columns. The M_K and $\bar{M}_K$ platform towers contain only the group (2a) particles. The M_ϕ platform tower contains just the group (2b) ϕ meson, but the ϕ also serves in turn as the ground state for the mass tripling J/ψ and Υ flavor-generating process described in Sec. 3.17.

The M^X tower excitations, when viewed all together, form an interesting overall pattern (Figs. 0.2.19 and 0.2.20 and Secs. 3.13 and 3.16). However each tower is fragmentary enough that it needs to be viewed in the context of all of the other towers. And if we did not include leptons and hadrons together in these towers, their symmetric construction would not be evident. The importance of these results is that they yield accurate estimates of the masses of all the tower particles in a manner that does not require arbitrary parameters, and they yield results which have not been forthcoming from other approaches to the elementary particle mass problem.

The phenomenological mass calculations of Chapter 3 are essentially devoid of theory. This can be regarded as a blessing, since we obtain useful results directly from the experimental data. But it can also be regarded, at least by elementary particle theorists, as a rather meaningless procedure — where is the dynamics? It depends which side of the particle fence a physicist is on. The real question is whether or not the elementary particle lifetime and mass measurements summarized in the present book are complete enough and accurate enough to reveal, by themselves, the answers we have been seeking about particle masses for the past century — specifically the proton-to-electron mass ratio. The judgment about the answer to this question is left up to the reader.

3.7 The M_π (π, η, η') Boson M^X Tower: The "Crown Jewels" of α-Quantized Masses

Elementary particle *phenomenology*, broadly speaking, is the study of the elementary particle data base for the purpose of determining how these data are organized. The phenomenologist has two general objectives in mind: (1) to see if the data fit some preconceived pattern — as suggested for example by a potential model; (2) to ascertain if there are any "surprises" lurking in the data — patterns that no one was expecting, but which seem to be of some significance. When we study the M_π platform state and the excitations that appear above it, we actually accomplish both of these objectives, and in a quite striking manner.

The preconceived pattern we are looking for here is an α^{-1}-quantized mass grid among related particles, and the motivation comes from the α-quantized lifetimes of Chapter 2. We have already had some suggestion of mass α-quantization from the "Rosetta-stone" results of Sec. 3.3. However, since the lifetime α-quantization extends through all 36 of the threshold-state particles — the ones with lifetimes longer than 1 zs, the mass quantization should really apply to most or all of these particles in order to provide a convincing confirmation of the lifetime results. The form that this α-quantization takes with respect to particle masses is not really spelled out by the lifetime systematics, so it is here that we may expect to find a surprise or two.

The M_π platform was formally defined in Eq. (3.5.5), and the M_π production channel was depicted symbolically in Fig. 0.2.16. The energy level diagram for the M_π platform state is shown in Fig. 3.7.1. This M_π platform generation process is not something that we would plausibly arrive at without the guidance of both experimental phenomenology (Chapters 2 and 3) and mathematical phenomenology (Chapter 4). In this scenario (top of Fig. 0.2.16), an electron–positron pair of spin-1/2 particles in a total spin-0 configuration is created out of the vacuum state. The electron and positron are acted upon by the coupling constant operator $\alpha = e^2/\hbar c$, and a pair of *spinless* bosonic mass quanta m_b and $\bar{m}_b$ (Eq. (3.4.1)) are created. Then the electron pair and the bosonic mass quanta combine together in a process which leads to charge fractionization (CF) (see Fig. 3.4.2) and consequent hadronic binding energies (HBE). As is shown in Fig. 3.4.2, a value of 2.7% for the HBE gives a precision fit to the average pion mass. However, when we apply this mass formalism simultaneously to the four meson states π, η, η' and ϕ, and use a common HBE for all of them, the value HBE = 2.6% gives the best average fit (see Tables 3.15.1 and 3.16.1). Thus we select HBE = 2.6% for the illustrative calculations of Secs. 3.7–3.12. The production process displayed in Figs. 0.2.16, 3.4.2 and 3.7.1 results in the formation of the platform states M_b and $\bar{M}_b$ (Eq. (3.5.3)) that obey the isotopic spin rules of the quark model, and that bind together hadronically and carry the quantum numbers of the excitation. The resulting $M_b\bar{M}_b$ bound state is manifested as the $\pi^\pm$ and π^0 pions. These platform masses can be split apart, and they then serve as platform states for the strange K mesons (Sec. 3.10). This scenario may seem contrived, in the sense that it starts out with a spin-1/2 electron–positron pair and ends up with a spin-0 bound state of two spinless mass quanta m_b, but it leads uniquely to a model that accommodates both leptons and hadrons in the same formalism, and that gives answers which are not otherwise forthcoming.

The first phenomenological piece of evidence for mass α-quantization is the factor-of-137 α-leap in energy (Figs. 3.3.2 and 3.3.3) that takes us from the 1 MeV mass of an electron pair to the ~137 MeV CI mass (the average energy) of the pion isospin doublet (Fig. 3.4.2). This is the

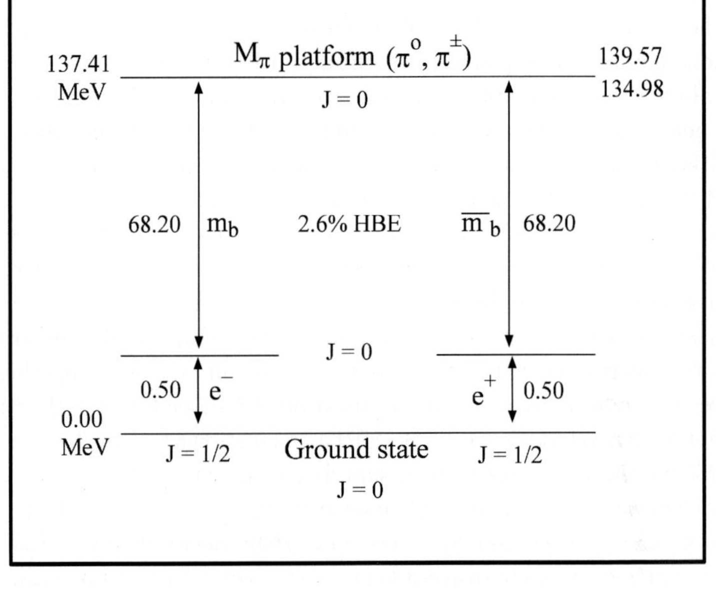

Fig. 3.7.1 The M_π platform state, which contains the $\pi^\pm$ and π^0 pions. In the creation of the M_π platform, an electron pair in a $J = 0$ spin state is operated on by the coupling constant α and produces a pair of $m_b = m_e/\alpha$ spinless boson excitation quanta (Eq. (3.4.1)). In this process, two things occur: (1) a charge fractionization (Fig. 3.4.2) produces a hadronic binding energy (HBE); (2) the m_b quanta merge with the electron quanta to form the hadronically bound M_π platform (Fig. 0.2.2). If we fit the calculated M_π platform mass (Fig. 3.7.1) to the charge-independent (CI) pion mass, we get a calculated HBE of 2.7% (Fig. 3.7.3). But if we also consider the η, η' and ϕ mesons and for simplicity assign the same HBE to all of them (Table 3.15.1), we obtain a consensus HBE of 2.6%, which is the value we employ for the figures of Secs. 3.7–3.12.

α-leap that we have just discussed in connection with Fig. 3.7.1. A second factor-of-137 energy leap on top of the first would take us to an energy of 137 MeV $\times$ 137 = 18.8 GeV, which is well above the mass range of observed particles (Fig. 3.3.1), and hence is not phenomenologically allowed until we consider the ultra-massive W and Z gauge bosons and top quark t, which we do in Sec. 3.20. This means that the first α-leap must have generated a type of α-quantized mass (the boson $m_b = 70$ MeV and its associated bound state $m_b\bar{m}_b \cong 137$ MeV) that can also appear in other threshold-state particles. Hence we expect to find an *additive* instead of *multiplicative* type of α-scaling for the M_π excitation tower, wherein the 137 MeV pionic quantum of energy (mass) serves as a basis state in the creation of other α-quantized pseudoscalar mesons. We are aided in our investigation here by the serendipitous fact that the mass of an electron pair — the generating "ground state" for pion α-production — is almost exactly 1 MeV. Thus if we simply plot a mass grid that is in units of 137 MeV, and then place the electron pair and the pion together with the higher-mass states on this mass grid, we can see by inspection if there is continuing evidence of α-quantization. The other two nonstrange PS mesons are the η and η' resonances (group (1a) in Table 3.6.1). Figure 3.7.2 displays the experimental mass levels for these three particles, plotted on a 137 MeV mass grid (also see Fig. 0.2.11). As can be seen, the α-quantization could hardly be more clear-cut: *the η and η' masses fit this 137 MeV mass grid to an absolute mass accuracy of 0.1%.* Furthermore, the decays of the η and η' mesons are back down to the pion platform state [10], which establishes a clear relationship between these particles. Figure 3.7.2 seems sufficient by itself to establish the α-quantization of hadronic particle masses, and thus the relevance of α to the hadronic domain.

The (π^+, π^0, π^-, η, η') "PS quintet" are the lowest-mass, lowest-spin, and most symmetric of the hadronic particles. This simplicity is manifested in the accuracy of their mass α-quantization (Fig. 3.7.2), and also in the accuracy of their lifetime α-quantization (Fig. 2.2.5), which extends over more than 12 orders of magnitude. The π^+, π^0, π^-, η and η' mesons are the "crown jewels" of hadronic α-quantization — the best examples we can offer.

We have unearthed the preconceived phenomenological pattern we were searching for — the 137 MeV α^{-1}-quantization of particle masses that is reciprocal to the 1/137 α-quantization of the lifetimes of these same particles. Now we should see if there are any surprises to be found? When we examine Fig. 3.7.2, we note that excited states do not appear every 137 MeV above the M_π platform level, as we might logically expect from the above discussion. Instead, the first level — the η meson — appears three pion masses above the pion platform state, and the second level — the η' meson — appears six pion masses above the platform state. This suggests that the mass unit $3M_\pi$ may be a natural platform excitation quantum. Since we have no *a priori* reason to expect this, it comes as a

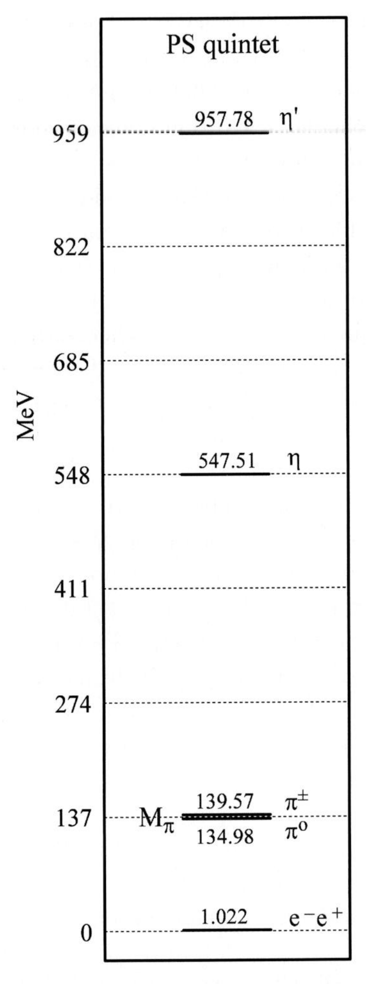

Fig. 3.7.2 The pseudoscalar η and η' mesons that are observed above the M_π platform state, shown plotted on a 137 MeV mass grid. This figure represents the most compelling experimental evidence for the α-quantization of elementary particle masses. All of these PS mesons match the α-quantized 137 MeV mass grid to an accuracy of 0.1%. Also, the extreme linearity of these PS meson masses brings into question the Standard Model use of *quadratic* masses in an attempt to achieve reasonable fits to the data (see Secs. 3.21 and 7.3). We might logically expect to find higher α-quantized PS mesons appearing at 137 MeV mass intervals above the M_π platform, but instead we see these mesons appearing at intervals of 3×137 MeV $=$ 411 MeV above the pions. This 411 MeV mass interval turns out to be the universal platform excitation quantum X (Sec. 3.9 and Figs. 0.2.18 and 0.2.19). The quantum X has an energy of about 420 MeV in unbound excitations (Fig. 3.8.3) and 410 MeV in hadronically bound excitations (Fig. 3.7.3).

surprise — something we did not expect to find. Let us denote this $3M_\pi$ mass unit as the quantum X. Since the pion is composed of two 70 MeV m_b mass quanta (plus the electron–positron mass), we will assume that the quantum X corresponds to six m_b mass units. The pion mass itself is an α-quantized mass unit with respect to the electron–positron ground state, so it follows that X is also an α-quantized mass unit. We make the following provisional mass assignment:

$$X = 6m_b = 420.15 \text{ MeV} . \tag{3.7.1}$$

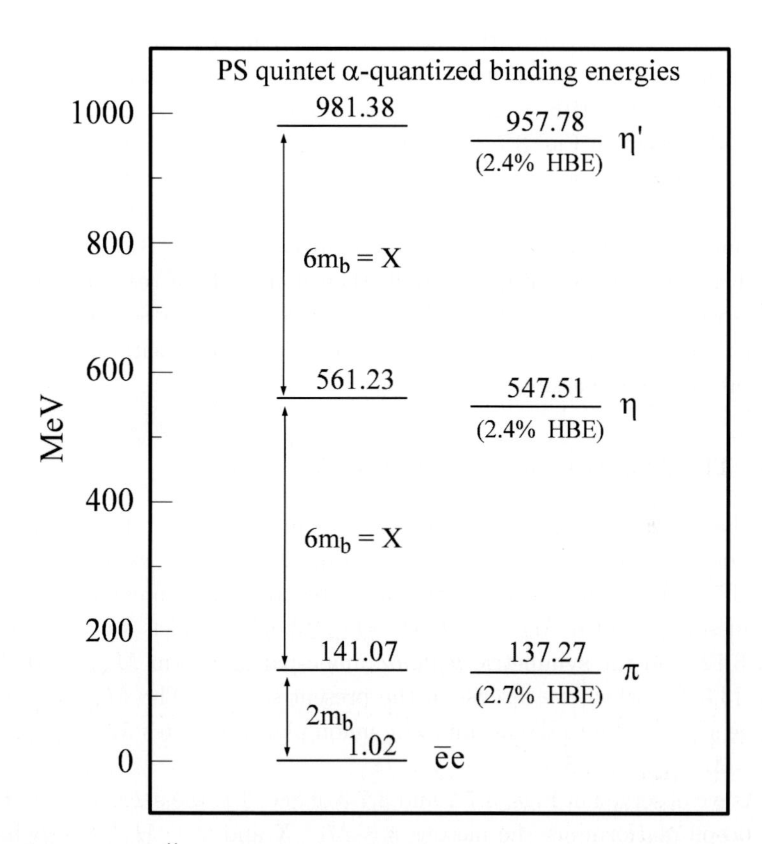

Fig. 3.7.3 The M^X α-quantization of the PS quintet masses in units of the boson α-mass $m_b \cong 70$ MeV. A comparison between the calculated α-mass values shown at the left in Fig. 3.7.3 and the experimental masses [Appendix B] shown at the right gives the calculated HBE hadronic binding energies displayed for each figure. These related PS mass levels serve to delineate the platform excitation quantum X for *hadronically bound* platforms M. A similar plot of excitation quanta X for *unbound* platforms M is displayed in Fig. 3.8.3.

Using this mass value, we plot theoretical mass levels above the 141.07 MeV unbound M_π platform mass (Eq. (3.4.5)), and then compare them with the experimental mass values, as displayed in Fig. 3.7.3. The mass difference between the unbound platform mass and the CI average pion mass of 137.27 MeV indicates a hadronic binding energy (HBE) of 2.7%, as we already demonstrated in Fig. 3.4.2. The difference between the calculated and experimental η masses in Fig. 3.7.3 gives HBE = 2.4%, and the η' mass values also give HBE = 2.4%. As a check on these results, we can compare the hadronically bound experimental $\eta' - \eta$ mass difference of 410.27 MeV [10] to the theoretical unbound 420.15 MeV excitation quantum X of Eq. (3.7.1) that creates the $\eta' - \eta$ mass difference. This comparison gives HBE = 2.4%, and is in agreement with the 2.4% HBE values displayed in Fig. 3.7.3, which were obtained using the α-quantized mass $X = 6m_b$. Hence Eq. (3.7.1) seems theoretically correct.

How significant is the mass quantum X? If it only appears in the PS meson quintet, it is probably just an interesting coincidence. But if it appears in other platform excitations as well, then it may be of real importance. When we move on to the other platform states defined in Sec. 3.6, we discover that the quantum X also dictates their low-mass excitations, so it is of more than just passing significance.

3.8 The $M_{\mu\mu}$ ($\mu\bar{\mu}, p\bar{p}, \tau\bar{\tau}$) Fermion M^X Tower

The basic *boson* elementary particle platform state is the symmetric *hadronic* pion platform $M_\pi = M_b\bar{M}_b$ (Eq. (3.5.5)) that we discussed in Sec. 3.7. The basic *fermion* platform states are the symmetric *hadronic* phi meson platform $M_\phi = M_f\bar{M}_f$ (Eq. (3.5.6)), which we discuss in Sec. 3.12, and the symmetric *leptonic* muon-pair platform $M_{\mu\mu} = M_f\bar{M}_f$ (Eq. (3.5.7)), which we discuss in the present section. The $M_{\mu\mu}$ platform can be separated into the asymmetric muon platform states $M_\mu = M_f$ and $\bar{M}_\mu = \bar{M}_f$ (Eq. (3.4.8)).

As we observed in Figs. 3.7.2 and 3.7.3 of Sec. 3.7, the excitations of the M_π boson platform are the mesons $\eta = M_\pi \cdot X$ and $\eta' = M_\pi \cdot XX$, where $X = 6m_b = 420$ MeV is the excitation quantum (Eq. (3.7.1)). If we now move to fermion platforms, where the quantum $m_f = 105$ MeV replaces $m_b = 70$ MeV as the fundamental α-quantized mass unit, will we still find the excitation mass $X = 420$ MeV serving as a characteristic platform excitation quantum? The answer turns out to be *yes*, as we demonstrate in the present section.

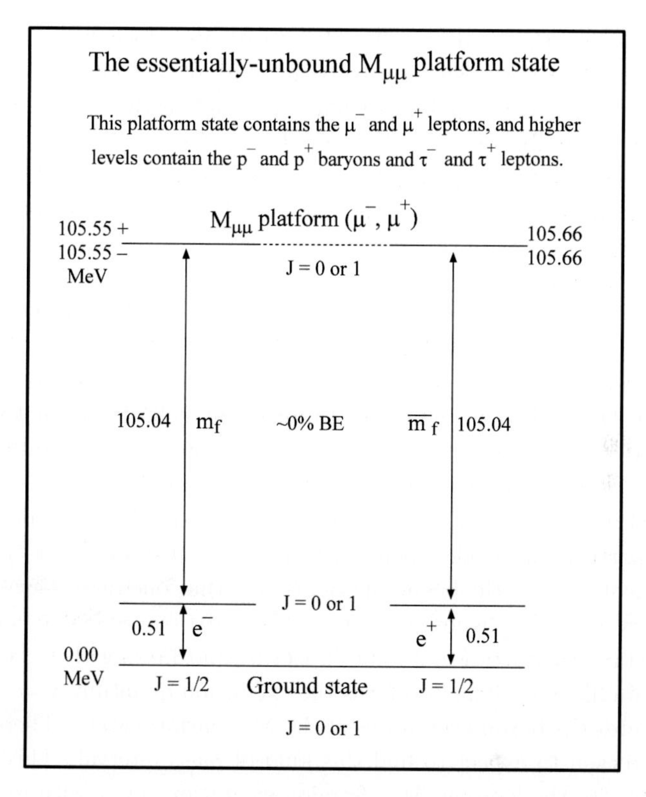

Fig. 3.8.1 The $M_{\mu\mu}$ platform state, which contains the μ^- and μ^+ muons. The coupling constant α operates on an electron pair to produce a pair of $m_f = 3m_e/2\alpha$ spinning fermion excitation quanta (Eq. (3.4.2)). The m_f quanta merge with the electrons to produce the $M_{\mu\mu}$ platform state (Fig. 0.2.16). Charge fractionization does not occur in this process, so there is no hadronic binding energy, and the particle and antiparticle components of the $M_{\mu\mu}$ platform operate almost independently, although they must be produced in a pairwise manner to conserve lepton number. Since HBE = 0, the excitations are right on the mass shell. The calculated mass of 105.55 MeV for each muon (see Fig. 3.4.2) is within 0.1% of the actual muon mass, which verifies the small $M_{\mu\mu}$ binding energy, and also demonstrates that the electron mass must be added to the α-quantized m_f excitation mass in order to get the correct muon mass. Higher-mass excitations of the $M_{\mu\mu}$ platform are displayed in Figs. 3.8.2 and 3.8.3.

Figure 3.8.1 displays the $M_{\mu\mu}$ platform state, which appears a factor of approximately $3/2$ higher in energy than the M_π platform state (Fig. 3.3.3). The $M_{\mu\mu}$ platform level corresponds to a muon–antimuon pair, just as the M_π platform level corresponds to a pion. Since the muon and antimuon do not bind together hadronically (HBE = 0), we expect the α-quantized excitations of the $M_{\mu\mu}$ platform to be in units of $3/2 \times 2m_b \cong 210$ MeV rather than $3/2 \times 137 \cong 205$ MeV, where 137 MeV is the hadronic (HBE = 2.6%) mass unit for the M_π platform (Fig. 3.7.2). (The 205 MeV mass interval also applies to the M_ϕ platform discussed in Sec. 3.12.) Since the muon and antimuon are only lightly bound together, we can conceptually divide up the $M_{\mu\mu}$ platform excitation tower into particle and antiparticle columns M_μ and $\bar{M}_\mu$ and consider each side separately. This corresponds to starting with a platform mass M_μ of about 105 MeV for each column of the tower, as indicated in Fig. 3.8.1, and then erecting 210 MeV mass grids on top of the platforms, as shown in Fig. 3.8.2, which displays the matching M_μ and $\bar{M}_\mu$ excitation columns that collectively make up the $M_{\mu\mu}$ platform excitation tower. When we fill up these excitation levels with the available threshold-state particles (group (1b) in Table 3.6.1), we end up with the $M_{\mu\mu}$ level diagram shown in Fig. 3.8.2, which contains the fermion particle-antiparticle pairs (μ^-, μ^+), (p^+, p^-) and ($\tau^-\tau^+$).

We mentioned at the beginning of Sec. 3.7 that one reason for studying elementary particle phenomenology is to see if there are any "surprises" in the data — features we did not expect to find lurking there. One such surprise we have unearthed so far is the excitation quantum $X = 420$ MeV that dictates the boson excitations of the M_π platform state. There was no *a priori* reason to expect to find this kind of mass interval. This surprise carries over to the leptonic $M_{\mu\mu}$ fermion excitation tower with its particle and antiparticle columns. As can be seen in Fig. 3.8.2, the p^+ proton appears 840 MeV above the μ^- muon, and the τ^- tau appears 840 MeV above the proton. These excitation intervals are each twice as large as the excitation intervals displayed in Fig. 3.7.3. Hence we have XX excitations occurring in the M_μ and $\bar{M}_\mu$ columns. Thus X is both a boson and a fermion excitation quantum. To see how this is possible, we note that instead of setting $X = 6m_b = 420.15$ MeV, as in Eq. (3.7.1), we can use

$$X = 4m_f = 420.15 \text{ MeV}. \tag{3.8.1}$$

Hence we obtain the same α-quantized $X = 420$ MeV mass unit in either the m_b boson representation of Sec. 3.7 or the m_f fermion representation of the present section. Figure 3.8.2 demonstrates that XX is a dominant fermion excitation, just as Fig. 3.7.2 demonstrates that X is a dominant (hadronically bound) boson excitation. Hence the phenomenological surprise that we discovered in the M_π platform excitations of Sec. 3.7 carries over to the $M_{\mu\mu}$ platform excitations of Sec. 3.8.

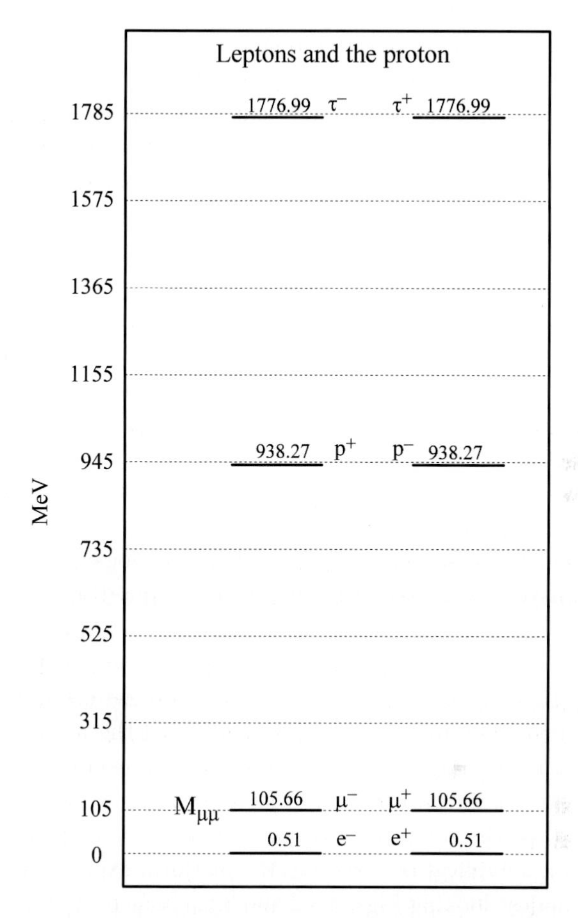

Fig. 3.8.2 The p^+, p^- baryon pair and τ^-, τ^+ lepton pair, displayed as spin-1/2 columnar excitations of the $M_{\mu\mu}$ leptonic μ^-, μ^+ platform state, with a 210 MeV mass grid erected above the $m_f = 105$ MeV columnar platform masses, and with the masses in each M_μ column treated separately. Since HBE $= 0$ for these separated excitation columns, their masses occur on the mass shell. There are three significant features to be noted in Fig. 3.8.2: (1) the precision 1% fit of the higher excitations to the 210 MeV mass grid; (2) the occurrence of the *proton* in a *lepton* energy level diagram, which is one of the most important results in the book; (3) the equal 840 MeV mass intervals between the μ, p and τ levels, which mirror the equal 411 MeV (HBE $= 2.6\%$) mass intervals between the π, η and η' levels in Fig. 3.7.2. The systematics displayed here suggests that the *proton* is initially created as a *lepton*, but its 9 m_f mass subunits convert (in charge-exchange processes with the associated antiproton) into 3 u and d quarks, each containing 3 m_f mass subunits. In this conversion process, the proton becomes hadronized, and can only de-hadronize in conjunction with the matching de-hadronization of an antiproton or antineutron. This hadronization process is not available to the 17 subunit τ or the 1 subunit μ.

The occurrence of the excitation quantum X (or XX) in the $M_{\mu\mu}$ excitation tower is not the only surprise contained there. The three $M_{\mu\mu}$ levels displayed in Fig. 3.8.2 are the $\mu^-\mu^+$ pair excitation that corresponds to the $M_{\mu\mu}$ platform level, the p^+p^- pair excitation that appears XX higher in each column of the $M_{\mu\mu}$ tower, and the $\tau^-\tau^+$ pair excitation that appears an additional XX higher in each column. These represent two new surprises: (a) we have leptons and hadrons mixed together in the same excitation sequence; (b) the proton–antiproton pair appears in this excitation tower with reversed signs. Surprise (a) is a result that violates the Standard Model separation of leptons and hadrons. Surprise (b) means that we have excited the positive proton, a *particle* state, from a negative electron, the *particle* "ground state," so how can the proton decay back down to this ground state? The answer to this question is simple, but profound: it cannot decay. We may have here the essence of proton stability. This possibility is discussed in Sec. 3.18.

In addition to these surprises — and in fact logically as a result of them — we have in Fig. 3.8.2 the answer to the question that has bedeviled particle physicists for the past century: namely, *how do we calculate the proton-to-electron mass ratio*? The answer, to an accuracy of 1%, is contained in the α-quantized levels of the $M_{\mu\mu}$ excitation tower, which are portrayed numerically in Fig. 3.8.3. By just doing the math, and without any use of adjustable constants, we reproduce the mass of the muon almost exactly, and we obtain close estimates for the masses of the proton and the tau lepton, using the same basic α-quantization systematics that appears (with hadronic modifications) in the M_π platform excitations of Sec. 3.7. The rather modest-looking Figs. 3.8.2 and 3.8.3 in fact represent the answer to a quest that started with the discovery of the electron in 1897 and the point proton in 1911.

The systematics displayed in Fig. 3.8.2 suggests that the proton is initially created as a *lepton*. The facts that three muon-like masses can readily convert into a fractionally charged u or d quark (Fig. 0.2.4), and that the nine muon-like masses in the proton can convert into three u and d quarks, facilitates the transformation to a stable fractionally charged three-quark structure within the proton. A matching antiproton must simultaneously be produced to provide the freedom for the necessary cross-channel charge exchanges in this proton–antiproton production process. Conversely, the proton requires an antiproton for its annihilation, which moves charges in the other direction. The τ lepton, with its 17 muon-like subquanta, is doomed to remain a lepton, since the number 17 does not contain the factor 3 that is necessary for u and d quark production. 17 is a prime

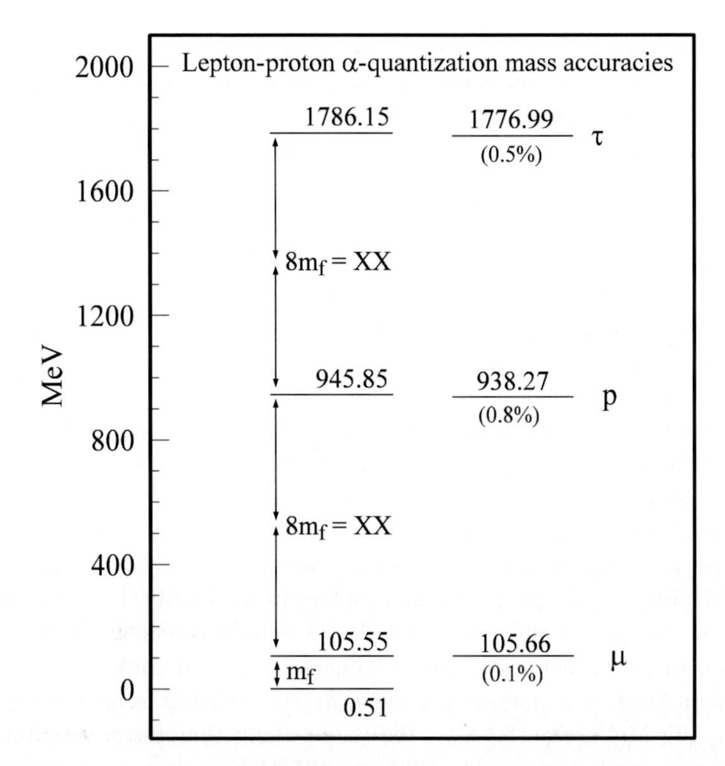

Fig. 3.8.3 The M^X α-quantization of the lepton and proton masses in units of the fermion α-mass $m_f \cong 105$ MeV. These α-quantizations are accurate to better than 1%. Since we have not incorporated isotopic-spin mass corrections into the m_b and m_f basis states, this level of accuracy is all that can be expected phenomenologically. If we define each 840 MeV mass interval displayed here as being composed of two X quanta (XX), then we have the same basic quantum X appearing in both boson and fermion platform excitations: $X = 6m_b = 4m_f$. Hence $X \cong 420$ MeV emerges as a universal platform excitation quantum. It is "supersymmetric" in the sense that it is the lowest possible excitation quantum which is isoergically symmetric with respect to particle–antiparticle substates in both the m_b and m_f representations (Fig. 0.2.18 and Sec. 3.9).

number, as is the number 1 in the muon, which also is doomed to remain a lepton.

Muons are formed out of m_f and $\bar{m}_f$ excitation quanta (Figs. 3.3.3 and 3.4.2), so it is logical to assume that the excitation quantum X in the $M_{\mu\mu}$ platform excitation tower corresponds to four m_f mass units. Since the muon mass is an α-quantized mass unit with respect to the electron–positron ground state, it follows that $X = 4m_f$ is also an α-quantized mass unit, which has precisely the same mass value as the boson mass quantum $X = 6m_b$. This coincidence in mass values is of course built in theoretically from the basic definitions of the α-masses in Eqs. (3.4.1) and (3.4.2). But the fact, as we will quantify in Sec. 3.16, that this coincidence seems to be maintained experimentally suggests that it is not accidental. In Sec. 3.9 we make a phenomenological analysis of the properties of the boson and fermion mass quantum X.

3.9 The "Supersymmetric" 420 MeV Excitation Quantum $X = 3m_b\bar{m}_b = 2m_f\bar{m}_f$

The α-quantization of the threshold-state elementary particle *lifetimes* that was described in Chapter 2 can be logically accounted-for by a corresponding α-quantization of the *masses* of these particles. An examination of the threshold-state masses showed that if we start with a 1 MeV electron–positron ground state, an "α-leap" by a factor of $\alpha^{-1} = 137$ brings us to the 137 MeV $\pi^{\pm}$, π^0, pion level, and enables us to identify the spin-0 boson "α-mass" $m_b = 70$ MeV (Eq. (3.4.1)). A second α-leap from the e^-e^+ ground state by a factor of $3/2\ \alpha^{-1}$ brings us to the 210 MeV $\mu^-\mu^+$ muon–antimuon level, and enables us to identify the spin-1/2 fermion "α-mass" $m_f = 105$ MeV (Eq. (3.4.2)). We might think that further excitations of the pion level would yield additional 137 MeV excitations, but instead we find $X = 3 \times 137 = 411$ MeV excitation units (Fig. 3.7.2). Correspondingly, we might think that further excitations of the muon–antimuon level would yield additional 210 MeV excitations, but instead we find $X = 2 \times 210 = 420$ MeV excitation units (Fig. 3.8.2). (The difference between the 411 MeV pion excitation units and the 420 MeV muon–antimuon excitation units can be attributed to a hadronic binding energy in the pion excitations.) Thus, after an initial α-leap from the e^-e^+ ground state to the pion or muon–antimuon level, subsequent excitations occur in X units of three pion masses or two muon–antimuon masses. Hence we denote the pion and muon–antimuon mass levels as "platform states" M upon which X-quantized excitations occur, thus yielding an "M^X" excitation scheme for long-lived threshold state particles.

The question then arises as to why the pion X-quanta seem to be basically equal to the muon–antimuon X-quanta. What do these X quanta have in common? The answer to this question may reside in the fact that the quantum X is "supersymmetric," in the sense that it is the lowest-mass excitation unit which can be particle–antiparticle symmetric when composed of either all boson α-mass excitation quanta m_b and $\bar{m}_b$ or all fermion α-mass quanta m_f and $\bar{m}_f$. The conditions for "supersymmetry" are that X must contain even numbers of m_b or m_f subquanta to be particle–antiparticle symmetric, and it must contain m_b or m_f quanta in the ratio of three to two to be isoergic in the two representations. Thus X is formally defined as follows:

$$X = 3(m_b + \bar{m}_b) = 2(m_f + \bar{m}_f) = 420.15 \text{ MeV}\,. \qquad (3.9.1)$$

These are the fully *symmetric* representations of the platform excitation quantum X. However, X evidently appears in different excitations with different forms. In hadronic channels X has a small (~2.6%) binding energy and exhibits an experimental mass of about 410 MeV, and in leptonic or unpaired hadron channels X has essentially zero binding energy and appears with a mass of about 420 MeV (Fig. 0.2.18). In the kaon, which is the excitation $K = M_K + X$ (Sec. 3.10), different configurations of X, including possible mixed m_b plus m_f modes, plausibly account for the properties of the $K^\pm$, K^0_L and K^0_S mesons, including their $K^0_L \to \pi\pi\pi$, $K^0_S \to \pi\pi$ and $K^\pm \to \pi\pi$ decay modes and their $K^0_L/K^\pm = 4$ (Fig. 0.1.6) and $K^0_S/K^\pm = \alpha$ (Fig. 2.3.1) lifetime ratios.

In its role as a symmetric excitation quantum, X corresponds to an increase in mass in a production channel, and hence does not represent a basic alteration of the channel quantum numbers. Figure 3.9.1 shows five sets of particles that exhibit X-spaced excitation intervals. The two top mass intervals, $\eta - \pi$ and $\eta' - \eta$, involve non-strange pseudoscalar mesons that are symmetrically formed from m_b excitation quanta (Sec. 3.7), and their experimental X values of 410.27 and 410.24 MeV reflect a 2.4% hadronic binding energy (Fig. 3.7.3). The next two mass intervals, $D^{*\pm} - D_1^{*\pm}$ and $D_s^{*\pm} - D_{s1}^{*\pm}$, involve charm and charm-strange mesons that are asymmetrically formed from m_f excitation quanta, and their experimental X values of 415.6 and 423.3 MeV indicate particles with essentially zero hadronic binding energy (see the discussion of the B_c meson binding energy in Sec. 3.17). The bottom mass interval, $\tau^\pm - \mu^\pm$, involves unpaired m_f excitation quanta, and the experimental $4X$ mass interval gives an average X value of 417.8 MeV, again denoting zero hadronic binding energy. In all five of these X excitations, the quantum numbers of the particles involved are unchanged, so the main function of the quantum X in each case is to simply increase the mass of the particle. (The proton level, which is half way between the μ and τ levels, represents a more complex interaction, as discussed in Sec. 3.18.)

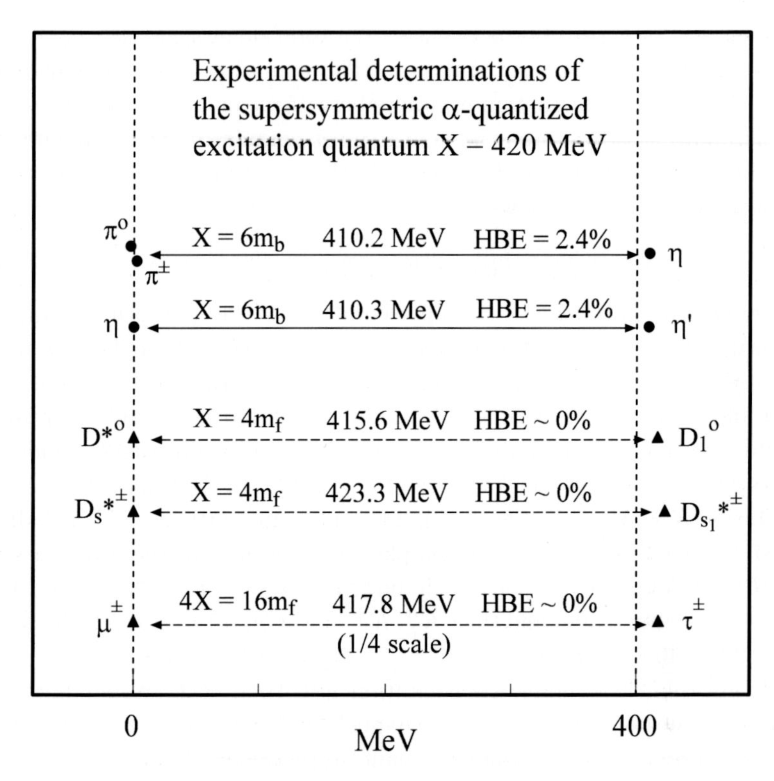

Fig. 3.9.1 Experimental values of the excitation mass $X = 420$ MeV as deduced from hadronically bound and unbound experimental mass intervals. The top two mass intervals are hadronically bound symmetric pseudoscalar meson excitations as displayed in Fig. 3.7.3, and they correspond to spinless m_b boson excitation quanta. The next two mass intervals represent unbound, unpaired charmed meson excitations as shown in Fig. 3.22.3, and they correspond to spin-1/2 m_f fermion excitation quanta. The bottom mass interval represents an unbound unpaired lepton excitation as displayed in Fig. 3.8.3, and it also corresponds to spin-1/2 m_f fermion excitation quanta. The mass difference of 10 MeV between hadronically bound and unbound excitation quanta X (Fig. 0.2.18) is not large, but it is quite clearly delineated in these very accurate mass measurements [Appendix B, Ref. 10].

There is one other area where the structure of the excitation quantum X may play an important role. We will demonstrate in Chapter 4 that the mass quanta m_b and m_f can be reproduced as spinless and relativistically spinning ($J = 0$ and $1/2$) modes of the same basic mass quantum, so the quantum X may facilitate transformations betweens these two types of particles, as for example in the decay modes $\pi \to \mu + \nu$ and $\tau \to \pi + \nu$, and thus serve as a "crossover" excitation. In particular, isoergic basis state transformations of the general form $m_b[m_b]\bar{m}_b \leftrightarrow m_f\bar{m}_f$ can take place, where the brackets denote the annihilation of a mass quantum m_b. Thus the symmetry of X in both the m_b and m_f representations (Eq. (3.9.1)) may be a significant characteristic.

We have one example where $m_b[m_b]\bar{m}_b \rightarrow m_f\bar{m}_f$ basis state transformations seem to occur in the same excitation tower. Figures 0.2.12 and 0.2.20 display an excitation-doubling sequence in the M_K column of the M_π platform that echoes a similar excitation-doubling sequence (Fig. 0.2.4) in the M_μ column of the $M_{\mu\mu}$ platform. The first particle produced in the M_K column is the spinless K meson that is composed (at least in part) of spinless m_b mass quanta. The second and third particles in the M_K column are the spin-1 $K^*(894)$ and $K^*(1717)$ mesons, which correspond to spin-1/2 $q\bar{s}$ excitations that require m_f mass quanta (Fig. 0.2.4). If all of these particles in fact start with the same $M_K = M_b$ ground state, as this excitation-doubling sequence implies, then transformations $m_b \rightarrow m_f$ necessarily take place. Furthermore, the $K^*(894)$ meson (the lowest-mass spin-1 kaon) has a mass value that is about 70 MeV higher (once binding energy effects are taken into account) than we would expect for a $q\bar{s}$ excitation (see Table 0.6.1 and Sec. 3.26), so it must in some manner have both m_f and m_b mass quanta incorporated into its structure. This same situation seems to occur in the Σ and Ω hyperons (Sec. 3.26).

The M_π and $M_{\mu\mu}$ platform excitations that we discussed in Secs. 3.7 and 3.8 represent symmetric particle–antiparticle configurations. We move on in Secs. 3.10 and 3.11 to the asymmetric M_K and M_μ platform excitations, where the quantum X again plays a significant role.

3.10 The Strange M_K (K, $\bar{K}$) Boson M^X Excitations and the $\eta' = K\bar{K}$ Bound State

The M_K and $\bar{M}_K$ kaon platform states are obtained by first creating the $M_\pi \equiv M_b\bar{M}_b$ pion platform state of Fig. 3.7.1, and then dividing it into its particle and antiparticle components, M_b and $\bar{M}_b$, which are then used to generate the K and $\bar{K}$ mesons. Thus we define $M_K \equiv M_b$ and $\bar{M}_K \equiv \bar{M}_b$ (Fig. 0.2.17). These kaon platform states are illustrated in Fig. 3.10.1. When the M_π compound platform state is separated into the unpaired M_K and $\bar{M}_K$ particle and antiparticle platforms, the ~2.6% hadronic binding energy vanishes, so the excitations of the M_K and $\bar{M}_K$ platforms occur right on the mass shell.

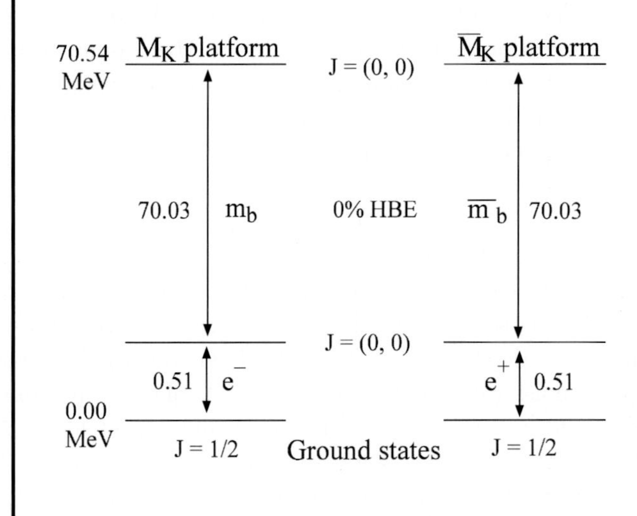

Fig. 3.10.1 The M_K platform states, which do not correspond to observed particles. They can be viewed as the separation of the M_π platform state of Fig. 3.7.1 into its particle and antiparticle excitation columns (Fig. 0.2.17). Since these separated states are not hadronically bound, the excitation masses are right on the mass shell. Each M_K platform is a spinless boson state that was generated from a spin-1/2 fermion, so it is clear that M_K and $\bar{M}_K$ platforms must be produced as matching pairs from $J = 0$ electron-positron ground states.

The M_π platform state corresponds to an observed particle — the pion, but the M_K platform state does not. The evidence for its existence comes from a higher level — the spin-0 kaon — that is erected on the spin-0 M_K platform. (In Figs. 0.2.12 and 0.2.20 we display two additional higher levels of this same platform, which correspond to spin-1 K^* mesons, and which require m_b to m_f basis state transformations in the X quanta.) In order to investigate the quantization of the higher levels of the M_π platform state (Sec. 3.7), we plotted these levels on a 137 MeV mass grid (Fig. 3.7.2). But since the excitations of the separated M_K and $\bar{M}_K$ platforms occur

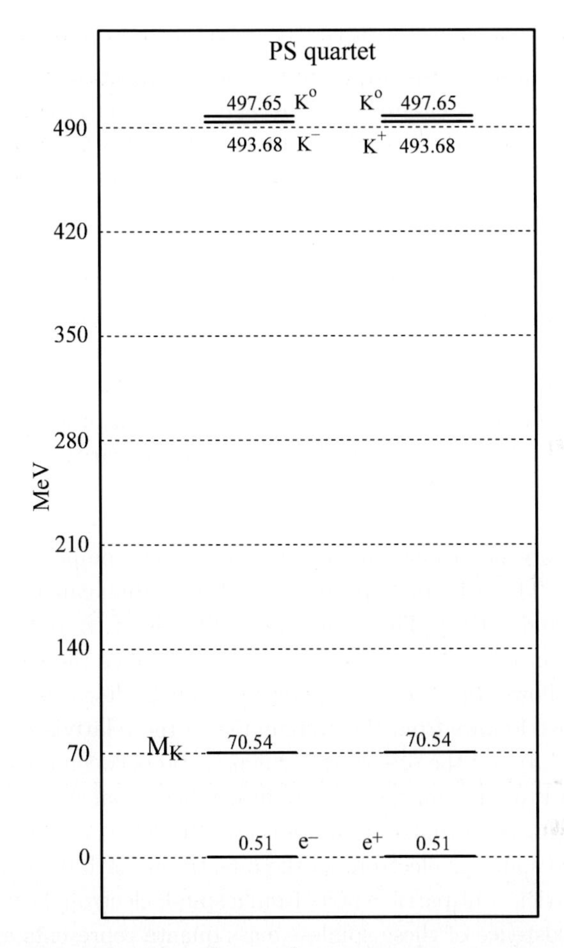

Fig. 3.10.2 The pseudoscalar $K^{\pm}$ and K^0 mesons that are observed above the M_K platform states of Fig. 3.10.1, shown plotted on a 70 MeV mass grid. Since these unpaired excitations do not have hadronic binding energies, the masses occur right on the mass shell. As can be seen, the same 420 MeV excitation quantization unit that was observed in the $M_{\mu\mu}$ platform levels of Fig. 3.8.2 is also occurring here. The asymmetric kaons carry the strangeness quantum numbers $S = \pm 1$.

on the mass shell, a 70 MeV mass grid naturally applies to them. This plot is shown in Fig. 3.10.2, extended up just to the mass of the kaon (Fig. 0.2.20 shows a further extension). As can be seen, the kaon appears about 420 MeV above its platform state, so we have the same excitation quantum X appearing in this asymmetric M_K platform excitation as we did in the symmetric M_π and $M_{\mu\mu}$ excitations of Secs. 3.7 and 3.8. This is another example of the α-quantized "M^X" excitation process, wherein a platform state M is created by an α-leap from an electron ground state, and then the α-quantized mass quantum X is added to the ground state mass so as to produce a higher-mass particle. The kaons form group (2a) in Table 3.6.1.

The K^+ and K^0 mesons carry the "strangeness" flavor quantum number $S = +1$, and the K^- and K^0 mesons carry $S = -1$. Their "strangeness" seems logically related to their particle and antiparticle asymmetries. In order to conserve strangeness, the K and $\bar{K}$ mesons must be generated in strangeness-conserving production channels (Sec. 3.11).

The mass systematics of the K meson excitation process is illustrated in Fig. 3.10.3. As can be seen, the masses add linearly, with no binding energy corrections. In fact, the experimental kaon levels are slightly above the calculated excitation mass, which can most likely be attributed to the fact that we are not using charge-dependent mass components in the m_b basis states. There is an important conclusion that can be drawn from Figs. 3.10.2 and 3.10.3. The K meson, unlike the π, η and η' mesons, is composed of an *odd* number of m_b mass quanta. Since the kaon is a spin-0 particle, it follows that the mass quantum m_b must itself be *spinless*. This conclusion also follows from the systematics of the relativistically spinning sphere (Sec. 4.3) and the observed 3/2 mass ratio between a muon pair and a pion, but it is interesting to see it confirmed here from the structure of the kaon. The spinless m_b mass quantum is created by the action of the generator α on the spin-1/2 electron, so the quanta m_b have to be produced in matching particle–antiparticle pairs from a spin-0 electron–positron ground state. The existence of these spinless mass quanta represents an extension of physics beyond the paradigm of the Standard Model, so it is important to bring out the manner in which the observed α-quantization of particle lifetimes and masses mandates their existence. The only elementary particles that seem to be composed entirely of spinless m_b and $\bar{m}_b$ bosons are the pseudoscalar mesons. Hence the PS mesons play a special role in elementary particle structure. Historically, they have for a long time presented difficulties for the Standard Model, as we discuss in Sec. 3.21.

The kaon states are created in the M^X production process with zero binding energies, as shown in Fig. 3.10.3. But they can, at least conceptually, bind together hadronically to form the η' meson. From Fig. 3.7.3, we see that $\eta' = M_\pi \cdot XX$, where $M_\pi \equiv M_b\bar{M}_b$, and from Fig. 3.10.3 we see that $K = M_K \cdot X$, where $M_K \equiv M_b$. Thus, structurally speaking, we have

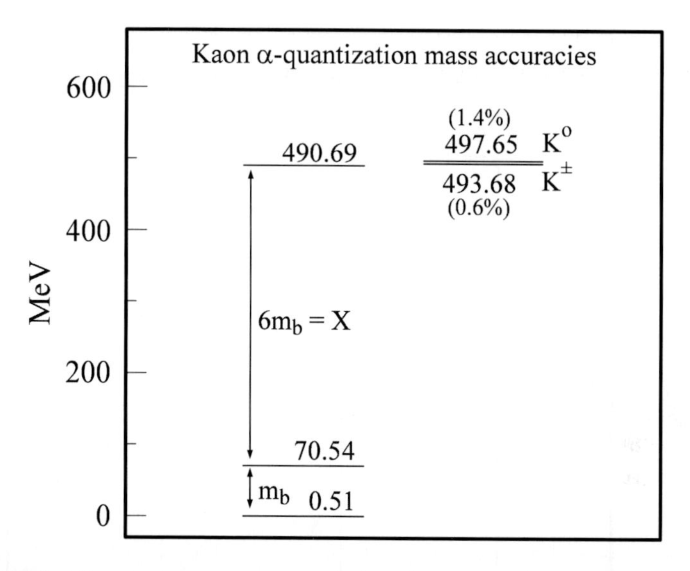

Fig. 3.10.3 The m_b boson structure of the K-meson excitations. The observed masses of the kaons are slightly larger than the calculated 490.69 MeV mass of the electron-based α-quantization process. The discrepancy of $\sim 1\%$ in the mass values can be attributed to the fact that we have not introduced a charge dependence in the masses of the m_b substates, as seems required in order to account for the observed mass splittings of both the pions (as shown in Fig. 3.4.1) and the kaons (as shown here). The key point to be noted with respect to Fig. 3.10.3 is that the spinless kaons each contain an odd number of subquanta m_b, so that the α-mass m_b must itself be spinless. The quarks of the Standard Model are all spin-1/2 entities, so these spinless m_b mass states represent physics beyond the Standard Model (see Sec. 3.21). The 3/2 mass relationship between the $J = 0$ m_b mass quanta of Eq. (3.4.1) and the $J = 1/2$ m_f mass quanta of Eq. (3.4.2) is established by the relativistically spinning sphere model of Sec. 4.3.

$\eta' = K\bar{K}$. This result is portrayed in Fig. 3.10.4, and it gives us a chance to evaluate hadronic binding energies directly from the experimental data, without having to resort to theoretical mass values. If we set $\eta' = K^+K^-$, we obtain HBE = 3.0%, and if we set $\eta' = K^0\bar{K}^0$, we obtain HBE = 3.8%. These values are comparable to the hadronic binding energies displayed in Fig. 3.7.3 and Table 3.15.1. We do not have a comprehensive formalism for dealing with the charge splittings of the masses in both the pion and the kaon isotopic spin multiplets. Thus we cannot achieve precision results in calculating HBE values. But what is important here is the range of HBE values involved, which are in the 2%–3% range. Calculated and measured hadronic binding energies are summarized in Sec. 3.15.

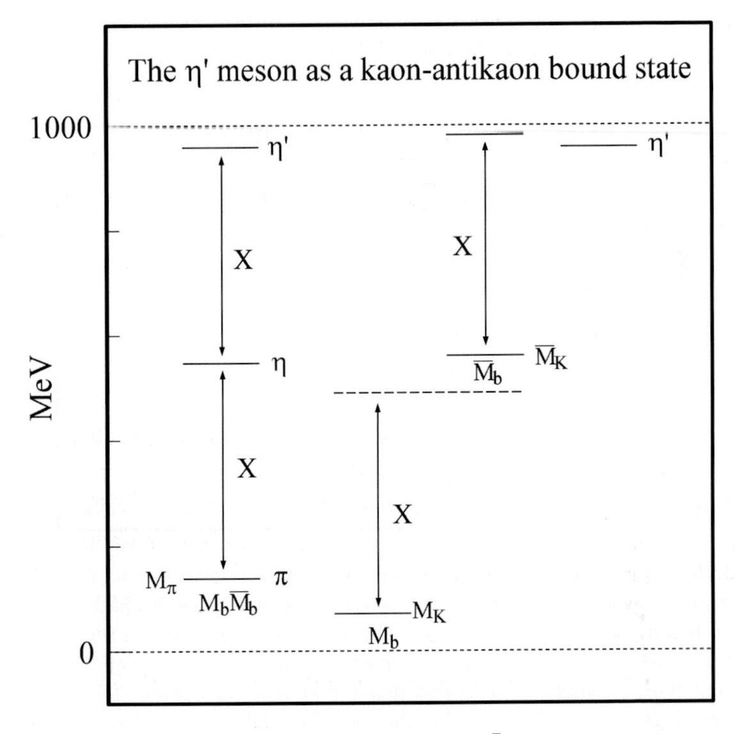

Fig. 3.10.4 The construction of the η' meson as a $K\bar{K}$ bound state. The left energy level diagram displays the η' as generated from the M_π pion platform, and the right diagram displays it as a spread-out $K\bar{K}$ generation process that is based on the M_K and $\bar{M}_K$ kaon platforms. The basis states in each case are the same, but they are shown here with the hadronic binding energy HBE $\cong$ 2.6% in the pion channel, and with HBE $\cong$ 0 in the $K\bar{K}$ channel, so that the difference between the calculated $K\bar{K}$ excitation mass and the experimental η' mass gives the HBE that must be applied once the K and $\bar{K}$ states are bound together. The essential point here is that if we assume that $\eta' = K\bar{K}$, and then use the experimental η' and kaon masses in this equation, we can determine the HBE directly from experiment. Assuming that $K\bar{K} = K^-K^+$ gives HBE = 3.0%, and assuming that $K\bar{K} = K^0\bar{K}^0$ gives HBE = 3.8%. When we similarly assume the relationship $\phi = s\bar{s}$ (Fig. 3.12.3), we obtain an HBE of 3.0%, so that $J = 0$ and $J = 1$ mesons reflect the same order-of-magnitude for the HBE values.

One final topic of interest here with respect to the pseudoscalar kaons is their decay systematics. In Fig. 2.3.1 we displayed the accurate 4:2:1 ratios of the K_L^0, $\pi^\pm$, $K^\pm$ lifetimes. These three particles contain 7, 2, 7 mass quanta m_b, respectively. The K_L^0 decay is into three pions, so it involves the annihilation of a single m_b substate in the decay process. The $\pi^\pm$ decay is into a muon and a neutrino, which involves the annihilation of a single m_b substate plus the conversion of the other spinless m_b bosonic substate into a spinning m_f fermionic substate. If we postulate that the K_L^0 decay is initiated by a single m_b substate, but the $\pi^{\pm}$ decay can be initiated by either of its two m_b substates, then we have an explanation for their factor-of-2 difference in lifetimes. The $K^{\pm}$ decay is into two pions, which involves the annihilation of three m_b substates. If we further postulate that the $K^{\pm}$ decay can be initiated by any one of four different m_b substates, then we have accounted for the 1:2:4 lifetime triad of these three particles. The K_S^0 decay is also into two pions, and the K_S^0 lifetime is one power of α shorter than the $K^{\pm}$ lifetime (Fig. 2.3.1). Thus the distribution of the charge states or the particle–antiparticle substates in these particles also affects the lifetime. These remarks do not constitute a theory of these decay relationships, but they point out the fact that by attributing a universal type of mass substructure to each of these particles, we can provide a mechanism for creating factor-of-2 lifetime relationships between them, even for particles as different as the pion and the kaon. The gluon clouds that are assumed to provide the particle masses in the Standard Model do not lend themselves as readily to discrete lifetime systematics, and they bear no obvious relationship to a global scaling of particle lifetimes in powers of α.

3.11 The Strange M_μ (s, $\bar{s}$) Fermion M^X Excitations: s and $\bar{s}$ Quarks

In Sec. 3.10 we discussed the asymmetric M_K platform state, which is created conceptually by dividing the symmetric M_π platform of Sec. 3.7 into its particle and antiparticle substates. In Sec. 3.11 we examine the asymmetric platform state M_μ, which is similarly created by the division of the symmetric $M_{\mu\mu}$ platform of Sec. 3.8. The M_K and M_μ platforms (and their corresponding antiplatforms) have somewhat reciprocal properties. In terms of generic basis states, we have $M_K = M_b$ $(J = 0)$ and $M_\mu = M_f$ $(J = 1/2)$. The M_μ platform corresponds to an observed particle — the muon, whereas the M_K platform does not correspond to a particle. The M_K platform excitation $M_K \cdot X$ is the kaon, which is an observed particle; the M_μ platform excitation $M_\mu \cdot X$ is the strange quark s, which has a fractional electric charge and is not a directly-observed entity. Both K and s carry the strangeness quantum number S. The K and the s give rise to the bound states $\eta' = K\bar{K}$ and $\phi = s\bar{s} \to K\bar{K}$. The M_μ platform state is illustrated in Fig. 3.11.1.

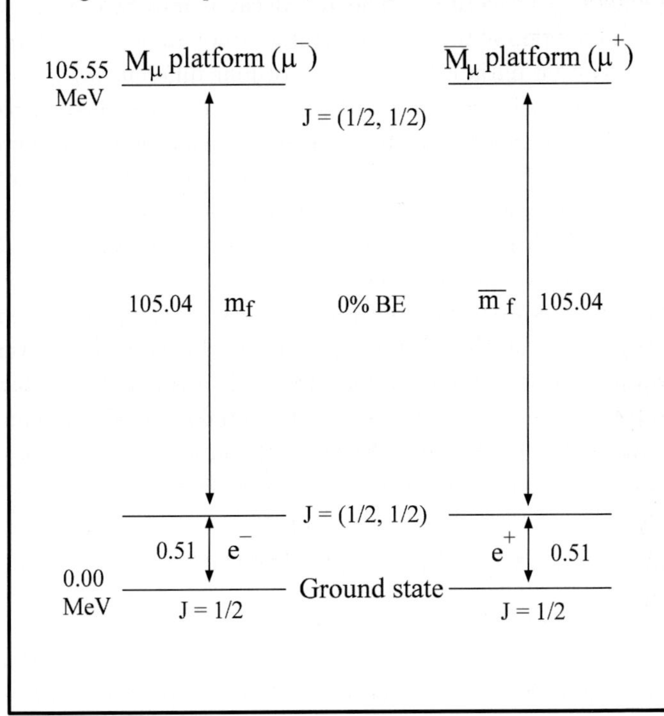

Fig. 3.11.1 The M_μ platform states, which each contain a μ meson. These represent the separation of the symmetric $M_{\mu\mu}$ platform of Fig. 3.8.1 into its particle and antiparticle components (Fig. 0.2.17). These spin 1/2 platforms and the spin-1/2 excitations that appear above them are composed of α-quantized m_f fermion masses which are combined with zero hadronic binding energy. The particles that appear in these excitation columns are displayed in Fig. 3.8.2.

The M_μ platform corresponds to the μ meson, as was demonstrated graphically in Fig. 3.4.2. If we add the 0.511 MeV mass of the electron ground state to the 105.038 MeV mass of the excitation quantum m_f, we obtain a calculated platform mass of 105.549 MeV, which compares within 0.1% accuracy to the measured muon mass of 105.658 MeV. When we further consider the facts that (a) the decay of the muon is back down to the electron, and (b) the electron and muon are closely related to one another within the formalism of quantum electrodynamics (QED), it seems clear that the muon represents an excited state of the electron. The calculation of the muon-to-electron mass ratio that we have just presented here may seem at first glance to be trivial or accidental, but it must be kept in mind that this mass ratio has been one of the major unsolved mysteries in particle physics ever since the discovery of the muon more than half a century ago.

The particles that are generated in the M^X excitation tower erected above the symmetric $M_{\mu\mu}$ platform state were portrayed in Fig. 3.8.2, where they were displayed as separate but parallel M_μ and $\bar{M}_\mu$ columnar excitations of $\mu^-\mu^+$, p^+p^- and $\tau^-\tau^+$ fermion pairs. These pairs are generated by successively adding excitation quanta XX in each column of the $M_{\mu\mu}$ platform, with no hadronic binding energy involved, as shown in Fig. 3.8.3. The calculated masses for all three of these levels are within 1% of the experimental masses, as is detailed in Fig. 3.8.3. One excitation that was not depicted in Fig. 3.8.2 was the addition of a single quantum X to each of the M_μ and $\bar{M}_\mu$ platforms. This single-X excitation is the fermion analogue of the single-X *boson* excitations $M_K \cdot X$ and $\bar{M}_K \cdot X$ of Sec. 3.10, which resulted in the formation of a K and $\bar{K}$ pair of "strange" kaons. Here we have the single-X *fermion* excitations $M_\mu \cdot X$ and $\bar{M}_\mu \cdot X$, which result in the formation of an s and $\bar{s}$ pair of "strange" quarks. Figure 3.11.2 illustrates the mass relationships between these two types of "strange" particle states.

The K meson is a spin-0 boson with an integer electric charge, and the s quark is a spin-1/2 fermion with a fractional electric charge, but they both carry the same strangeness quantum number S. In a process known historically as "associated production" [54], where strangeness is conserved, a typical reaction is a negatively charged pion striking a proton with enough energy to produce a neutral kaon and a lambda hyperon:

$$\pi^- + p \rightarrow \Lambda^0 + K^0 .$$

The π^- and p are non-strange ($S = 0$), the Λ^0 carries strangeness $S = -1$ (in its s quark), and the K^0 carries strangeness $S = +1$. Thus overall strangeness $S = 0$ is preserved in the production process. When the Λ^0 and K^0 decay, they have only non-strange final states available. Hence the decays must violate strangeness, and they proceed very slowly. According to the Standard Model, the strangeness of the kaon is due the fact that it contains a strange antiquark $\bar{s}$. Thus the fact that both the Λ^0 and the K^0 carry the *same* strangeness quantum number (but with opposite signs) is

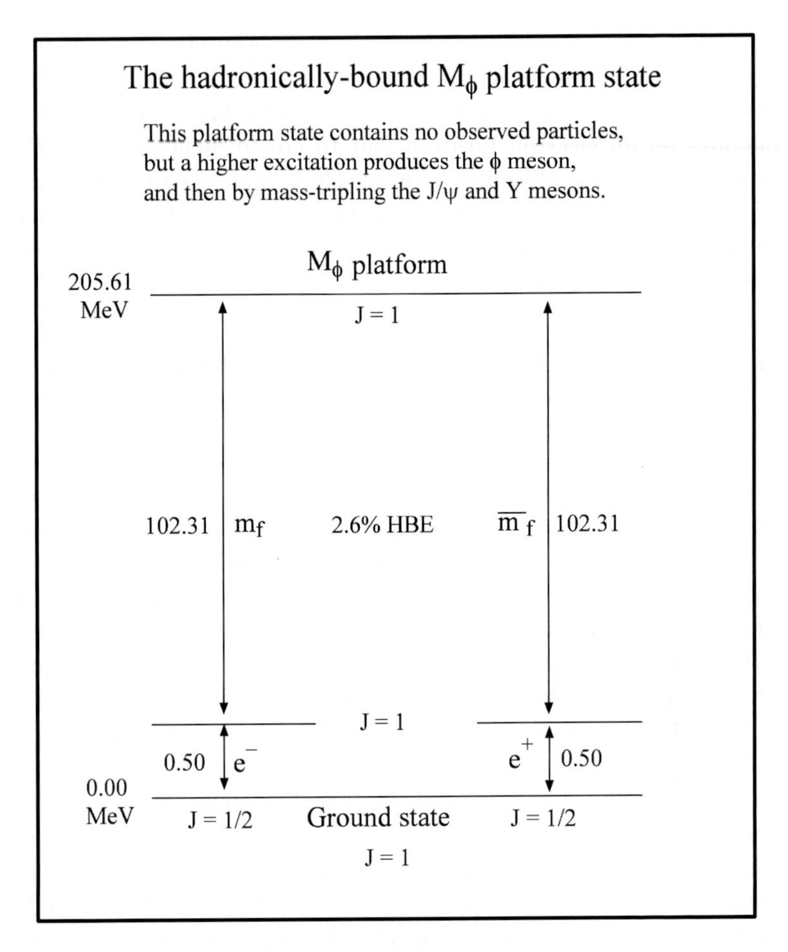

Fig. 3.12.1 The M_ϕ platform state, which does not correspond to an observed particle. This symmetric platform has the same fermionic α-excitation structure as the symmetric $M_{\mu\mu}$ platform, but it has the particle and antiparticle components hadronically bound together in a spin-1 configuration (Fig. 0.2.16). It is identified by the spin-1 ϕ vector meson that appears as an XX excitation of the platform state (Figs. 3.12.2 and 3.12.3). The ϕ meson represents the threshold for producing $s\bar{s}$ quark pairs, just as the J/ψ and Υ vector mesons represent the thresholds for producing $c\bar{c}$ and $b\bar{b}$ quark pairs (Sec. 3.17).

Fig. 3.12.3 is a comparison of the $\phi = s\bar{s}$ and $\eta' = K\bar{K}$ bound states, using an HBE of 3% for each of them. The α-quantized mass value of 525.7 MeV is used for an unbound s quark, and the CI charge-averaged experimental mass is used for the K meson. As can be seen in Fig. 3.12.3, both the ϕ

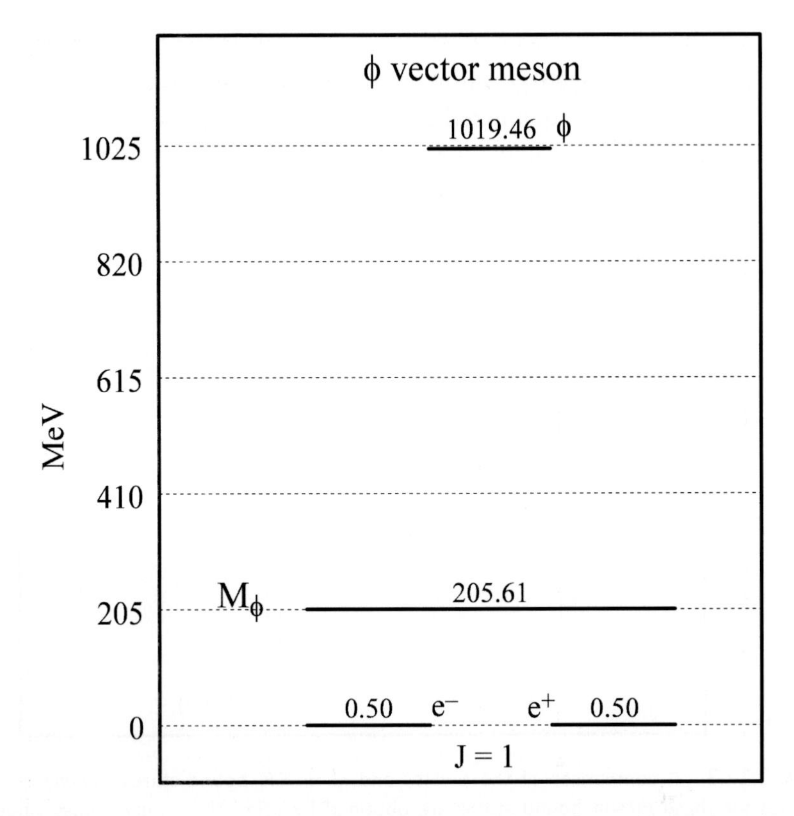

Fig. 3.12.2 The ϕ meson that appears above the M_ϕ platform, shown plotted on a 205 MeV mass grid that applies to hadronically bound *fermion* states. This 205 MeV mass grid is analogous to the 137 MeV mass grid that applies to hadronically bound *boson* states, as displayed for the M_π platform in Fig. 3.7.2. The 205 MeV grid contrasts with the 210 MeV mass grid erected for the particles in the hadronically *unbound* fermion particle and antiparticle columns of the $M_{\mu\mu}$ platform in Fig. 3.8.2. As can be seen, the ϕ meson falls accurately on this grid, at an excitation energy of $2X \cong 820$ MeV above the M_ϕ platform.

and η' bound states exhibit the same general magnitude for the hadronic binding energy (see Fig. 3.15.1). The interesting feature of the $\eta' = K\bar{K}$ bound state calculation is that it is based on the masses of the observed η' and K mesons, so we obtain a hadronic binding energy value that is independent of theory. Some textbooks describe the ϕ, whose principal decay mode is into a $K\bar{K}$ pair, as having "hidden strangeness." We demonstrate in Sec. 3.17 the manner in which the $\phi = s\bar{s}$ vector meson serves as the mass generator for the $J/\psi = c\bar{c}$ and $\Upsilon = b\bar{b}$ vector mesons.

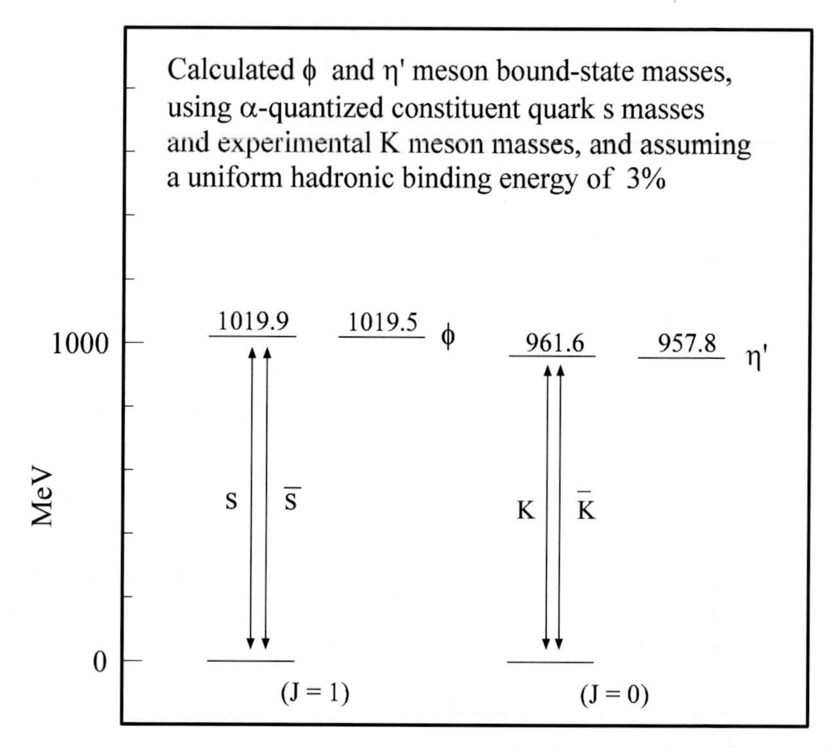

Fig. 3.12.3 A comparison of the $\phi = s\bar{s}$ and $\eta' = K\bar{K}$ bound states. Theoretical values for these meson bound states are obtained by using the s-quark mass value shown in Fig. 3.11.2 for the ϕ meson, the average K mass (the kaon CI mass in Sec. 3.14) for the η' meson, and a hadronic binding energy of 3% for each meson. As can be seen in Fig. 3.12.3, this procedure gives accurate calculated mass values. Furthermore, the dominant decay mode of the ϕ meson is into a $K\bar{K}$ pair [10]. The main point of this comparison between the ϕ and η' mesons is to provide evidence for the viewpoint that they really are $\phi = s\bar{s}$ and $\eta' = K\bar{K}$ bound states, respectively. This result does not emerge in any transparent manner from the current-quark mass values of the Standard Model.

Starting with the Rosetta-stone particles of Sec. 3.3, we have defined a set of α-quantized platform states M, and we have also discovered an excitation quantum X that serves as the mass unit for erecting excitation towers on the platforms M. Then we described in detail the particles that are generated in this M^X excitation process. We now move on in the next few sections to gather together the "M^X octet" of particles that occur in the various M_X excitation towers, and to display the calculated mass values we can deduce for them.

3.13 The Fundamental "M_X Octet" of Threshold-State Particles

In Sec. 3.6 we listed the 36 long-lived elementary particles that exhibit α-quantized lifetimes (Chapter 2). We then added in the stable electron and proton, and also the ϕ meson, whose decay into a $K\bar{K}$ pair of equal strangeness leads to an anomalously short lifetime. We subtracted the Ξ_b baryon. This list is displayed in Table 3.6.1, where the 38 particles are sorted into "generations" that correspond roughly to their "hierarchy" in the mass generation process. The particles in the first generation are the non-strange bosons (PS mesons) in group (1a) and the fermions (leptons and nucleons) in group (1b) of Table 3.6.1. The second generation particles are the strange bosons and fermions, which are separated into a low-mass tier of spin-0 (PS mesons) and spin-1 (phi meson) bosons that comprise groups (2a) and (2b), respectively, and a higher-mass tier of hyperons that comprise group (2c). The third and fourth generations M^X are the charm and bottom particle states, respectively. The particles that appear in the excitation towers of Secs. 3.7–3.12 are those in groups (1a), (1b), (2a) and (2b), which are the ones of greatest interest from the standpoint of studying the α-quantization of elementary particle masses. In the calculation of particle masses we use the charge-independent (CI) masses $\pi = 1/2\ (\pi^{\pm} + \pi^0)$ and $K = 1/2\ (K^0 + K^{\pm})$ to represent the pions and kaons (Sec. 3.14), and we use the proton itself to represent the proton-neutron "nucleon" pair. These group 1(a), 1(b), 2(a) and 2(b) particle states form the "M^X octet", with the electron mass serving as the reference mass for the octet.

Before listing the M^X octet particles, we first briefly recapitulate the M^X production process. When a symmetric $m_b\bar{m}_b$ or $m_f\bar{m}_f$ two-electron α-leap is generated from an electron–positron pair, the resulting energy level acts as an M_π, M_ϕ or $M_{\mu\mu}$ platform mass upon which further excitations can be made (Eqs. (3.5.5)–(3.5.7)). Asymmetric one-electron α-leaps can also be identified that exist as components of the symmetric production process, and they lead to the asymmetric platform masses M_K and M_μ (Eqs. (3.5.9) and (3.5.8)). We might logically expect that subsequent excitations of the M_K and M_μ platform masses would be in single α-leaps of $m_b = 70$ MeV or $m_f = 105$ MeV, respectively, and that excitations of the M_π, M_ϕ and $M_{\mu\mu}$ platform masses would be in double α-leaps of $m_b\bar{m}_b = 140$ MeV or $m_f\bar{m}_f = 210$ MeV. Instead, what we find is that the subsequent excitations of all five of these platform masses are in units of the supersymmetric "multiple-α-mass" quantum

$X = 3m_b\bar{m}_b = 2m_f\bar{m}_f = 420$ MeV (Eq. (3.9.1)). Thus we have a two-step "M^X" production process: (a) an initial α-leap from an electron ground state to a platform mass M; (b) the addition of excitation quanta X to the mass M.

The M^X (a) and (b) steps for the filled asymmetric *one-electron* M^X platform excitation levels, which have HBE = 0, are

$$\begin{aligned} &\text{(a) } m_e(1 + 1/\alpha) = M_K\,, \\ &\text{(b) } M_K + nX = \text{spin-0 particle}\,, \quad n = 1\,; \end{aligned} \tag{3.13.1}$$

$$\begin{aligned} &\text{(a) } m_e(1 + 3/2\alpha) = M_\mu\,, \\ &\text{(b) } M_\mu + nX = \text{spin-1/2 particle}\,, \quad n = 0, 1, 2, 4\,, \end{aligned} \tag{3.13.2}$$

where m_e is the electron mass. The M^X (a) and (b) steps for the filled symmetric *electron-pair* M^X hadronic platform levels that have HBE $\cong$ 2.6% are

$$\begin{aligned} &\text{(a) } m_{e-\bar{e}}(1 + 1/\alpha) = M_\pi\,, \\ &\text{(b) } M_\pi + nX = \text{spin-0 particle}\,, \quad n = 0, 1, 2\,; \end{aligned} \tag{3.13.3}$$

$$\begin{aligned} &\text{(a) } m_{e-\bar{e}}(1 + 3/2\alpha) = M_\phi\,, \\ &\text{(b) } M_\phi + nXX = \text{spin-1 particle}\,, \quad n = 1\,, \end{aligned} \tag{3.13.4}$$

where $m_{e-\bar{e}}$ is the electron–positron mass. Equation (3.13.1) is for the kaon and antikaon mesons; (3.13.2) is for the muon and antimuon leptons, the s and $\bar{s}$ quarks, the proton and antiproton nucleons, and the tau and antitau leptons; (3.13.3) is for the pi, eta and eta$'$ mesons; and (3.13.4) is for the phi(1020) meson. These M^X excitations generate the "M^X octet", which includes the following eight fundamental elementary particle masses:

$$\pi\,,\ \eta\,,\ \eta'\,,\ K\,,\ \mu\,,\ p\,,\ \tau\,,\ \phi \tag{3.13.5}$$

and the related antimasses $\bar{K}$, $\bar{\mu}$, $\bar{p}$, $\bar{\tau}$. These are the basic low-mass threshold states. (The u, $\bar{u}$, d and $\bar{d}$ fermion quarks correspond to $n = 1/2$ levels in Eqs. (3.13.2) and (3.13.4), which are not allowed as direct electron excitations in this X-quantized M^X formalism.)

As a notation for describing M_X production in detail, we attach a superscript n to each platform M, and we denote the hadronic binding energy (HBE) used. This gives

$$M_\pi^n \equiv (M_\pi + nX)(1 - \text{HBE})\,; \quad M_\phi^n \equiv (M_\phi + nXX)(1 - \text{HBE})\,; \tag{3.13.6}$$

$$M_K^n \equiv (M_K + nX)\,; \quad M_\mu^n \equiv (M_\mu + nX)\,; \tag{3.13.7}$$

This notation is used in Table 3.16.1 for the precision mass calculations of Sec. 3.16.

Table 3.13.1 The "M^X octet" excitation matrix. The eight M^X threshold-state particles of Eq. (3.13.5) are shown separated into hadronically-bound and unbound excitations, and into nonstrange or leptonic and strange excitations, using the M^X notation of Eqs. (3.13.1)–(3.13.4). (Two related excitations are shown in parentheses.) Isotopic-spin-averaged masses are used for the pions and kaons (Sec. 3.14), and the proton mass is used for the proton–neutron doublet. As can be seen in Table 3.13.1, leptons and hadrons blend into one comprehensive α-quantized pattern. Without combining these two particle types, we would not obtain this M^X structure. Also, we would not obtain an accurate parameter-free value for the proton-to-electron mass ratio.

	Hadronically-bound states	Unbound states
Nonstrange and leptonic states	$M_\pi^{0,1,2} = \pi, \eta, \eta'$	$M_\mu^{0,2,4} = \mu, p, \tau$
Strange states	$M_\phi^1 = \phi = s\bar{s}$ $(\eta' = K\bar{K})$	$M_K^1 = K$ $(M_\mu^1 = s)$

Table 3.13.1 displays the binding energy and flavor symmetry properties of the M_X octet particles. The eight particles are arrayed vertically as hadronically bound and unbound states, and horizontally as (non-strange or leptonic) and strange states. The M^X mass excitation levels shown in Table 3.13.1 for these particles follow the platform notations of Eqs. (3.13.6) and (3.13.7). As can be seen, each element in this M^X binding energy and flavor matrix is fragmentary, but together they form a complementary pattern that certainly is not random, and they yield mass ratios which are not obtainable elsewhere.

Before we present the final α-generated mass values for the M^X octet particles, which we do in Sec. 3.16, we first review the available information on isotopic spin charge splittings of the masses (Sec. 3.14), and on hadronic binding energies (Sec. 3.15).

3.14 Isotopic Spin Mass Splittings and Charge-Independent (CI) Particle Masses

Some elementary particles exist in single charge states, whereas others have two or more charge states (isotopic spins) with different mass values in those states. In the M^X-octet mass matrix defined in Table 3.13.1 and

Eqs. (3.13.6) and (3.13.7), we used the same averaged mass value for the two isotopic spin states of the pion, and also of the kaon. This limits the accuracy of the mass calculations. If we are going to accurately reproduce a particle from a set of substates, then the masses of these substates, or at least the mass of the overall structure, must depend on the charge of the particle. However, when we examine the experimental variations in the mass splittings that are associated with the charges, it becomes apparent that this is not an easy task to accomplish.

The lowest-mass — and thus in some sense the simplest — isotopic spin multiplets occur in the pseudoscalar mesons. The pions are the lowest-mass multiplet, and they have an intriguing charge splitting of the mass. The measured $\pi^{\pm} - \pi^0$ mass difference is 4.5936 ± 0.0005 MeV, or 3.3%. This is the largest percentage charge splitting of any of the elementary particles except the $W^{\pm}$ and Z^0. The pion charge splitting seems to be related to the mass of the electron, as we pointed out in Sec. 3.4. Figure 3.4.1 is a plot of the $\pi^{\pm} - \pi^0$ mass difference, shown in comparison to a mass interval of nine electron masses — 4.5990 MeV. These two mass intervals agree to an accuracy of 0.1%. In view of the factors of three and nine that are observed in the particle masses, and particularly in the particle lifetimes (Figs. 2.9.1, 2.10.2 and 2.10.3), the mass agreement displayed in Fig. 3.4.1 is probably not accidental. Since the mass of the pion is derived from the mass of the electron by making an α-leap from a spin-0 electron–positron pair to an $m_b\bar{m}_b$ bosonic α-mass pair, the mass of the pion is directly related to the mass of the electron. Thus both the pion mass and the pion charge splitting seem to depend on the mass of the electron.

We represent the pion here as $\pi = m_b\bar{m}_b$, so one way of reproducing the $\pi^{\pm} - \pi^0$ mass difference is by assigning charge-dependent masses to m_b and $\bar{m}_b$. But if we now move on to the kaon isotopic spin multiplet, where we use these same m_b and $\bar{m}_b$ masses, we find a $K^0 - K^{\pm}$ mass difference of 3.995 ± 0.034 MeV. This mass interval is different, and the sign of the mass difference has been reversed! Thus we cannot automatically reproduce the kaon charge splitting by using the same m_b and $\bar{m}_b$ charge-dependent masses that we use for the pion. We could qualitatively reproduce the reversed sign of the kaon charge splitting by juggling the number of charged and uncharged mass quanta m_b we assign to the kaon states, but do we then have to consider coulomb interactions between the various charges? Also, are we dealing here with integer charges or with fractional quark charges? When we move on to the mass differences observed in the proton and neutron isospin doublet, in the hyperon isospin multiplets, and in the

c and b meson isospin multiplets, the charge-splitting mechanism becomes even more obscure.

Another way to handle the charge splitting is to keep the basic mass quanta m_b and $\bar{m}_b$ independent of charge effects, and instead attribute the mass splitting to Coulomb charge effects among the fractionally charged quark states. If we reproduce the pion with the charge states $\pi^{\pm} = (\pm 2/3, \pm 1/3)$ and $\pi^0 = (\pm 1/3, \mp 1/3)$, then the Coulomb repulsion between the charges in the $\pi^{\pm}$ raises its mass, and the Coulomb attraction between the charges in the π^0 lowers its mass, so a charge spacing can be deduced that reproduces the pion mass splitting. But this same mechanism will not work in any simple manner for the charge-reversed kaon mass splitting.

In lieu of a comprehensive theory for handling these charge splittings, we use *charge-independent* (CI) average masses as the best representation of the "intrinsic" experimental mass values for the pion and kaon isotopic spin multiplets, and also for the higher-mass threshold-state particles discussed in Sec. 3.23. In addition to the pions and kaons, the other isotopic spin multiplet in the M^X octet is the proton–neutron doublet. Since we have focused strongly in the present book on the determination of the proton-to-electron mass ratio, we treat the proton mass directly in the mass calculations of Sec. 3.16, instead of averaging it in with the nearby mass of the neutron.

The isotopin spin mass splitting is 3.3% in the pion, 0.8% in the kaon, and 0.14% in the nucleon. In most particle isospin multiplets, it is a few tenths of a percent. Thus if we use CI masses in phenomenological mass calculations for a collection of particle states, we should expect the average *absolute percent deviation* (APD) of the masses to be on the order of a few tenths of a percent. This applies to the individual particle masses, but probably also to the CI averaged masses, because we cannot make an accurate separation between charge-associated mass effects and hadronic binding energy effects. In the M^X octet mass calculations shown in Table 3.16.1, the average APD value is about 0.4%, which is what we might expect when using CI mass values.

When we expand these α-quantized masses to encompass the supermassive $W^{\pm}$, Z^0 isotopic-spin doublet and the top quark t, we discover that the CI average mass value $\overline{WZ} = 85,795$ MeV of the widely-separated W and Z vector bosons seems to be of real physical significance. This conclusion follows from the following three facts: (1) multiplying the experimental Υ_{1S} vector meson mass by 9 (two mass-triplings) gives a value of 85,143 MeV, which is within 0.8% of the experimental $\overline{WZ}$ CI mass (Fig. 3.19.1); (2) the sum of the W and Z masses, which is twice the $\overline{WZ}$ mass, is 171,591 MeV, which is almost identical to the top quark t mass of 172,500 MeV; (3) when we go to second-order α-leaps, which we do in Sec. 3.20, the agreement of the second-order "α-quarks" with the $\overline{WZ}$ mass and with the top quark t mass is even closer than the values cited here. Hence isotopic-spin-averaged CI masses may represent more than just a convenient way of fitting charge-dependent isotopic spin multiplets with charge-independent mass units, and may eventually guide us to a proper understanding of the charge-state contributions to particle masses.

3.15 Hadronic Binding Energy (HBE) Systematics

In the constituent-quark model, the mass of a particle is determined by (1) the intrinsic quark masses, (2) small isotopic spin effects, and (3) small binding energy effects. The isotopic spin effects, which involve the charge splitting of the masses among multiplets with the same basic quark structure, are generally on the order of a few tenths of a percent except for the lightest particles. Specifically, they typically amount to a few MeV, independently of the total mass of the particle. Binding energy effects are larger than isotopic spin effects — a few percent at the lower masses — but they remain relatively constant in magnitude as the mass increases, which means that they decrease percentagewise at higher energies, and then they seem to almost vanish at energies above 4 GeV. Binding energies not only directly affect the mass values, but they can also provide some indirect information about the internal charge structures of the particles if we make the assumption that they are associated with the gluon fields produced by fractionally charged quarks. In some sets of lower-mass particles (e.g., π, η, η'), the binding energy scales with the mass, so it is essentially a fixed (small) percentage of the mass. In at least one set of particles, the ϕ, J/ψ, Υ paired-quark threshold states, which have particularly simple and well-defined charge structures, the hadronic binding energy does not seem to scale with the mass, but rather with the charge states of the quark pairs.

A key point in the discussion of particle hadronic binding energies is to identify what causes them. An important clue is provided by the leptons, and in particular the electron. We know that the interactions of the electron with other particles are purely electromagnetic: electrons have no hadronic forces. We also know that the charge on the electron is very localized, and in fact exhibits no measurable size effects at all, down to at least 10^{-17} cm. If we represent the electron as being just its charge, and if we represent the charge as having a classical electric field structure, then the calculated self-energy of the electron is much larger than its observed mass. Hence the mass of the electron is certainly not all electromagnetic — a point that has been emphasized for example by Abraham Pais [55]. Thus the electron must possess a non-electromagnetic "mechanical" mass, and this mass must be non-interacting (neutrino-like). As we have demonstrated in the present studies, the mass of the electron is directly linked to the masses of the hadrons, so the hadrons must similarly possess non-interacting masses. This suggests that the hadronic binding energy derives from the hadron fractional electric charges and their associated gluon fields, and not from the hadron masses.

When an integer electric charge is broken apart into 1/3 and 2/3 fractions, which effectively occurs in the SM quarks, the charge fractions can move apart — as evidenced for example by Bjorken scattering in the proton — but they never emerge as fractional charges on final-state particles. This indicates that the fragmented parts of an electric charge are not free, but are bonded to one another by an unbreakable gluon-like cluster of forces which seem to play a significant role in hadronic binding energies. We can formally define a gluon cluster as follows:

A "gluon cluster" (GC) is a localized set of fractional electric charges formed from an initial integer charge, and held together by unbreakable gluon-like bonds.

The phenomenological significance of the GC is that it appears to serve as a center for producing hadronic forces, as set forth in the following postulate:

Hadronic binding energy (HBE) is a GC–GC interaction.

A proton has a single GC, and it does not have an appreciable hadronic self-binding energy: it appears right on the $M_{\mu\mu}$ excitation tower mass shell, as discussed in Sec. 3.8 and displayed in Figs. 3.8.2 and 3.8.3. The proton GC interacts hadronically with the GCs of other protons, antiprotons and antineutrons. The pion, like the proton, is a particle with an integral overall electric charge, but the pion charge, unlike the proton charge, is the summation of two (possibly intertwined) GCs, one centered on the M_b substate and the other centered on the $\bar{M}_b$ substate (see Fig. 0.2.16 and Eq. (3.5.5)). Hence the pion has a substantial (~2.7%) HBE. An integrally charged lepton does not have a GC, and it interacts only electromagnetically. Thus a muon–antimuon pair, for example, appears in the $M_{\mu\mu}$ excitation tower as two essentially separate particles — one in each column of the tower (Fig. 3.8.2).

The above discussion of GCs seems rather obvious. Where it is not quite so obvious is when we consider hyperons such as the Λ and Ξ. These are typically formed by collisions between a proton and a pion or kaon. The Λ contains one strange quark s, which is 210 MeV more massive than a $q \equiv u$ or d quark. The Ξ contains two strange quarks s, and thus should be 210 MeV larger in mass than the Λ, which it is. But the calculated Λ and Ξ masses are each about 3.5% larger than the experimental masses. This is not the result we obtained in the proton, where the total mass was accurately given (to better than 1%) as the sum of the three q quarks. One way of accounting for this mass deficiency in the experimental Λ and Ξ masses is to say that the proton excitation process in the creation of a Λ or Ξ contains both particle and antiparticle components, thus giving rise to a second GC gluon cloud. Hence the resulting Λ or Ξ excitation has a substantial hadronic binding energy, which empirically must be about 3.5% in order to bring the calculated and experimental mass values into line. This is a rather indirect line of reasoning, but it introduces a method for addressing some of the complexities of the particle excitation process.

Another notable feature of hadronic binding energies is the existence of some particle groups that have HBE's which are proportional to the particle masses (and hence equal percentage-wise), plus one group — the φ, J/ψ and Υ vector mesons — that has decreasing HBE's which essentially vanish at the higher energies. To illustrate this, we present in Table 3.15.1 some experimental and theoretical HBE values Δ for sets of threshold-state particles. These Δ values are displayed graphically in Fig. 3.15.1. The "experimental" Δ values are obtained directly from measured mass values. The "theoretical" Δ values are obtained by comparing "unbound" particle masses (as given in Table C1 of Appendix C) to measured masses, which are compiled in Appendix B. The deviation Δ between the two mass values is the HBE. As can be seen in Table 3.15.1, the π, η and η' PS meson binding energies are proportional to the masses, and they are in reasonable agreement with the HBE that we calculate for the η' under the assumption that it is a $K\bar{K}$ bound state. The Λ and Ξ hyperon HBE's are also proportional to the masses (where we assume that they represent GC–GC interactions), and they agree quite well with the measured binding energy of the $\bar{p}n$ bound state [53].

The vector mesons, by way of contrast to the PS mesons, baryons and hyperons, have HBE Δ values that do not vary proportionally with the

Table 3.15.1 Calculations of hadronic binding energies (HBE) Δ for some representative threshold-state particles. The "theoretical" values for Δ are obtained by comparing calculated unbound-quark mass values as defined in the equations of Table C1 in Appendix C with the measured mass values shown here. The experimental masses for these particle are listed in MeV, together with the HBE $= \Delta$ values expressed in percent (relative to the calculated values). A few "experimental" values for Δ are shown, as obtained directly from measured mass values. The PS mesons and hyperons and the baryon-antibaryon state have values for Δ that are roughly proportional to the masses, whereas the vector mesons and B_c meson have values for Δ that scale with the quark charges, and hence tend to vanish at high energies (see Fig. 3.15.1 and Sec. 3.17).

Theoretical Δ values (pseudoscalar mesons)		
$\bar{\pi} = 137.27$ MeV	$\Delta = 3.80$ MeV $= 2.69\%$	
$\eta = 547.51$ MeV	$\Delta = 13.72$ MeV $= 2.44\%$	
$\eta' = 957.78$ MeV	$\Delta = 23.60$ MeV $= 2.40\%$	
Experimental Δ values		
$\eta' = K^+K^-$	$\Delta = 29.57$ MeV $= 2.99\%$	
$\eta' = K^0\bar{K}^0$	$\Delta = 37.52$ MeV $= 3.77\%$	
Theoretical Δ values (hyperons)		
$\Lambda = 1115.68$ MeV	$\Delta = 40.25$ MeV $= 3.48\%$	
$\bar{\Xi} = 1318.07$ MeV	$\Delta = 47.93$ MeV $= 3.51\%$	
Experimental Δ value (baryon-antibaryon)		
$\bar{p}n = 1794.5$ MeV	$\Delta = 83.3$ MeV $= 4.44\%$	[53]
Theoretical Δ values (vector mesons, D and B_c)		
$\phi = 1019.46$ MeV	$\Delta = 31.94$ MeV $= 3.04\%$	
$\bar{D} = 1866.9$ MeV	$\Delta = 24.80$ MeV $= 1.31\%$	
$J/\psi = 3096.92$ MeV	$\Delta = 55.24$ MeV $= 1.75\%$	
$B_c = 6826$ MeV	$\Delta = 17.29$ MeV $= 0.27\%$	
$\Upsilon = 9460.30$ MeV	$\Delta = -5.87$ MeV $= -0.06\%$	

masses, but instead, roughly speaking, vary according to the charge states of the quark pairs, and thus are approximately constant, which makes them tend toward zero (percentagewise) at high energy (see Fig. 3.17.2). The ϕ and Υ resonances each have $(-1/3, +1/3)$ charge pairs. The J/ψ has a $(+2/3, -2/3)$ charge pair, and it exhibits a somewhat larger HBE than we

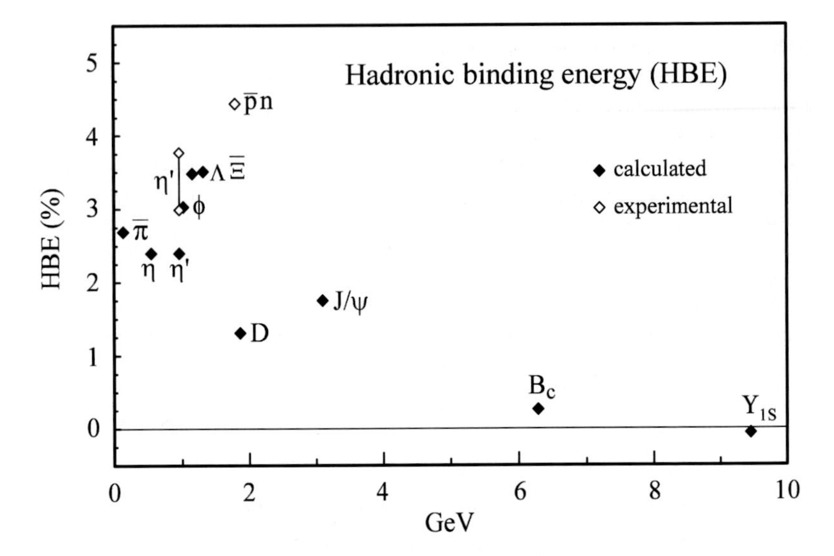

Fig. 3.15.1 A plot of hadronic binding energies (HBE) for long-lived threshold-state paired-quark particles. The calculated values are obtained from quark masses in the cases where they exceed the experimental masses (see Table C1). The experimental values are for the $\bar{p}n$ annihilation into pions (Ref. [53]), and for the postulated $\eta' \equiv K\bar{K}$ bound state with a range of values that is limited by the charged and neutral K masses. As can be seen, the HBE value are in the 2%–4% range for particle masses up to about 4 GeV and then fall off sharply for mass values above 4 GeV. These results are extended in Fig. 3.15.2.

might expect from the trend of the ϕ and Υ HBEs. The $B_c^{\pm}$ [10] has a $(\pm 2/3, \pm 1/3)$ charge pair and a small Δ value. These calculated HBE's in MeV (Table C1) are shown together in Eq. (3.15.1)

$$\Delta_{\phi} = 31.94\,, \quad \Delta_{J/\psi} = 55.24\,, \quad \Delta_{B_c} = 17.29\,, \quad \Delta_{\Upsilon} = -5.87\,. \quad (3.15.1)$$

In addition to the hadronic binding energies that are shown in Table 3.15.1, there are other HBE values for Δ that we can obtain from platform particles, as shown in Table 3.15.2. The particle masses calculated in Table 3.15.2 are all somewhat below the experimental masses, which indicates that the excitation process is more complicated than just adding together the basic q, s, c, b quark masses. To see how this works out phenomenologically, we have increased the K^* and Σ masses by one excitation quantum m_b each (see Sec. 3.26), and we have increased the B_s, Ω and charmed baryon masses by two excitation quanta m_b (one pion mass) each. The B meson and Λ_b baryon masses have each been increased by

Table 3.15.2 Calculations of hadronic binding energies Δ for some platform states that were not shown in Table 3.15.1. The particles included here all have directly-calculated quark masses that are slightly below the experimental values shown here (see Fig. 0.2.9, and Table C2 in Appendix C). Thus additional m_b excitation quanta have been included, as indicated in the table (see Sec. 3.26). The Δ values that this procedure gives are plotted as X's in Fig. 3.15.2, which shows that they are in general agreement with the Δ values of Table 3.15.1.

Theoretical Δ values (mesons)

$K^* = q\bar{s} + m_b = 893.83$ MeV $\quad \Delta = 17.01$ MeV $= 1.87\%$

$B = q\bar{b} + 4m_b = 5279.2$ MeV $\quad \Delta = 43.75$ MeV $= 0.83\%$

$B_s = s\bar{b} + 2m_b = 5367.5$ MeV $\quad \Delta = 25.47$ MeV $= 0.47\%$

Theoretical Δ values (hyperons)

$\bar{\Sigma} = qqs + m_b = 1193.15$ MeV $\quad \Delta = 32.80$ MeV $= 2.68\%$

$\Omega = sss + 2m_b = 1672.45$ MeV $\quad \Delta = 44.19$ MeV $= 2.57\%$

Theoretical Δ values (charmed and bottom baryons)

$\Lambda_c = qqc + 2m_b = 2286.46$ MeV $\quad \Delta = 59.90$ MeV $= 2.55\%$

$\bar{\Xi}_c = qsc + 2m_b = 2469.45$ MeV $\quad \Delta = 86.98$ MeV $= 3.40\%$

$\bar{\bar{\Xi}}_{cc} = qcc + 2m_b = 3518.9$ MeV $\quad \Delta = 87.91$ MeV $= 2.44\%$

$\Omega_c = ssc + 2m_b = 2697.5$ MeV $\quad \Delta = 70.03$ MeV $= 2.53\%$

$\Lambda_b = qqb + 4m_b = 5624.0$ MeV $\quad \Delta = 13.55$ MeV $= 0.24\%$

four excitation quanta m_b. This procedure yields the Δ values shown in Table 3.15.2 and also displayed in Table C2 of Appendix C. These new Δ values are illustrated in Fig. 3.15.2, which is just Fig. 3.15.1 with the new Δ values indicated as X's. As can be seen, the agreement between the two sets of Δ values is pretty good. This does not establish a "theory" for these excitation mechanisms, but it serves as a phenomenological guide for further investigation of the particle excitation mechanisms.

The hadron binding energies Δ that are discussed here have no counterpart in the SM mass calculations. The reason is that, unlike the SM *current-quark* masses, the α-quantized m_b and m_f masses that are deduced from the reciprocal lifetime and mass relationships of the present paper are *constituent-quark* masses, wherein the mass of the particle comes mainly from the masses of the quarks that make up the particle. If the quarks have enormous binding energies, as was originally assumed in QCD [36], then the quark masses can have a wide range of energies. But if the quarks

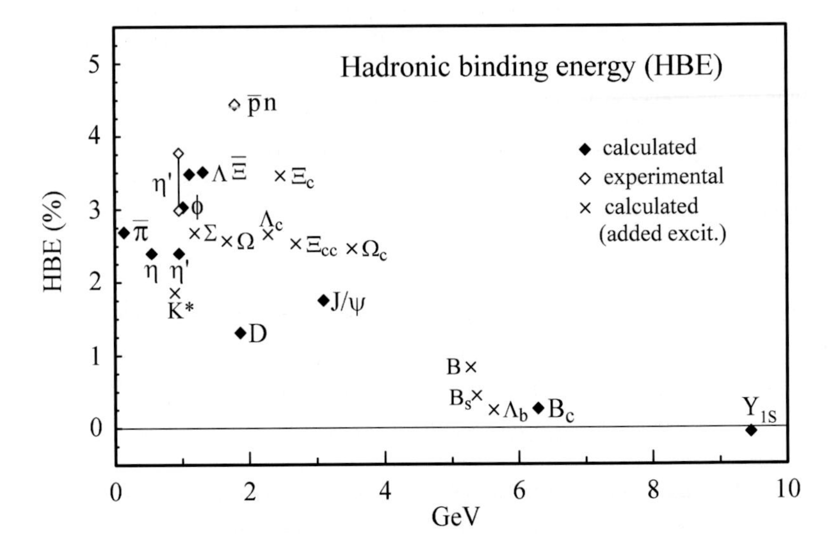

Fig. 3.15.2 A plot of hadronic bind energies for the particles of Fig. 3.15.1 plus additional particles (labeled with ×'s) for which pionic excitation quanta have been added to the basic muonic quark configurations to bring their calculated masses up to the experimental values (see Table C2). These added particles reinforce the two-tier HBE binding energy systematics displayed in Fig. 3.15.1, wherein the HBE is in the 2%–4% range below 4 GeV, and is almost negligible above 4 GeV. It seems to essentially vanish completely for mass values above 9 GeV (Sec. 3.20).

have small binding energies (a few percent), then their masses are delimited by the particle masses, and we can investigate small deviations from the theoretical masses. In the current-quark approach, gluon fields are assumed to make large contributions to the mass.

In addition to the hadronic binding energies displayed in Table 3.15.1, there is one other experimental value we can cite that sets the scale, at least at low mass values, for the effective overall HBE. The M^X excitation diagram that is displayed schematically in Fig. 0.2.19 and also shown in Fig. 3.16.1 contains five independent values of the quantum X in unbound configurations with an average experimental mass of 419.3 MeV, and four values of X in hadronically bound configurations with an average experimental mass of 408.6 MeV. The ratio of these two experimentally determined mass averages corresponds to a hadronic binding energy of 2.6%, which is in line with the HBE values shown in Tables 3.15.1 and 3.15.2 and displayed in Figs. 3.15.1 and 3.15.2.

The sharp decrease in hadronic binding energies above 4 GeV that is observed in Figs. 3.15.1 and 3.15.2 can be attributed to the decreasing radii of the Compton-sized particle states with increasing mass as compared to the intrinsic (unstretched) "length" of the gluon "rubber-band" forces that hold the particle together. This subject is discussed in more detail in Sec. 3.17.

In the next section we analyze the M^X octet excitations of Fig. 0.2.19 in detail.

3.16 Almost-Parameter-Free M^X Octet Mass Calculations

The mass calculations that we present here represent one of the main goals we have been working toward in these studies of the α-quantization of elementary particle lifetimes and masses. The eight particle masses of the M^X octet (plus their antiparticles) are among the most important particles to consider from the standpoint of mass generation, and their mass values have not emerged from other approaches to the mass problem.

The M^X octet mass values are phenomenologically obtained here in a manner that is essentially independent of theory. Furthermore, the mass calculations depend on just one adjustable parameter — the hadronic binding energy, which is a relatively small correction, and whose range is bounded by experimental limits. The α-quantized mass units are obtained from the operation of the particle generator $\alpha = e^2/\hbar c$ on the electron mass, and the way they are combined together is dictated by the fits to the experimental data.

The motivation for using this α-quantized particle mass formalism was obtained from the global lifetime analysis of Chapter 2, which demonstrated that the lifetimes of the long-lived ($\tau > 10^{-21}$ s) threshold state particles are α-dependent. In evaluating the significance of this lifetime α-dependence, two questions had to be addressed: (1) is it *comprehensive*? (2) Is it *accurate*? The answer to both of these questions is in the affirmative. The lifetime α-dependence only applies to the threshold-state particles, but in that domain it is comprehensive: all 36 threshold particles fit onto the α-spaced lifetime grid, as displayed for example in Fig. 2.6.1; there are no "rogue" lifetimes. And when "corrections" are applied to remove the factor-of-2 "hyperfine" structure (Figs. 2.5.5 and 2.5.6), the global accuracy of the lifetime α-quantization is impressive (Fig. 2.6.2), especially when allowances are made for the factor-of-3 "flavor" structure (Fig. 2.9.1). Another striking feature of α-quantized threshold-state particle lifetimes is the α^4 lifetime spacing between unpaired-quark and paired-quark decays, which is a spacing of more than eight orders of magnitude. This α^4 spacing applies to essentially all of the threshold particles, as demonstrated in Figs. 2.8.7 and 2.8.8. The $\pi^\pm$, π^0, η and η' mesons, in particular, have highly accurate α-spaced lifetime intervals that span almost thirteen orders of magnitude (Fig. 2.2.5).

Elementary particle lifetimes and mass widths are conjugate quantities, so the α-dependence of the lifetimes carries over to the widths via the uncertainty principle. Furthermore, we logically expect the mass widths (mass stability) to be related to the masses themselves (since mass regularities must give rise to width regularities). Thus it follows that the α-quantization of elementary particle *masses* ought to be as comprehensive and accurate as the α-quantization of the corresponding *lifetimes*. In Chapter 3 we have already implicitly addressed the problem of comprehensiveness. Our focus in Sec. 3.16 is quite narrow. We study the masses of the pseudoscalar mesons, leptons, proton, and $\phi = s\bar{s}$ vector meson, which can all be accounted-for by the M^X production mechanism of Sec. 3.13. But when we add in the M^T flavor-tripling mechanism of Sec. 3.17, we reproduce the $J/\psi = c\bar{c}$ and $\Upsilon = b\bar{b}$ vector mesons. And when we extend these results to encompass the rest of the 36 threshold particles (Sec. 3.23), the α-quantized constituent-quark basis states qualitatively reproduce the observed masses. Thus the mass α-quantization of the threshold-state particles is *comprehensive*. The remaining question is that of mass *accuracies*. Interestingly, the most parameter-free and accurate results occur at the low and high ends of the mass scale. At the low end, accurate and parameter-free mass calculations occur for the M^X octet states of Table 3.13.1, as we demonstrate in the present section. At the high end of the mass scale, the single α-leap B_c and Υ_{1S} mesons below 12 GeV, and the double α-leap $\overline{WZ}$ and top quark t particle states above 12 GeV, are all given with precision directly from the constituent-quark masses. But complexities set in for the intermediate-mass hyperon and meson threshold particles (Sec. 3.23), which involve a mixture of α-mass basis states and possibly a more involved hadronic binding energy formalism.

The M^X octet states — π, η, η', K, μ, p, τ, ϕ — are the simplest particle excitations, and should have the most straightforward masses to calculate. Their M^X production channels are plotted together in the mass diagram of Fig. 3.16.1, where the platform mass M is set equal to zero in each channel. As can be seen in Fig. 3.16.1, and also in Fig. 0.2.19, the excitation towers which are erected above the various platforms M are formed out of mass units X. The numerical values for the X masses in

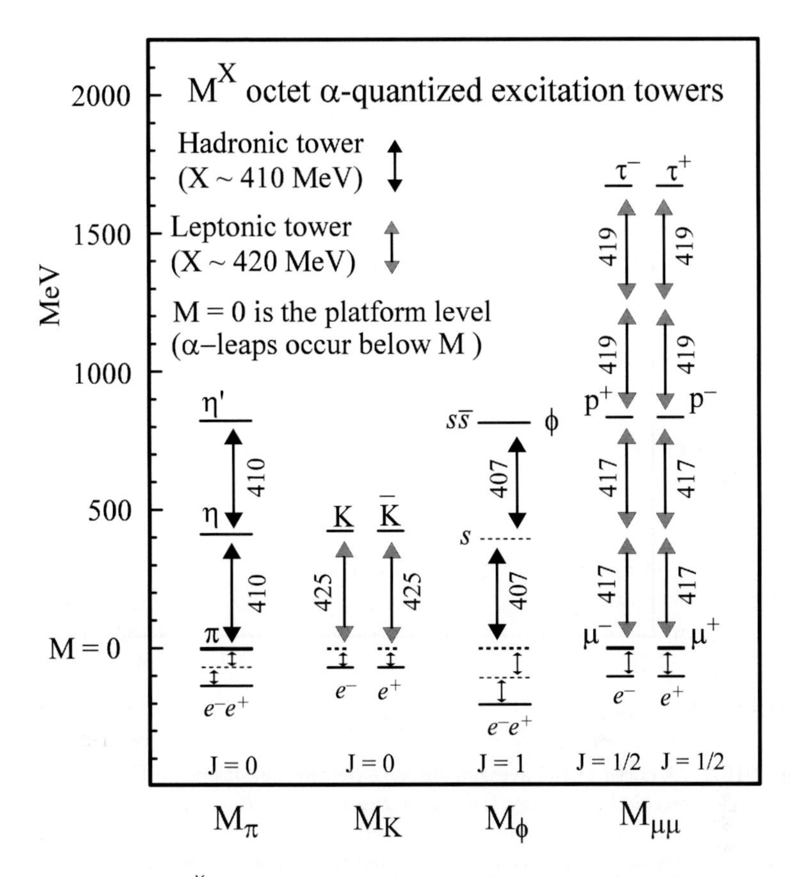

Fig. 3.16.1 The M^X octet platform excitations. Quantized α-leaps from electron–positron ground states create the platforms M_π, M_K, M_ϕ and $M_{\mu\mu}$ and subsequent additions of mass units X create the excitation towers. The displayed X-energies represent the experimental excitation level mass intervals. The averaged observed X-energy is 419.3 MeV in the M_K and $M_{\mu\mu}$ platform towers, where HBE $= 0$, and 408.6 MeV in the M_π and M_ϕ towers, where HBE $\neq 0$. The ratio of these values gives an experimentally determined X-quantum HBE of 2.55% (see Fig. 3.16.2). Table 3.16.1 displays the mass calculations for the π, η, η', K, μ, p, τ, ϕ "M^X octet", where HBE $=$ 2.6% is used. The average calculated mass accuracy for these eight particles is 0.4%, and the average mass error is 3.7 MeV, where 70, 105 and 420 MeV additive mass units are employed. There are three main points that should be noted with respect to this M^X-octet diagram: (1) leptons and baryons are interleaved together in one unified mass pattern; (2) the mass generator for these masses is the fine structure constant α, which operates on the electron ground state; (3) these accurate mass values, which require no freely adjustable parameters except the small HBE, do not emerge from theories that divide leptons and hadrons into unrelated mass domains.

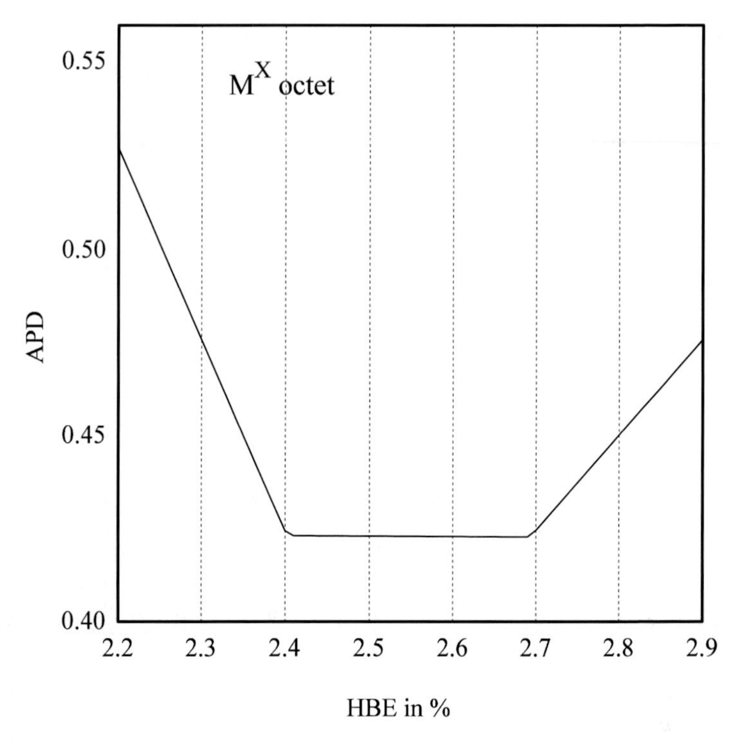

Fig. 3.16.2 A curve of the average absolute percent deviation (APD) as a function of the hadronic binding energy (HBE) for the π, η, η', K, μ, p, τ, ϕ meson octet particles of Sec. 3.13. Only the π, η, η' and ϕ are functions of the HBE. As can be seen, the average APD has a flat minimum from HBE = 2.4% to 2.7%, with an APD value of 0.42% at the minimum (see Table 3.16.1). The discontinuous nature of the APD curve is due to the discreteness of the data, and not to coarse zoning in the HBE scan.

Fig. 3.16.1 are obtained from the experimental differences between the various energy levels, with theoretical values used for the unfilled M_K and M_ϕ platform energies. The five independent *leptonic* X excitations (gray-tipped arrows) have an average X-mass of 419.3 MeV, as compared to the theoretical unbound X-mass of 420.2 MeV (Eq. (3.9.1) and Fig. 0.2.18). The four independent *hadronic* X excitations (black-tipped arrows) have an average X-mass of 408.6 MeV, as compared to the theoretical bound X-mass of 409.2 MeV (Fig. 0.2.18). The ratio of these averaged experimental X-masses indicates a hadronic binding energy of 2.55%. This small but significant hadronic binding energy exhibits itself in a clear-cut manner in these leptonic and hadronic production channels.

We can examine the effect that the hadronic binding energy has on the M^X octet mass calculations by applying a series of HBE values to the data set. Of these eight particles, only the π, η, η' and ϕ depend on the HBE. A plot of the average absolute percent deviation (APD) values for the mass calculations of the M_X octet as a function of the HBE is shown in Fig. 3.16.2. As can be seen, the APD curve has a flat minimum that extends from HBE = 2.4% to 2.7%. This is in agreement with the experimental leptonic-to-hadronic X-mass ratio cited above which yielded the value HBE = 2.55%. For uniformity, we generally select 2.6% as the HBE to apply in discussing all of these threshold-state particle excitations.

Mass calculations for the M^X octet mesons of Table 3.13.1 and Fig. 3.16.1 are displayed in Table 3.16.1. These calculations in a sense represent the culmination of the first-order α-leap studies in the present book. In making them, we use CI average values for the experimental pion and kaon masses (Sec. 3.14). The only adjustable parameter in the mass calculations is the hadronic binding energy, and a uniform HBE of 2.6% is employed for all four of the hadronically bound states. The average *absolute percent deviation* (APD) of the calculated masses from the experimental values is 0.42%.

The average *absolute mass deviation* (AMD) of the calculated masses from the experimental values for the M^X octet particles in Table 3.16.1 is 3.7 MeV. This mass error is much smaller than the m_b = 70 MeV and m_f = 105 MeV α-masses used for the platform states M, and the 420 MeV X-mass used for the platform excitations. Hence the accurate experimental α-quantization of this M^X octet is apparent. Of equal importance is the manner in which the leptons and baryons are seamlessly interleaved, thus demonstrating the universality of the mass generation operator $\alpha = e^2/\hbar c$.

The phenomenologically important point with respect to Fig. 3.16.1 is that the excitation quantum X applies to all six of the particles which occur as mass levels on the four excitation towers. The number of X quanta that are needed for each case is of course dictated by experiment, and not by theory. But with such a large building block, the choice of how many X quanta to employ is unmistakable. The M^X phenomenology is simple enough that we can obtain the mass ratios of these particles with respect to that of the electron almost by inspection.

The mass calculations displayed in Table 3.16.1 are simple, but they are not simplistic. They yield the electron–proton mass ratio, which is perhaps the oldest unattained task in elementary particle physics. They also yield the electron–muon and electron–tau mass ratios, which, according to the most recent books that have been published on elementary particles,

Table 3.16.1 Mass calculations for the M^X octet particles of Table 3.13.1. The experimental pion and kaon masses are averaged over isotopic spin states (Appendix B). The theoretical masses are calculated from Eqs. (3.13.6) and (3.13.7). The hadonic binding energy (HBE) of 2.6% (Fig. 3.16.2) is the only adjustable parameter in these calculations. The α-leap platforms M_π, M_φ, M_μ, M_K are defined in Eqs. (3.5.5), (3.5.6), (3.5.8) and (3.5.9). Two of the M^X octet particles (π and μ) appear right on the platforms, and the other six (η, η', K, p, τ and ϕ) appear at accurate X-quantized mass intervals above the platforms (Fig. 3.16.1), where the unbound supersymmetric excitation quantum $X = 420.15$ MeV is defined in Eq. (3.9.1). The average *absolute percent deviation* (APD) for these eight basic particle masses is 0.42%, and the average *absolute mass deviation* (AMD) is 3.67 MeV.

Particle	Spin	Platform state	HBE	Exp. mass (MeV)	Calc. mass (MeV)	AMD (MeV)	APD
e	1/2	(input)					
π	0	M_π^0	2.6%	137.27	137.40	0.13	0.10%
η	0	M_π^1	2.6%	547.51	546.64	0.87	0.16%
η'	0	M_π^2	2.6%	957.78	955.86	1.92	0.20%
K	0	M_K^1	0	495.66	490.69	4.97	1.00%
μ	1/2	M_μ^0	0	105.66	105.55	0.11	0.10%
p	1/2	M_μ^2	0	938.27	945.85	7.58	0.81%
τ	1/2	M_μ^4	0	1776.99	1786.16	9.16	0.52%
ϕ	1	M_ϕ^1	2.6%	1019.46	1024.06	4.60	0.45%

remain a complete mystery. And they answer I. I. Rabi's rhetorical question about the muon [56]: "*Who ordered that?*" The muon mass is the basic building block for all of the fermions in the universe that are heavier than the electron.

3.17 The M_ϕ (ϕ, J/ψ, Υ) = ($s\bar{s}$, $c\bar{c}$, $b\bar{b}$) Vector Meson M^T Mass-Tripling Tower

In the past few sections we have been discussing α-quantized masses that were created by the M^X production process, wherein an α-leap from an electron pair produces a platform mass M, and then mass elements are added to M in units of the supersymmetric excitation quantum X. This M^X process accounts for the creation of the M^X octet of particles described in Tables 3.13.1 and 3.16.1. In the present section we describe a different mode of mass (and flavor) α-quantized production, a mode that involves successive mass triplings (TR) of the ϕ meson threshold state. We will demonstrate in Sec. 3.18 that this M^T mode also involves a charge exchange (CX) at each tripling step.

The mass-tripling production mechanism ties together the SM quark-pair threshold states $s\bar{s}$, $c\bar{c}$ and $b\bar{b}$. The threshold state for $s\bar{s}$ pair production is the $\phi(1020)$ vector meson, which has a hadronic binding energy of about 3% (Sec. 3.12). The spin-1/2 s quark and the spin-0 K meson both carry the strangeness quantum number S. The dominant ϕ meson decay mode is $\phi \rightarrow K\bar{K}$, where the $K\bar{K}$ final state is lower in mass than the $\phi \equiv s\bar{s}$ initial state. Thus this ϕ decay can proceed without changing strangeness, which gives it a shorter lifetime than those of the J/ψ and Υ threshold states. The ϕ lifetime of 1.5×10^{-22} s is shorter than the zeptosecond threshold-state boundary of 10^{-21} s, so the ϕ does not appear in the lifetime systematics of Chapter 2 as one of the 36 threshold-state particles displayed in Fig. 2.8.8, although it is in fact an important threshold resonance, as we demonstrate here.

The ground state for the M^T mass-tripling excitation tower is an electron–positron pair in an overall spin-1 configuration (Fig. 0.2.16). The ϕ, J/ψ and Υ vector mesons that are created from this ground state are all spin 1 resonances. The first excitation step is an α-leap to the M_ϕ platform, which is unoccupied. This is followed by an XX excitation that produces the $\phi(1020) = s\bar{s}$ meson (Fig. 3.12.1). This is the lowest-mass flavor threshold state. (The $u\bar{u}$ and $d\bar{d}$ flavor thresholds at 630 MeV require $(1/2\ X)(1/2\ X)$ M_ϕ platform excitations, which are not allowed in the M^X formalism, and are not observed experimentally.) The creation of the second and third flavor threshold states — the $J/\psi(3097) = c\bar{c}$ and $\Upsilon(9460) = b\bar{b}$ mesons — is accomplished by successive mass triplings in which the entire $s\bar{s} = MXX$ excitation (the platform mass M_ϕ plus the excitation XX) is tripled twice to create first the $c\bar{c}$ pair and then the $b\bar{b}$ pair. The resulting mass ratio of three between c-quark and b-quark states (Fig. 2.9.2) echoes the lifetime ratio of three between states with unpaired c and b quarks (Fig. 2.9.1). This M^T mass tripling flavor sequence is depicted graphically in Fig. 3.17.1.

The hadronic binding energies of the $\phi(1020) = s\bar{s}$, $J/\psi(3097) = c\bar{c}$ and $\Upsilon(9460) = b\bar{b}$ mesons in this M^T mass tripling process are of considerable phenomenological interest, especially when we extrapolate these results to even higher energies (Sec. 3.19). The numerical values of the α-quantized mass units that we use here to calculate HBEs were defined in Secs. 3.4 and

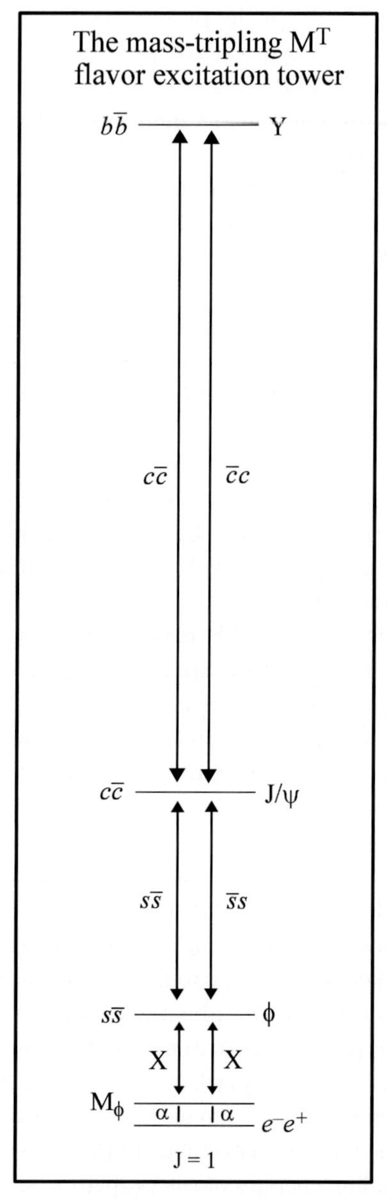

Fig. 3.17.1 The s, c, b excitation tower. An initial α-leap from a spin-1 electron–positron pair generates the (unoccupied) M_ϕ platform state. Then an XX excitation creates the ϕ meson in the form of an $s\bar{s}$ quark pair (Fig. 3.12.3). A subsequent $(s\bar{s})$ $(\bar{s}s)$ mass-tripling (M^T) excitation generates the $J/\psi \equiv c\bar{c}$ meson, and an additional $(c\bar{c})$ $(\bar{c}c)$ M^T excitation generates the $\Upsilon \equiv b\bar{b}$ threshold. Other features of this s, c, b excitation tower are displayed in Fig. 3.18.1.

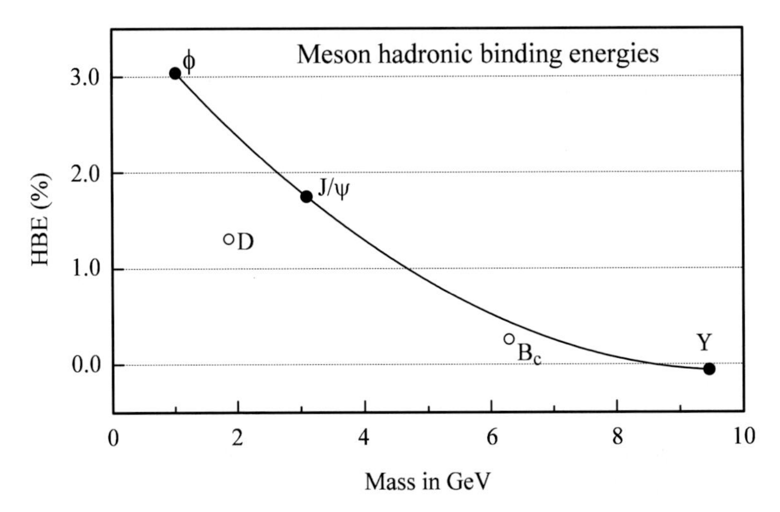

Fig. 3.17.2 Hadronic binding energies for the $\phi \equiv s\bar{s}$, $J/\psi \equiv c\bar{c}$ and $\Upsilon \equiv b\bar{b}$ resonances. These binding energies are obtained by starting with the low-energy α-masses and platform states defined in Secs. 3.4 and 3.5, extrapolating these masses all the way up to the Υ meson, and comparing the calculated unbound masses of these states with the experimental masses (Table 3.15.1 and Eq. (3.15.1)). These three binding energies define the curve in Fig. 3.17.2, which is an eyeball fit to the data. As can be seen, the binding energy (percentagewise) goes essentially to zero at the energy of the Υ. Also shown are the binding energies of the $D \equiv q\bar{c}$ meson $q \equiv (u, d)$ and the $B_c \equiv b\bar{c}$ meson (whose mass has recently been accurately measured [56]). As discussed in the text, the decrease in HBE values shown in Fig. 3.17.2 can plausibly be regarded as a manifestation of "asymptotic freedom".

3.5, and were deduced from the masses of the lowest elementary particle states — the electron, pion and muon "Rosetta stones" of Sec. 3.3. We use multiples of these same low-energy masses to reproduce other particle masses all the way up to the $\Upsilon(9460)$ meson, which is almost two orders of magnitude higher in energy. It then turns out that the hadronic binding energy that we need to empirically apply to these $s\bar{s} \to c\bar{c} \to b\bar{b}$ M^T basis states in order to reproduce their experimental values systematically decreases from 3.04% to 1.75% to −0.06%, as is displayed in the HBE curve in Fig. 3.17.2, and listed in Table 3.15.1. This suggests that the HBE is proportional to the charge states (which are not proportional to the masses), so that the HBE essentially vanishes (percentagewise) at the higher mass values. Two more high-mass data points we can add in here come from the $D = q\bar{c}$ meson, ($q \equiv u$, d), whose averaged HBE is 1.31%, and the $B_c = b\bar{c}$ meson (whose mass has recently been measured with high accuracy as 6286 ± 5 MeV [10]), with an HBE of 0.27% (Table 3.15.1). Since the D and B_c mesons feature unpaired quarks, we might expect their binding energies to be relatively weaker than for balanced $c\bar{c}$ and $b\bar{b}$ excitations. However, if the HBE comes just from the fractional electric charges, then the mismatch in masses may not matter. It should be noted that other unpaired D and B mesons have fluctuating binding energies which suggest that the quark basis state assignments for these particles may be more complex.

The hadronic binding energies displayed in Fig. 3.17.2 indicate that the HBE essentially vanishes at the higher energies in the M^T mass-tripling excitation scheme. The magnitude of the binding energy effect is displayed graphically in Fig. 3.17.3, which shows the ϕ, J/ψ and Υ vector meson masses plotted in comparison to a mass-tripled straight line (in this log–log plot) that passes through the Υ meson mass. As can be seen, the J/ψ and ϕ masses lie slightly below the M^T line, but the overall mass tripling is quite accurate. In Sec. 3.19 we extrapolate this M^T generation process up to the very-high-energy region of the W and Z gauge vector bosons. Section 3.20 contains a quite different method — a second α-quantized mass leap from the u and d quarks — to move up into the mass region of the W and Z bosons and the top quark t.

An explanation for the decrease in hadronic binding energy with increasing mass that is observed in Fig. 3.17.2 (and also in Figs. 3.15.1 and 3.15.2) can be deduced from the decreasing "sizes" of the elementary particle excitations at higher mass values, assuming that these particles and their constituent subunits have Compton-sized radii, $R_c \approx \hbar/mc$. If the particle HBE arises from the gluon field of the fractional charges in the particle, then the HBE will vanish in magnitude when the charge-separation distance of the fractional charges falls below the natural (unstretched) "length" of the gluon field. From the Compton radii of the particle states displayed in Fig. 3.17.2, we can estimate this "unstretched gluon length" to be roughly one fermi (1 fm $= 10^{-13}$ cm). A recent toy model calculation of the proton mass [57] assumes a "gluon length" of about 2 fm. We do not really know the sizes of (e.g.) the J/ψ and Υ excitation clusters, so these observations are just qualitative, but it is interesting to view the HBE curve in Fig. 3.17.2 as a portrayal of "asymptotic (unstretched) freedom", wherein the gluon forces vanish at very short charge-separation distances.

A question arises as to whether the basic mass-scaling mass units should be adjusted slightly to eliminate the falloff in HBE values at the higher energies shown in Figs. 3.17.2 and 3.17.3. The M^T extrapolation shown

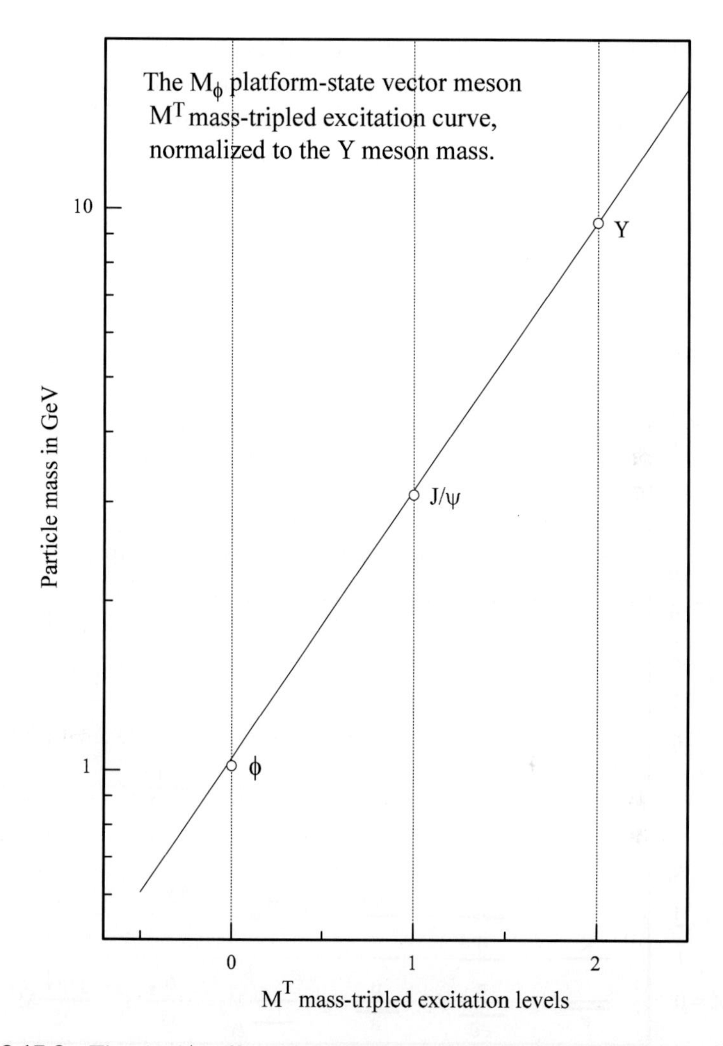

Fig. 3.17.3 The ϕ, J/ψ, Υ masses, shown plotted against a mass-tripling curve that passes through the Υ mass. Small hadronic binding energy effects lower the ϕ and J/ψ masses below the M^T curve, but the overall mass-tripling systematics is quite accurate. In Sec. 3.19 we extend this M^T curve up to the region of the W and Z gauge bosons.

here is based on the use of an s-quark mass of 525.70 MeV (Eq. (3.15.1)), which is $s = m_e + 5m_f$, where m_f is the α-quantized fermion mass of Eq. (3.4.2) and m_e is the electron mass. When the s-quark mass is raised (e.g.) to 527.75 MeV, which is five times the mass $(m_f + m_e)$, a more uniform HBE curve is obtained. However, if we also pass this slight mass increase on to the other calculated particle masses, in particular those of the proton and τ lepton, which have no internal HBE, the discrepancy between the calculated and experimental mass values is increased. Thus the requirement of making a precision fit to all of the M^X octet particles, as displayed in Table 3.16.1, rather strictly delimits the allowed range of the α-quantized basis states.

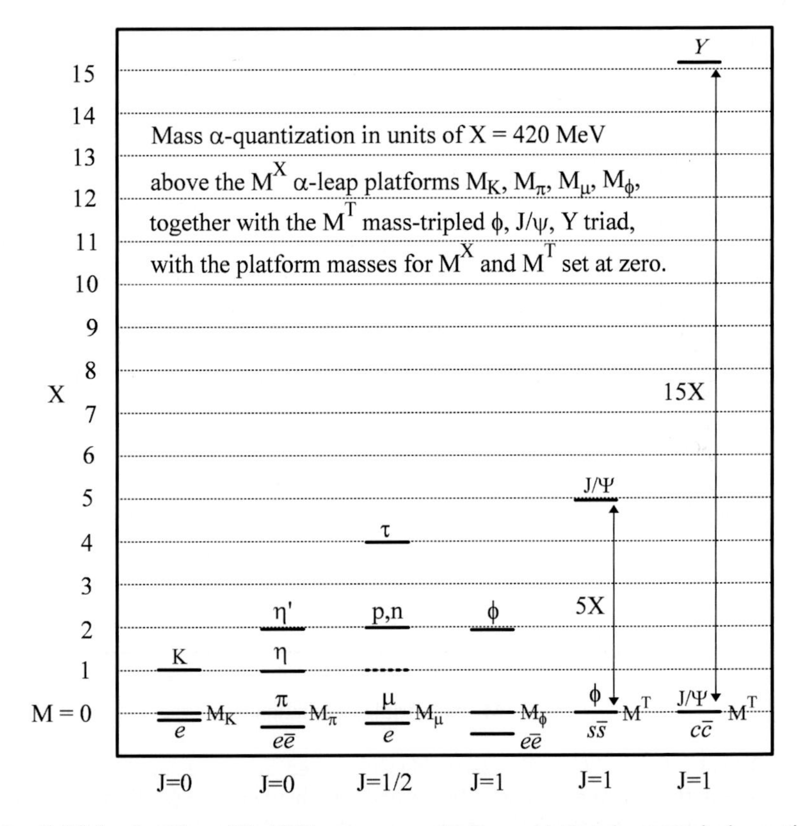

Fig. 3.17.4 An $X = 420$ MeV mass quantization grid that is erected above the (zero) masses of the M_K, M_π, M_μ and M_ϕ platform states, which are the "ground states" for the M^X production process. Also included are the $J/\psi - \phi$ and $\Upsilon - J/\psi$ mass intervals in the M^T production process, which are plotted above "zero-mass" ϕ and J/ψ "ground states," respectively. As can be seen, all of the M^X octet mass intervals are in units of X, as are the two mass-tripled intervals at the right of the figure. But the M^T mass intervals of $5X$ and $15X$ differ in character from the M^X mass intervals of $1X$, $2X$ and $4X$. Thus mass tripling represents a distinct excitation process.

As a final result in this section, it is interesting to display the M^T mass-tripling process on an excitation mass grid that uses X-units of mass. This is shown in Fig. 3.17.4, which is a combined plot of the M^X octet excitations of Fig. 3.16.1 and the M^T excitations of Fig. 3.17.1. In this plot the masses of the octet platforms M are set equal to zero, and the masses of the initial M^T states at each tripling stage are also set equal to zero. The particles that appear above these excitation "ground states" are shown against a mass grid that is in units of $X = 420$ MeV. As can be seen, all of the observed M^X and M^T levels fall accurately on this grid except the unbound Υ meson, which is slightly above the grid. The observed M^X octet X-intervals are all in units of $1X$, $2X$ or $4X$, but the M^T X-intervals exhibit a different spacing. The M_ϕ platform has a mass, apart from HBE effects, of 0.5 X. The $s\bar{s}$ excitation requires another $2X$ of mass, so the $\phi = s\bar{s}$ threshold state represents 2.5 X-quanta. If we triple this amount, we get 7.5 X-quanta for the $c\bar{c}$ threshold, which is an increase of 5 X-units. Tripling this again gives 22.5 X-quanta for the $b\bar{b}$ threshold, which is a further increase of 15 X-units. Thus these M^T steps each require an *odd* number of X quanta — $5X$ and $15X$. Hence the M^X and M^T particle generation mechanisms are both X-quantized, but are phenomenologically of quite different natures. The M^T tripling process is not just a different way of describing M^X excitation processes, but represents a different type of hadronically bound excitation.

In the absence of a comprehensive theory for this α-quantized systematics, both the M^X and M^T excitations represent empirical results that occur within the framework of electron-based, α-generated mass units. Combined together, they accurately reproduce the main elementary particle threshold states, including both leptons and hadrons, in one comprehensive α-quantized formalism.

3.18 Charge Exchange (CX) and Fragmentation (CF) Excitations and Proton Stability

In addition to providing accurate mass values, the M^T and M^X excitation towers have another interesting property: they delineate the charge-exchange (CX) transfers that take place during tower excitations, and they at least symbolically suggest the stages where charge fragmentation (CF) occurs. These CX and CF mechanisms provide the rationale for the positive charge on the proton (which is produced in the *negatively* charged particle channel of the $M_{\mu\mu}$ excitation tower), and also for the stability of the proton.

The M^T mass tripling tower is erected on the M_ϕ platform excitation, and it contains the $\phi = s\bar{s}$, $J/\psi = c\bar{c}$ and $\Upsilon = b\bar{b}$ flavor threshold states. This M^T tower was displayed in Fig. 3.17.1, where the *mass values* were emphasized. It is shown here schematically in Fig. 3.18.1, where the emphasis is now on the *charge states*. The M^T excitation process starts with the spin-1 M_ϕ tower, where an XX excitation creates an $s\bar{s}$ quark pair (Fig. 3.12.3) with fractional electric charges $e = (-1/3, +1/3)$ that result from a CF charge fragmentation process. Then the $s\bar{s}$ mass is tripled, creating a $c\bar{c}$ quark pair with fractional charges $e = (+2/3, -1/3)$. These altered quark charges indicate that a cross-channel CX charge transfer has occurred during the mass-tripling process. Finally the $c\bar{c}$ mass is tripled, creating a $b\bar{b}$ quark pair with fractional charges $e = (-1/3, +1/3)$ which indicate that a *reversed* CX charge transfer e has occurred. Thus these flavored-quark excitations involve the CX exchange of a unit e of electric charge during each M^T mass-tripling step. The fractional charges on the quark pairs give rise to gluon fields with associated hadronic binding energies that confine the quark pairs (or at least the quark charges) together with unbreakable bonds.

The M^X excitation tower is erected on the $M_{\mu\mu}$ platform excitation displayed in Fig. 3.18.2, where the $M_{\mu\mu}$ platform is shown divided into separate M_μ and $\bar{M}_\mu$ platform channels. Their respective M^X excitation channels contain the μ, p, τ and $\bar{\mu}$, $\bar{p}$, $\bar{\tau}$ fermions, as generated in matching particle and antiparticle excitations. The $\mu\bar{\mu}$ pair is created by the initial α-leap from an electron–positron pair to the $M_{\mu\mu}$ platform. Then separate XX excitations occur in each of the M_μ and $\bar{M}_\mu$ channels. These XX excitations are accompanied by two CX cross-channel charge exchanges e, which result in the creation of a $p\bar{p}$ pair with reversed charges. Finally, additional XX excitations occur in each channel which are accompanied by two *reversed* CX cross-channel charge exchanges $-e$, and which produce a $\tau\bar{\tau}$ meson pair with the same charge signs as the $\mu\bar{\mu}$ meson pair.

The CX transfer is interesting in its own right, particularly as a way of accounting-for the alternating charges on the s, c, and b quarks in the M^T excitation tower. However, its real interest is in its implications for the proton. The proton is, at least conceptually, created (leptonically) with a mass equal to *nine* muon masses. It then decomposes into three equal-mass quark substates — u, u, d, which each have *three-muon* masses. In this decomposition process the original mass divides into three parts,

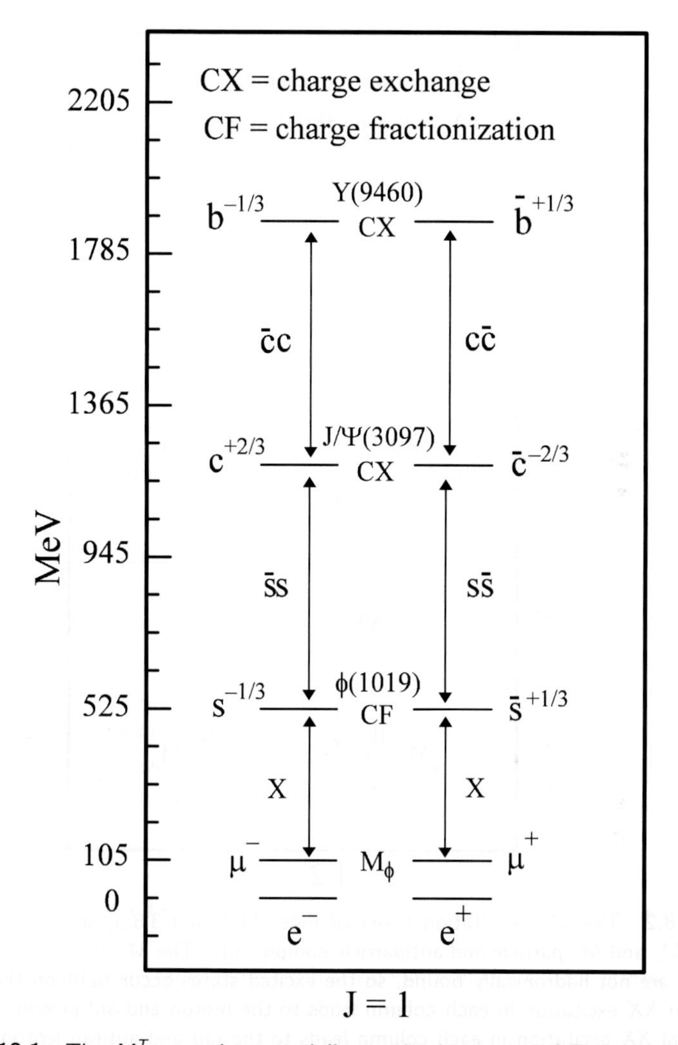

Fig. 3.18.1 The M^T s, c, b mass-tripling excitation tower of Fig. 3.17.1, shown schematically separated into its particle and antiparticle components. An XX excitation of the M_ϕ platform state leads to the generation of a $\phi \equiv s\bar{s}$ quark pair (see Fig. 3.12.3), where charge fragmentation (CF) creates $e = (-1/3, +1/3)$ charges on the s and $\bar{s}$ quarks, respectively. The resulting quark gluon field produces a hadronic binding energy of 3% (Table 3.15.1). Then two successive mass-triplings of the ϕ accurately reproduce the $J/\psi \equiv c\bar{c}$ and $\Upsilon \equiv b\bar{b}$ masses (Fig. 3.17.3). These sequential M^T excitations are accompanied by a CX charge exchange e for the J/ψ and then a reverse CX charge exchange $-e$ for the Υ, which create the charges $e = (+2/3, -2/3)$ on the c and $\bar{c}$ quarks, and $e = (-1/3, +1/3)$ on the b and $\bar{b}$ quarks.

triad that is displayed graphically in Figs. 3.17.1 and 3.17.3. These are all hadronically bound resonances, but their binding energies decrease with increasing mass, and essentially vanish at the mass of the Υ, as displayed in Fig. 3.17.2 and tabulated in Table 3.15.1.

The elementary particle mass spectrum extends up to 11 GeV with the Υ family of vector mesons, then has a large energy gap that extends up to 80 GeV, where the $W^{\pm}$ vector meson appears, followed by the Z^0 vector meson at 91 GeV. These two particles form an isotopic spin doublet. In the SM formalism, they are denoted as *gauge bosons*, and are placed in the same family group as the massless spin-1 photon, the spin-1 gluon, and the mysterious Higgs boson. Thus their masses are not expected to be directly related to the particle or quark masses that appear in hadrons and leptons, including in particular the mass of the supermassive t quark, which shows up at 173 GeV. However, as we demonstrate here and in Sec. 3.20, accurate masses relationships between the W, Z doublet and these other particles do in fact exist.

The ϕ, J/ψ and Υ vector mesons are successively generated by the M^T mass-tripling process, and the W and Z vector mesons are the only particle states that appear above them. Thus the natural phenomenological approach is to extend the M^T mass-tripling curve of Fig. 3.17.3 up to higher energies and see where it intersects the W and Z masses. Guided by the binding energy systematics displayed in Fig. 3.17.2, we can ignore binding energies at the higher masses. In order to obtain a numerical estimate for this M^T mass extrapolation, we select the isotopic-spin-averaged mass $m_{WZ} = 1/2\ (m_W + m_Z) = 85.7953$ GeV to represent the W, Z doublet, and we divide it by the experimental mass $m_\Upsilon = 9.4603$ GeV of the Υ_{1S} vector meson. This gives

$$m_{WZ}/m_\Upsilon = 9.069 \cong (3.011)^2\,. \tag{3.19.1}$$

The average mass m_{WZ} is almost exactly equal to two mass-triplings of the m_Υ! Pushing this phenomenological mass analysis to the limit, we note that the mass ratio shown in Eq. (3.19.1) is slightly larger than 9, which indicates that the trend toward decreasing hadronic binding energies shown in Fig. 3.17.2 continues on up into the higher energy domain of the W and Z. (Of course, it must be kept in mind that we are basing these results on the average mass value of two particles whose masses differ by 11 GeV.)

The extension of the M^T mass-tripling curve of Fig. 3.17.3 to higher energies is displayed in Fig. 3.19.1. In this plot the W and Z vector mesons (open circles) are each assigned the abscissa $M^T = 4$, with their ordinates

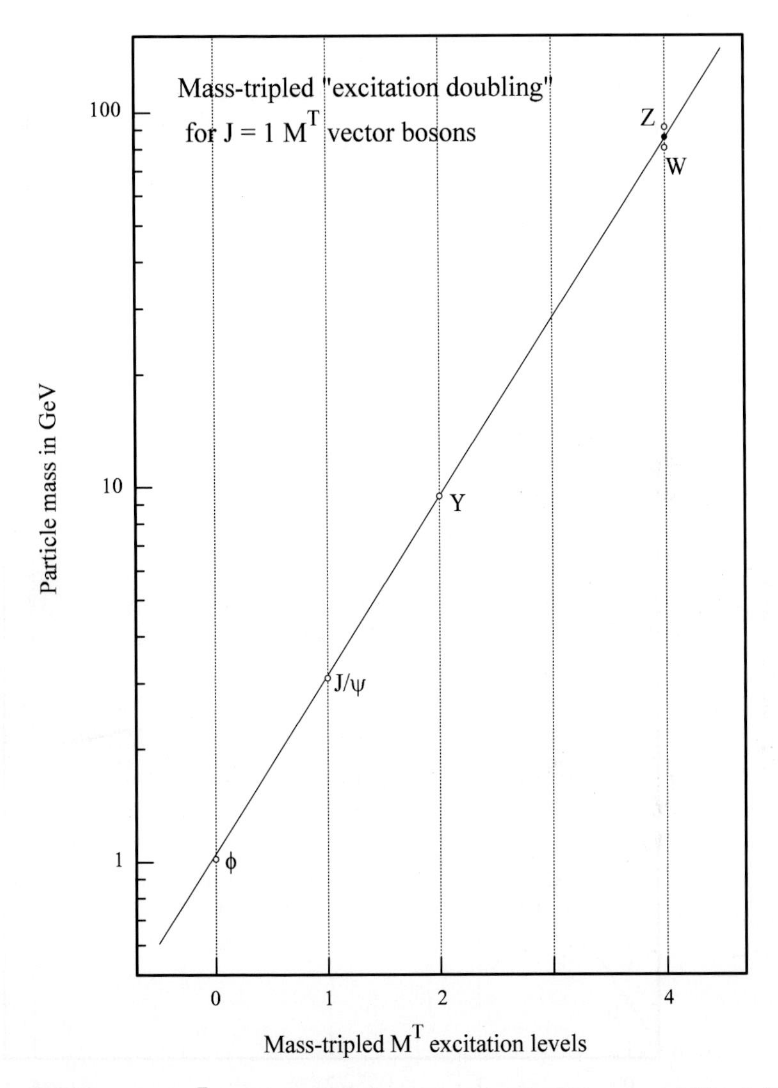

Fig. 3.19.1 The M^T mass-tripled vector meson excitation curve of Fig. 3.17.3, shown extrapolated up to the region of the W and Z vector mesons. Open circles denote the W and Z masses, and the closed circle represents their average mass m_{WZ}. The M^T mass-tripling curve is normalized to the experimental Υ meson mass. As can be seen, the M^T curve coincides with the mass m_{WZ} precisely at the abscissa $M^T = 4$. The $M^T = 3$ level is not occupied, which is analogous to the situation that is displayed for the M^X platform excitations in Fig. 3.19.2 and Table 3.19.1. However, a quite different mass excitation process — an α-enhancement of the u and d quarks, which represents a second mass α-leap — is employed in Sec. 3.20 to reproduce not only the W and Z average mass m_{WZ}, but also the isotopic spins of the W and Z and the mass of the top quark t, which this M^T process does not encompass.

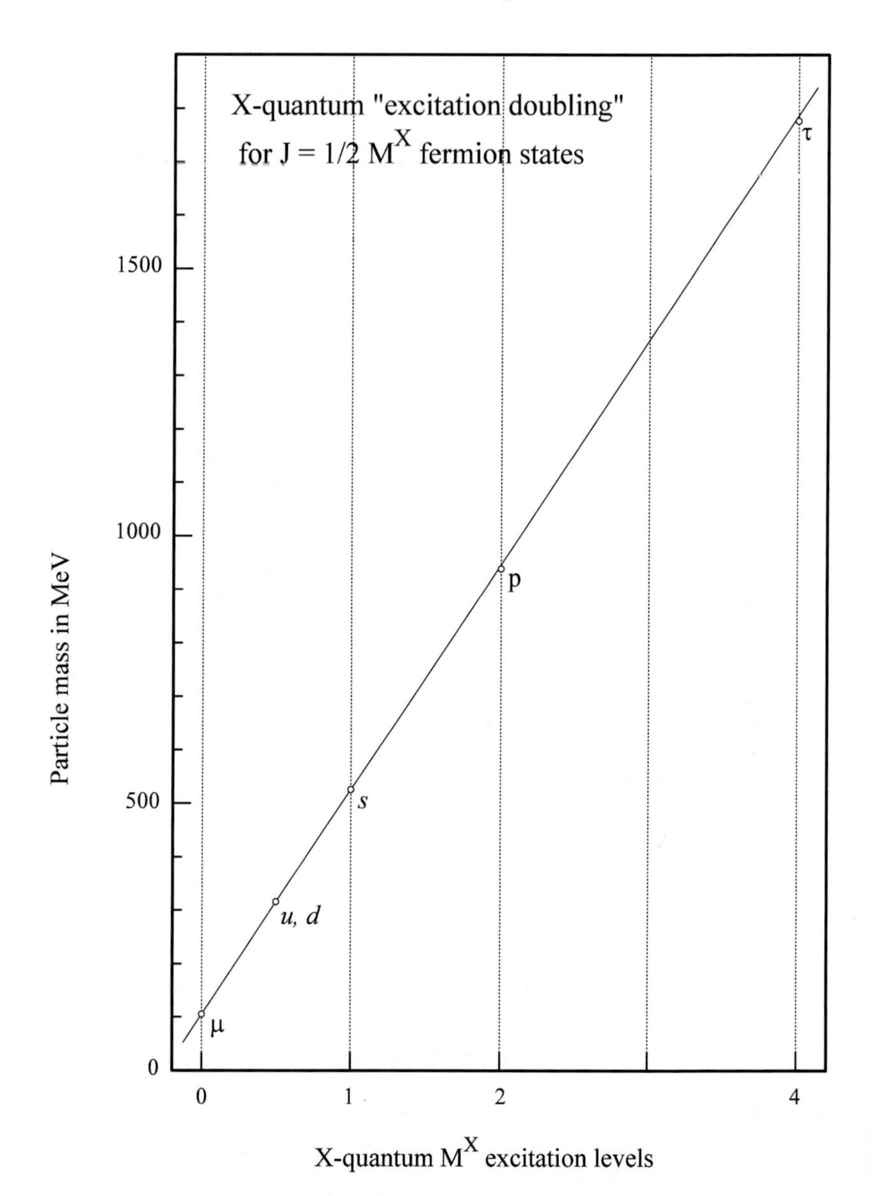

Fig. 3.19.2 The M^X excitation levels μ, p and τ of the X-quantized M_μ particle platform state (Fig. 3.8.3), displayed together with the masses of the (not directly observed) u, d, and s quarks. These mass levels suggest an "excitation doubling" process that leaves the $X = 3$ level unoccupied, as also occurs for the M^T excitations of Fig. 3.19.1. The M^T and M^X platform excitations are summarized in Table 3.19.1.

representing their mass values. The filled circle between them is their average mass value m_{WZ}. As can be seen, the W and Z masses evenly straddle the extrapolated M^T mass-tripling curve at the value $M^T = 4$, and their average value lies essentially right on the curve: the mass of the Υ meson tripled twice is 85.143 GeV, which differs by 0.76% from the m_{WZ} mass of 85.795 GeV. The closeness of this extrapolated mass ratio, when combined with the extreme experimental accuracy of the Υ, W and Z mass values (Appendix B), the fact that these are all vector mesons, and the additional fact that the W and Z are the only particles discovered anywhere nearby in this vast high energy mass region, suggests that this result may not be accidental. On the global mass scale displayed in Fig. 3.19.1, the 11 GeV mass splitting of the W and Z does not seem as large as the bare numbers suggest.

There is one more feature of Fig. 3.19.1 that is of phenomenological interest. The occupied M^T mass levels in this M_ϕ platform tower are the $M^T = 0$, 1, 2 and 4 levels: the $M^T = 3$ level is unoccupied. An analogous situation occurs with the M^X threshold-state excitations of the $M_{\mu\mu}$ platform tower, which are summarized in Secs. 3.13 and 3.16. Figure 3.19.2 shows the occupied X-spaced excitation levels in the M_μ particle column of the $M_{\mu\mu}$ tower. The muon itself occupies the $X = 0$ level, and the s quark, proton and τ occupy the $X = 1$, 2 and 4 levels. (The u and d quarks occupy the $X = 1/2$ level, but because of the "$X =$ integer only" rule they are not excited directly as pairs.) The pertinent fact in comparison to the M^T excitation levels of Fig. 3.19.1 is that the $X = 3$ level is again unoccupied. These M^T and M^X excitations of the M_ϕ and $M_{\mu\mu}$ platforms, respectively, are summarized in Table 3.19.1. If we postulate that cascade excitations occur as an "excitation-doubling" process, wherein a basic excitation (M^T or X) is doubled (M^TM^T or XX) and then doubled again ($M^TM^TM^TM^T$ or XXXX), the excitation levels $M^TM^TM^T$ and XXX are excluded, which is what we find in Figs. 3.19.1 and 3.19.2 and display in Table 3.19.1.

We have just laid out the phenomenological evidence that an extrapolation of the ϕ, J/ψ, Υ mass-tripled M^T excitation series to two more mass triplings reproduces the average mass m_{WZ} of the W, Z doublet. However, there are two problems raised by this extrapolation: (1) the ϕ, J/ψ and Υ vector mesons are all isotopic-spin singlets with zero charge states only, so there is no reason to expect that their further extension upwards would yield the charged configuration of the $W^\pm$ vector meson; (2) the successive

Table 3.19.1 The "excitation-doubling" systematics of the M_ϕ and $M_{\mu\mu}$ platform excitation towers. The M_ϕ mass-triplings M^T, M^TM^T and $M^TM^TM^TM^T$ are observed, but $M^TM^TM^T$ is not (Fig. 3.19.1). Similarly, the $M_{\mu\mu}$ excitations X, XX and XXXX are observed, but XXX is not (Fig. 3.19.2).

α-quantized excitation	Observable
Hadronically bound vector meson platform excitations:	
$e^-e^+(1+3/2\alpha) \to M_\phi(J=1)$	———
$M_\phi + X \cdot X$	$s\bar{s} \equiv \phi$
$s\bar{s} \times M^T \cdot M^T$	$c\bar{c} \equiv J/\psi$
$s\bar{s} \times M^TM^T \cdot M^TM^T$	$b\bar{b} \equiv \Upsilon$
$s\bar{s} \times M^TM^TM^T \cdot M^TM^TM^T$	———
$s\bar{s} \times M^TM^TM^TM^T \cdot M^TM^TM^TM^T$	W, Z
Hadronically unbound fermion pair platform excitations:	
$e^-e^+(1+3/2\alpha) \to M_{\mu\mu}(J=0 \text{ or } 1)$	μ^-, μ^+
$\mu^-, \mu^+ + (\frac{1}{2}X) \cdot (\frac{1}{2}X)$	$(u,\bar{u})$ $(d,\bar{d})^*$
$\mu^-, \mu^+ + X \cdot X$	$(s,\bar{s})^*$
$\mu^-, \mu^+ + XX \cdot XX$	p, $\bar{p}$
$\mu^-, \mu^+ + XXX \cdot XXX$	———
$\mu^-, \mu^+ + XXXX \cdot XXXX$	τ^-, τ^+

*Not observed directly.

M^T mass triplings that generate the $c = 3s$ and $b = 3c$ constituent-quark masses do not extend upward so as to encompass the 173 GeV top quark t mass. We can remedy both of these defiencies by invoking a second α^{-1}-leap of the already α^{-1}-quantized elementary particle fermion basis states, which we do in the next section.

3.20 The Second-Order α^{-2} Fermion Mass Leap to the W and Z Bosons and Top Quark t

The supermassive $W^\pm$ and Z^0 vector bosons and top quark t represent the last stage of our phenomenological investigation into the α-quantization of elementary particle lifetimes and masses. They are currently at the frontier in the elementary particle mass domain (until the next mega-accelerator, the Large Hadron Collider — LHC — at CERN, comes online). This investigation started with particle lifetimes, where it was discovered that the long-lived threshold-state particles with lifetimes $\tau > 10^{-21}$ s are arranged in α-spaced lifetime groups that reflect their SM u, d, s, c, b quark substructures. This lifetime grouping extends over 11 powers of α, or 23 orders of magnitude. Then the masses of these threshold-state particles were studied, which revealed a reciprocal α^{-1} mass quantization that starts with the electron ground state and proceeds by a single α^{-1}-leap to the generation of two fundamental particle mass quanta — the $m_b = m_e/\alpha = 70$ MeV boson α-mass, and the $m_f = (3/2)(m_e/\alpha) = 105$ MeV fermion α-mass. These two α-enhanced mass quanta then serve as the constituent-quark "bricks" for constructing all of the threshold-state elementary particles up to 11 GeV. Hence we have a lifetime α-quantization that extends over many powers of α, and a reciprocal mass α^{-1} quantization that extends over only one power of α^{-1}, but whose effect is spread throughout the entire mass region up to 11 GeV. Looking at α-quantization from this perspective, what do the supermassive W, Z and t particle states represent?

In Sec. 3.19 we extrapolated the low-energy mass results up to the supermassive region, where we discovered that we could accurately reproduce the average $\overline{WZ}$ mass, but not the charge states of this isotopic spin doublet, and we could not account for the mass of the t quark. In Sec. 3.20 we start with the *ansatz* that these W, Z and t particle states represent a second α^{-1} quantization step. This ansatz leads to the identification of an α-enhanced pair of massive u and d quarks — $u^{\alpha} \equiv u/\alpha$ and $d^{\alpha} \equiv d/\alpha$ — that reproduce the W and Z average mass and account for their isotopic spin states (but not the mass splitting), and that precisely reproduce the mass of the top quark t.

The accurate measurement of the experimental mass of the top quark is a real *tour de force* — one of the marvels of present-day physics. It is the only quark mass to be directly determined. As such, it has tremendous *phenomenological* significance. For example, if it turns out to be a constituent-quark mass, then it indicates that all of the other quark masses (to which it is directly related, at least in the present formalism) are also constituent-quark masses. The top quark mass also has *theoretical* significance, in that it verifies QCD perturbation theory expansions whose systematics led to its prediction and eventual discovery.

The prediction of the top quark mass came from experiments on parity violation in the production of the Z^0 from electron–positron annihilation, as is described in interesting detail in Bruce Schumm's book *Deep Down Things* [58]. The probability of producing the Z^0 depends on the handedness of the initial-state electron and positron, which is a violation of parity conservation. The theoretical calculation of this process involves contributions from the twelve lepton and quark pairs in the final state of the reaction. These contributions involve the masses of the final-state particles. All of the masses were known except that of the top quark. The experimental and theoretical results were refined down to the point where their agreement depended sensitively on the choice made for the top quark mass. By varying the mass in the theoretical calculation, it was discovered that to have experiment and theory in exact agreement one had to have a top quark mass of about 170 GeV, which was an astounding prediction: the b quark, the third-generation companion of the t quark in the quark hierarchy, has a mass of less that 5 GeV. Guided by this result, the experimenters at Fermilab, the only installation with a high enough available beam energy, soon discovered the t quark right at the predicted location. But this is not the end of the story. When the experimental t quark mass is added to the theoretical calculations, there is still a small discrepancy between experiment and theory. This can be attributed to the fact that there is one remaining particle that can contribute to the electron–positron scattering — the Higgs boson. In fact, the experimental and theoretical calculations are so precise that the same game can be played for the Higgs as was played for the top quark — assume different values for the Higgs mass and see which one brings experiment and theory into complete agreement. The SM results indicate that the Higgs mass is, with 95% confidence, no larger than about 200 GeV. LEP electron–positron experiments at CERN were able to push the lower limit of the Higgs mass up to about 110 GeV. Running at very end of their energy range, they found a few events which indicated a possible Higgs particle with a mass of 115 GeV. The decision was then made to shut down LEP and convert it into the higher energy LHC, which is projected to be on line in 2007.

From our present point of view, the interesting fact with respect to the Higgs particle is that the calculation of its mass depends on the precise value of the top quark mass, which makes a much larger contribution to the electron–positron scattering process. This has spurred efforts at Fermilab to obtain a precision mass value for the t, which is produced in matching $t\bar{t}$ pairs. Precision mass values are obtained by exclusive measurements which focus on a particular decay mode that can be followed through energy-wise in complete detail. In the case of the t, the decay mode $t^{+2/3} \rightarrow W^{+1} + b^{-1/3}$ is used at Fermilab, where the superscripts indicate the charge of each particle state. The subsequent W decay is essentially instantaneous and occurs at the same location where the initial $t\bar{t}$ pair was created. The b quark travels an appreciable distance before it decays, so the entire t decay involves two separated decay centers. This is a characteristic signature for a t event, and it is obtained experimentally in coincidence with a matching two-center decay from the $\bar{t}$ quark.

Another salient feature of the Fermilab $t\bar{t}$ production process is that it requires a head-on collision between two quarks — one in the proton and one in the antiproton. At these high energies of 1 TeV (tera-electron-volt) per colliding particle, the proton and antiproton interact as if they each consist of three separated scattering centers (which are also observed, for example, in the high energy Bjorken scattering of electrons on protons). In order to produce a $t\bar{t}$ pair, the proton collision process must be focused on a single pair of quarks, so that they obtain most of the interaction energy, with only a small residual amount of energy being transferred to the remaining four "spectator" quarks. This type of head-on collision process, with complete t and $\bar{t}$ decay channel measurements being obtained in coincidence, occurs once in every 10^{10} p–$\bar{p}$ collisions. Thus the running time for these experiments is measured in years, with teams of hundreds of workers operating each of the CDF and D0 experimental stations at the Fermilab Tevatron, which are located at points where the counter-rotating proton and antiproton beams intersect.

The CDF and D0 groups each take top quark mass data, and they have been posting periodic "best concensus values" for the measured mass of the t on the Los Alamos-Cornell "arXiv" data base. The most recent value [59] is

$$m_t = 172.5 \pm 2.3 \text{ GeV} . \tag{3.20.1}$$

This result combines data from Run-I (1992–1996) and Run-II (2001-present). Continuing measurements will push the error bars down even farther. It is interesting that we obtain the best value for the mass of an individual SM quark by going to experiments at the highest energies. The mass values of the W and Z gauge bosons are [10]

$$m_{W^\pm} = 80.403 \pm 0.029 \text{ GeV} , \quad m_{Z^0} = 91.1876 \pm 0.0021 \text{ GeV} . \tag{3.20.2}$$

As can be seen, the masses of all three of these ultra-mass states have been measured very accurately. In the SM hierarchy of particle states, the top quark t is the sixth member of the three-generation (u, d), (s, c), (b, t) set of quarks. As such, we might logically expect to final mass relationships between these six quarks — particularly if we adopt the constituent-quark point of view. The W and Z particle states, however, are designated at gauge bosons that are grouped together in the Standard Model hierarchy with the photon, gluon and Higgs particle, and their masses have no direct tie-in with the top quark mass, except through the weak-interaction perturbation calculations mentioned above [58]. Hence weak interaction theorists have had no reason to attempt to relate the W and Z masses to the t mass.

The role of a phenomenologist is to examine *what* is out there without worrying particularly (at least initially) *why* it is there. Thus to the trained phenomenological eye, Eqs. (3.20.1) and (3.20.2) lead to an interesting mass relationship (see Fig. 0.2.21):

$$m_W + m_Z = 171.59 \pm 0.03 \text{ GeV}, \quad m_t = 172.5 \pm 2.3 \text{ GeV}. \qquad (3.20.3)$$

This agreement is well within the experimental errors, and it raises the question as to whether these supposedly disparate entities in fact share a common set of basis states. From the SM current-quark point of view, all of the quark masses are treated as independent adjustable parameters that do *not* have related values. Hence is not surprising that when the mass relationship in Eq. (3.20.3) was pointed out to a group of physicists who have spent years studying electroweak theory and carrying out experiments in that area, they did not seem to be aware of its existence, and they did not feel it could be a significant result.

We demonstrated in Sec. 3.19 that the m_{WZ} average mass is accurately reproduced by an upward extrapolation of the s, c, b flavor triad masses (Fig. 3.19.1), but the extrapolation of the isotopic-spin singlets ϕ, J/ψ and Υ does not account for the fact that the $W^{\pm}$ and Z^0 form an isotopic-spin mass doublet, and it does not encompass the mass of the t quark. Equation (3.20.3) indicates that if we want to extend the constituent-quark mass concept up into this high-mass region, we need to devise a mass scheme that includes all three of these particle states. The key to this mass extension is provided by the creation process for the top quark t. The proton–antiproton collisions that produce $t\bar{t}$ pairs are actually collisions between the individual $q \equiv (u, d)$ quarks in these particles. Thus the proton and antiproton q quarks logically serve as "ground states" for generating the t. The mass leap upward from the 315 MeV q quarks to the 172,500 MeV t quark is more than two orders of magnitude, and there are no observed particles in the 12,000 MeV to 80,000 MeV mass region between them. This suggests that a very large mass increase of a q quark occurs in a single step, which in turn indicates that this process may be an α-enhancement of the q quark mass, just as the initial q quarks were created by α-enhancements of the electron ground-state mass. The equation for the α-enhanced q quark mass with respect to the initial electron mass is

$$m_q = m_e(1 + 9/2\alpha) = 315.625 \text{ MeV} . \tag{3.20.4}$$

The equation for the α-enhancement of a q quark mass up to a much-higher-mass q^α quark, which is two α-leaps above the electron ground state, is

$$m_{q^\alpha} = m_e(1 + 9/2\alpha^2) = 43{,}182.5 \text{ MeV} . \tag{3.20.5}$$

With this q^α α-quark as a new high-mass basis state, the W and Z bosons are reproduced as a $q^\alpha\bar{q}^\alpha$ quark-antiquark pair, whose calculated isotopic-spin-averaged mass value is

$$(m_{WZ})_{\text{calc}} \equiv (\overline{WX})_{\text{calc}} = q^\alpha\bar{q}^\alpha = m_e(2 + 9/\alpha^2) = 86.365 \text{ GeV} , \tag{3.20.6}$$

which is within 0.66% of the experimental mass

$$(\overline{WX})_{\text{exp}} = 85.795 \text{ GeV} . \tag{3.20.7}$$

Furthermore, the $q^\alpha\bar{q}^\alpha$ α-quark pair has the charge freedom to reproduce the $W^\pm$ and Z^0 isotopic-spin states. The W–Z to t quark mass relationship $m_W + m_Z = m_t$ (Eq. (3.20.3)) then leads to the quark assignment $t \leftrightarrow 4q^\alpha$. Thus the calculated t mass is (in terms of an electron ground state)

$$(m_t)_{\text{calc}} = 4m_{q^\alpha} = m_e(1 + 18/\alpha^2) = 172.73 \text{ GeV} , \tag{3.20.8}$$

which accurately matches the experimental value $(m_t)_{\text{exp}} = 172.5 \pm 2.3$ GeV.

The phenomenological justification of these q^α α-quark assignments for the W, Z and t is provided by inter-comparing the quark "excitation towers" for the W, Z and t — the highest-mass elementary particle states — and the π, η and η' — the lowest-mass hadron states. These "platform" excitations are displayed in Figs. 3.20.1 and 3.20.2 (also see Fig. 0.2.22). It is informative to make a step-by-step comparison of these two excitation processes, which is shown in Table 3.20.1. As can be seen, the basis states for these two excitation towers feature radically different (but related) mass scales; however, the excitation patterns are identical. Thus we are not in the awkward phenomenological position of having to devise a completely new type of systematics for each particle family we encounter.

If these were the only particle states we could reproduce by using α-enhanced electron masses, the results shown here could conceivably be regarded as accidental. However, keeping in mind the very accurate α-enhanced masses that are displayed in Table 0.10.1 and Figs. 0.2.23 and 0.2.24, which span the entire spectrum of threshold-state particles, there

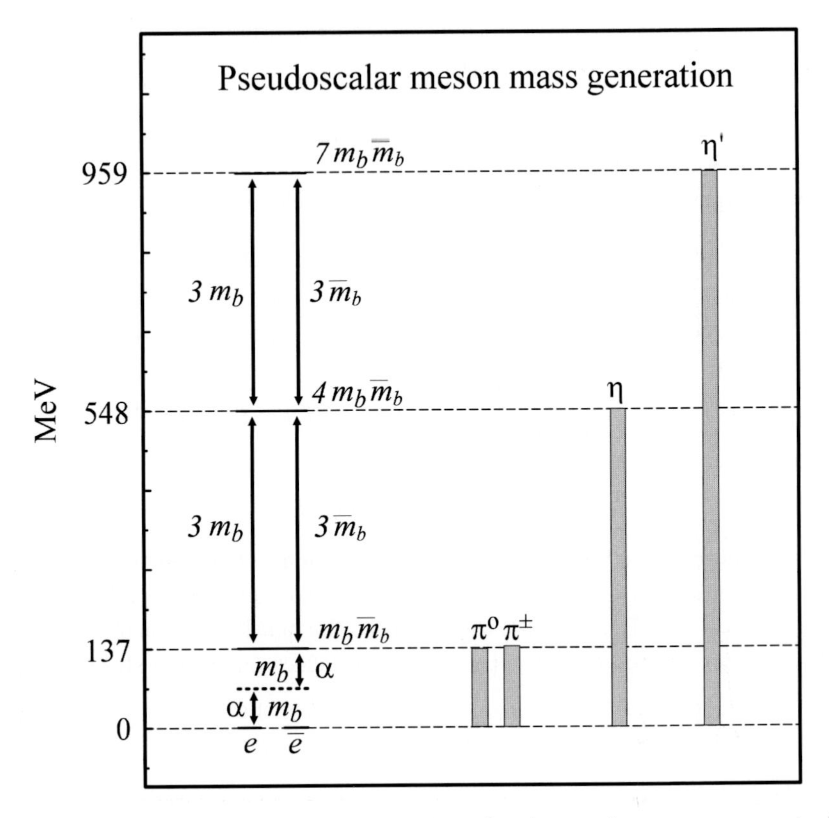

Fig. 3.20.1 The pattern of mass excitations for the very-low-mass π, η and η' pseudoscalar mesons, shown displayed against a 137 MeV mass grid. As can be seen, the experimental mass values (the gray columns) fall right on the 137 MeV-quantized grid. (The 137 MeV mass grid — instead of 140 MeV — reflects a hadronic binding energy of about 2.6% for these particle states, as pictured in Fig. 3.16.2.) This excitation pattern is repeated in Fig. 3.20.2 for the very-high-mass W, Z and t particle states. Table 3.20.1 summarizes these results.

seems to be no way that these inter-related constituent-quark masses can be considered as irrelevant to the elementary particle paradigm. In the final analysis, physics is all about numbers, and if physicists do not pay careful attention to the numbers that they themselves have generated, they may lose their bearings in the complexity and richness of the physical world in which we live.

As a way of summarizing the present α-quantized constituent-quark mass systematics, we display in Appendix C two tables of α-enhanced mass values for the threshold elementary particle states that have been identi-

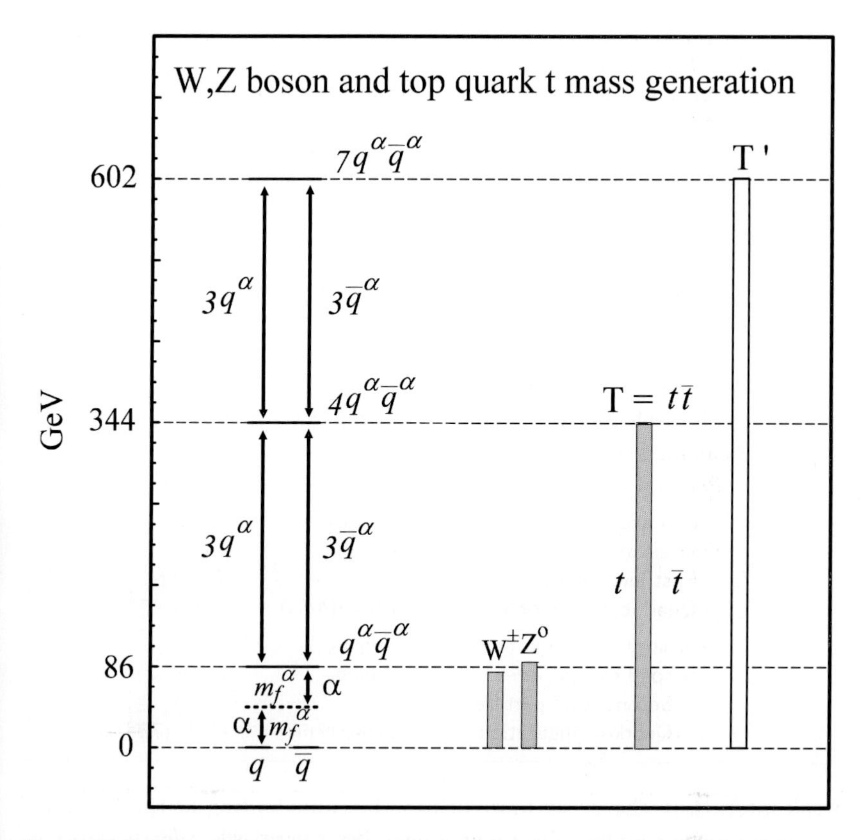

Fig. 3.20.2 The pattern of mass excitations for the ultra-high W and Z bosons and top quark t, shown displayed against an 86 GeV mass grid. The experimental isotopic-spin-averaged $\overline{WZ}$ mass and t quark mass (the gray columns) are in precise agreement (to better than 1%) with this 86 GeV mass quantization. The excitation pattern shown here is the same as that displayed in Fig. 3.20.1 for the pseudoscalar mesons, even though the two mass scales differ by a factor of about 628 in energy.

fied to date. The first table gives the straightforward calculated quark mass values as compared to the experimental particle mass values, and the second table shows plausible α-quantized excitation quanta that we have to add to the calculated mass values in the cases where they fall below the experimental values.

The workings of the mysterious and as yet undiscovered Higgs particle, which Lederman characterizes as the "God particle" [4], have been well described, e.g. by Schumm [60] and Veltman [8, 9, 12]. The Higgs enters into

Table 3.20.1 The matching platform excitation towers for the very-low-mass (π, η, η') pseudoscalar mesons and the ultra-high-mass (W, Z, t) gauge boson and top quark particle states. These are based on the $e\bar{e}$ and $q\bar{q}$ ground states, respectively, and have parallel excitation systematics. The calculated values for the isotopic-spin-averaged masses of these states are all within 1% of the measured mass values (when the low-mass 2.6% HBE is included).

Basis states:

m_e, $m_b = m_e/\alpha$, $q \equiv (u, d)$, $q^\alpha = q/\alpha \equiv (u^\alpha, d^\alpha)$
$m_q = m_e + (9/2)m_b$,
$m_{q^\alpha} = m_e + (9/2\alpha)m_b$,

Ground state	$e\bar{e}$	$q\bar{q}$
Mass enhancement	α-leap	α-leap
Platform state	$m_b\bar{m}_b$	$q^\alpha\bar{q}^\alpha$
Platform particles	π^0, $\pi^\pm$	$W^\pm$, Z^0
Platform excitation quanta	$3m_b\bar{m}_b$	$3q^\alpha\bar{q}^\alpha$
First excited level	$4m_b\bar{m}_b$	$4q^\alpha\bar{q}^\alpha$
First level particle	η	$T \equiv t\bar{t}$
Quark configuration	$(4m_b)(4\bar{m}_b)$	$(4q^\alpha)(4\bar{q}^\alpha)$
First level excitation quanta	$3m_b\bar{m}_b$	$3q^\alpha\bar{q}^\alpha$
Second excited level	$7m_b\bar{m}_b$	$7q^\alpha\bar{q}^\alpha$
Second level particle	η'	$T'(?)$
Quark configuration	$(7m_b)(7\bar{m}_b)$	$(7q^\alpha)(7\bar{q}^\alpha)$

weak interaction theory on more-or-less the same footing as the W and Z bosons and the t quark, all of which have been predicted and then discovered with the correct masses and correct behavior. But the Higgs has also been ascribed one additional property: theoretically, it couples to the other particles with an interaction strength that is proportional to the masses of these particles, so it has been proposed as the actual "generator" of these masses. However, from the viewpoint of the present studies, the fine structure constant $\alpha = e^2/\hbar c$ couples to the electron and generates the elementary particles. Thus the Higgs is no longer needed for the purpose of mass generation. However, it may be still needed for all of the other purposes it serves in weak interaction theory. Hence it is worthwhile to keep searching for it in the upper reaches of the elementary particle mass domain.

After this sojourn into the highest energy region of particle physics, we return in the next section to the lowest energy region, which is occupied by the pseudoscalar mesons, who continue to resist their SM quark assignments.

3.21 The PS Lifetime and Mass Nonet: Physics Outside of the Standard Model

The subject of the pseudoscalar meson nonet has been accorded a separate listing in Chapter 3 because we want to make a simple but far-reaching statement about the PS mesons. This statement is the following:

> *From the point-of-view of experimental masses, the η and η' mesons are on a completely equivalent footing.*

This is a companion statement to the one we made in Sec. 2.4, which is:

> *From the point-of-view of experimental lifetimes, the η and η' mesons are on a completely equivalent footing.*

In the reciprocal lifetime and mass studies of the present paper, the nine PS pseudoscalar mesons — the "PS octet" of the Standard Model plus the η' meson — exhibit the most accurate examples of α-quantization. They are the "crown jewels" — the best we can offer. And the simplest and most accurate results are displayed by the non-strange ($\pi^{\pm}$, π^0, η, η') subset of PS mesons — the "PS quintet" of Secs. 2.2 and 3.7. Figure 3.21.1 illustrates the stunning accuracy of the *lifetime* ratios of this PS quintet when expressed in powers of the scaling factor $\alpha \cong 1/137$. Figure 3.21.2 shows the equally stunning reciprocal accuracy of the *masses* of these same particles when plotted on a 137 MeV mass grid. The accuracy of these results is not surprising, in the sense that these PS mesons are the lowest-mass spinless states in the elementary particle zoo, and logically have simple and relatively stable structures. If we characterize Figs. 3.21.1 and 3.21.2 in nuclear physics terms, the η and η' mesons appear as the first and second excited states, respectively, above the ground-state π^0 meson. This idea is reinforced by the observed mass generation systematics

$$\pi^0 + \{\pi\pi\pi\} \to \eta \quad \text{and} \quad \eta + \{\pi\pi\pi\} \to \eta' \,,$$

and the corresponding decay modes

$$\eta \to \pi^0 + [\pi]\pi\pi \quad \text{and} \quad \eta' \to \eta + [\pi]\pi\pi \,,$$

where $\{\pi\pi\pi\}$ denotes the excitation quantum X, and $[\pi]$ denotes the annihilation of a mass quantum π. Thus the (π^+, π^-, π^0, η, η') quintet of pseudoscalar mesons forms an interrelated family of particles.

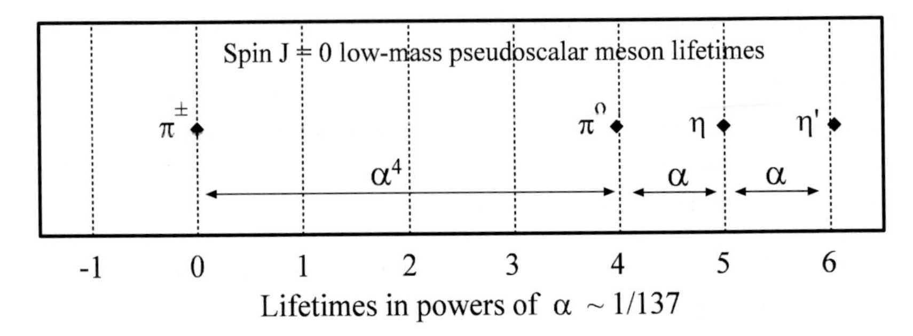

Fig. 3.21.1 The scaling of the non-strange pseudoscalar meson *lifetimes* in powers of α (see Figs. 2.2.3–2.2.5). This scaling is accurately maintained over more than 12 orders of magnitude.

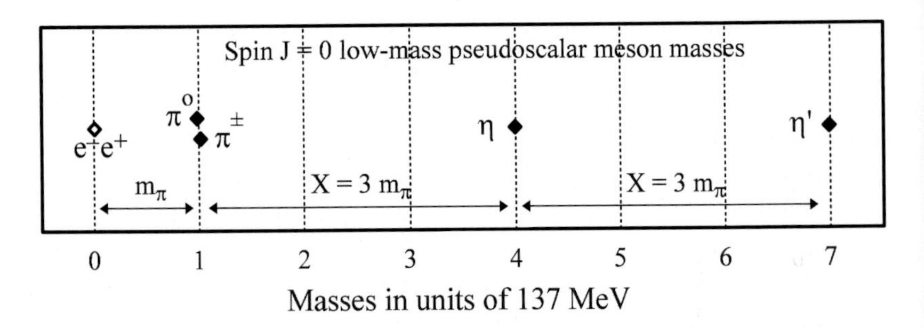

Fig. 3.21.2 The reciprocal scaling of the non-strange pseudoscalar meson *masses* in powers of α^{-1} (see Fig. 3.7.2). The accuracy of the linear mass intervals in fitting to the 137 MeV mass grid is on the order of 0.1%.

The lifetime and mass α-quantization of the non-strange PS quintet of mesons is displayed *pictorially* in Figs. 3.21.1 and 3.21.2. It is also instructive to consider the accuracy of this α-quantization from a *numerical* viewpoint. Consider first the lifetimes. The $\pi^{\pm}$, π^0, η and η' PS quintet lifetimes are spread out over more than 12 orders of magnitude. With the $\pi^{\pm}$ meson used as the reference lifetime, the π^0, η and η' mesons have the lifetime ratios S^{x_1}, S^{x_2} and S^{x_3}, where $S \cong \alpha^{-1}$ and $x_i \cong 4$, 5 and 6. Taking a range of S-values, obtaining precise fits for the x_i at each value of S, and calculating the absolute deviations of the x_i's from integers, we obtain an *absolute deviation from an integer* (ADI) minimum at $S = 136.09$ (Fig. 2.2.3). Similarly, using the experimental lifetimes as weighted by their uncertainties in a quadratic $\chi^2(S)$ sum, we obtain a χ^2 minimum at

$S = 138.96$ (Fig. 2.2.4). Each of these independent determinations of S is within 1.4% of the value $1/\alpha = 137.036$, and their average S value is within 0.4%. These results could hardly be more accurate, considering the tremendous span of lifetimes included here. This lifetime α-dependence also extends to the "strange" PS quartet of K mesons, as depicted in Figs. 2.3.2 and 2.3.3. In particular, the $K^{\pm}$ and K_S^0 mesons, which are related to one another by their $\pi\pi$ decay modes (Fig. 2.3.1), have an experimental lifetime ratio of 138.3, which is very close to the value of α^{-1}. Thus the nonet of PS mesons fit with numerical accuracy into the α-spaced lifetime grid.

The α-quantization of the non-strange PS meson *masses* in 137 MeV mass units is also impressive. Lacking a formalism for handling the pion isotopic-spin mass splitting, the best guess we can make for the "intrinsic" pion mass $\bar{\pi}$ is its average value of 137.27 MeV. The experimental masses of the η and η' mesons are 547.51 and 957.78 MeV, respectively. Thus the $\eta - \bar{\pi}$ and $\eta' - \eta$ mass intervals are 410.24 and 410.27 MeV, respectively, which are each approximately equal to $3 \times 137 = 411$ MeV. These $\eta - \bar{\pi}$ and $\eta' - \eta$ mass intervals represent *mass linearity* at an accuracy level of about 0.1%.

The 0.1% accuracy of the α-quantized $\eta - \bar{\pi}$ and $\eta' - \eta$ *mass* intervals, when combined with the 0.4% average accuracy of the *lifetime* ADI and χ^2 fits to the α-grid, leads to an important phenomenological conclusion. Empirically, the η and η' mesons occur as the *first excited state* and *second excited state*, respectively, of the π meson, and they are on a completely equivalent footing from this standpoint. This conclusion seems borne out by their dominant $\eta' \rightarrow \eta + \pi\pi$ and $\eta \rightarrow \pi + \pi\pi$ decay channels. However, that is *not* the way the η and η' are fitted into the Standard model, as we now discuss.

The pseudoscalar QCD octet in the Standard Model includes the π^+, π^0, π^-, K^+, K^0, $\bar{K}^0$, K^- and η mesons. The η' is thrown in with this octet as an associated singlet that has its own adjustable mass parameter. A strong early argument in favor of the QCD approach was the accuracy of the linear mass intervals for related spin-1/2 baryons and hyperons. However, linear mass intervals turned out not to work nearly as well for the PS mesons, so quadratic mass relationships were invoked (RPP2006 includes both linear and quadratic fits). The argument advanced for quadratic mass relationships is that mass is a self-energy phenomenon, and boson self-energies such are those for the PS mesons are calculated by the Klein–Gordon equation, which gives quadratic masses [61]. However, the results of the present paper suggest that elementary particle masses are not a self-energy phenomenon, but instead come from an α-coupling to the mass of the electron, which undercuts the use of Klein–Gordon quadratic masses. Furthermore, as described above, the observed $\eta - \bar{\pi}$ and $\eta' - \eta$ mass intervals, are in fact very linear. (It took a number of years until the η' mass was accurately measured.) A more serious conceptual difference between these two approaches is that the η and η' mesons occur as the first and second "excited states" of the pion in the reciprocal α-quantized mass and lifetime data displayed in Figs. 3.7.1 and 3.7.2, and hence should not be handled differently from one another in the theoretical mass formalism. However, in the QCD mass formulation the η and η' occur in different SU(3) groups that are linked by a rather awkward coupling constant [10, 62].

The bottom line in this discussion is that from the experimental standpoint of lifetimes and masses, the five non-strange PS mesons and the four strange PS kaons constitute a pseudoscalar *nonet*, and not an *octet*. At the time that the SM PS octet was originally set forth, the properties of the η' were not well established, and it was for a while denoted as the X-particle. Thus to add it as a tacked-on singlet seemed permissible. However, this is no longer experimentally justified. A nonet is not an octet.

Historically, the pseudoscalar mesons have created many problems for the Standard Model. In particular, they have steered the SM theorists into an awkward situation with respect to the issue of current-quark and constituent-quark masses. This topic is addressed in Sec. 7.3. In physics it is the experimentalists with their well-measured data who usually end up making the decisive determinations in these matters.

3.22 Examples of Reciprocal α-Quantized Lifetimes and Masses

The simplest and therefore probably the most informative hadronic elementary particle states are the low-mass spin-0 non-strange pseudoscalar mesons $\pi^{\pm}$, π^0, η and η' that we discussed in Sec. 3.21. In the lifetime studies, these states show the most accurate α-quantization, as displayed in Fig. 3.21.1. Two important aspects of this figure are that (a) all of the particles in this category are included, and (b) the accuracy of the α-scaling is maintained over six powers of 1/137, or almost 13 orders of magnitude. The lifetime results are interesting in themselves, but of equal interest are the implications they carry with respect to the masses of these same particles, which are displayed in Fig. 3.21.2. When plotted on a linear mass grid in units of 137 MeV, the $\pi^{\pm}$, π^0, η and η' masses fall squarely on this grid (see Fig. 3.7.2). In order to account for the relevance of using a 137 MeV mass grid, we included another low-mass non-strange particle system — the $J = 0$ electron–positron pair — in Fig. 3.21.2. The mass of the e^-e^+ pair is 1.022 MeV, and it sets the mass scale for the other particles. The precise value of the π/e^-e^+ mass ratio is somewhat blurred by the charge-splitting of the pion masses (Fig. 3.4.1) and by the hadronic binding energy of the pion (Fig. 3.4.2), but it is close to the value $\alpha^{-1} = 137.036$. Hence the e^-e^+ to π mass excitation represents the factor-of-137 mass "α-leap" that is reciprocal to the 1/137 lifetime "α-leaps" displayed in Fig. 3.21.1. The ~137 MeV mass quantum m_π generated by this α-leap serves as the mass unit for the $X = 3m_\pi$ excitation quantum that generates the η and η' higher-mass excitations, as shown in Fig. 3.21.2. The α-dependent lifetimes and masses of Figs. 3.21.1 and 3.21.2 demonstrate the concept of reciprocal lifetime and mass α-quantizations in a transparent, complete, and accurate manner.

The interest in Figs. 3.21.1 and 3.21.2 is due not only to the displayed dependence on α of the non-strange PS meson lifetimes and masses, but also to the *patterns* of this α-dependence. In the lifetimes we see a basic reference lifetime followed by a lifetime that is a factor of α^4 shorter, and then lifetimes which are additional factors of α shorter. In the masses we see a basic spin-0 "ground state" mass followed by a "platform state" mass that is one m_π α-mass larger, and then masses which are additional $X \equiv 3m_\pi$ mass units larger. These reciprocal α-dependences arise in a system which is based on the spin $J = 0$ e^-e^+ ground state. A similar pattern of reciprocal α-quantized lifetimes and masses is also observed in another system that has a quite different spin 0 "platform state" — the $J = 0$ D meson and its excited states. Figures 3.22.1 and 3.22.2 display these two systems, which are placed together for comparison purposes. The lifetimes are shown in Fig. 3.22.1, with all lifetimes expressed as ratios to the $\pi^{\pm}$ reference lifetime. The D-meson lifetimes use the logarithmic x_i scale at the top of the figure, which is shifted by two powers of α with respect to the pion lifetime scale at the bottom of the figure. The "platform state" $D^{\pm}$ and D^0 lifetimes are separated from one another by about a factor of 2 (Figs. 2.5.3 and 2.5.4), and the D^0 lifetime (which is in the central c quark lifetime group, as shown in Fig. 2.9.1) exhibits the characteristic factor of 3 c-quark displacement from the $\pi^{\pm}$-based α^{x_i} lifetime grid (Fig. 2.9.1). The $D^{*\pm}$ lifetime is about a factor of α^4 shorter than the D lifetimes (the D^{*0} lifetime has not been accurately measured), and the D_1^0 and $D_1^{\pm}$ lifetimes are another factor of α

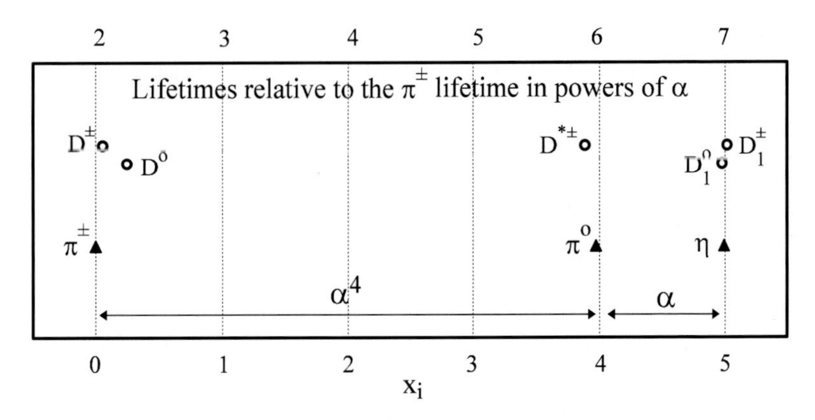

Fig. 3.22.1 Matching α-quantized lifetime patterns for the low-mass non-strange pseudoscalar mesons (bottom) and the higher-mass charmed D mesons (top). Although the lifetimes of these two meson families are shifted relative to one another by two powers of α (top and bottom abscissa scales), the overall lifetime patterns are essentially the same. The fact that these two lifetime patterns encompass a total of seven powers of α (fifteen orders of magnitude) shows that this lifetime α-scaling is a global phenomenon which can only be analyzed on a comprehensive global scale. Of even more significance is the fact that the masses of these two systems of particles exhibit a reciprocal α^{-1} scaling, as is displayed in Figs. 3.22.2 and 3.22.3.

shorter. Thus these D-meson lifetime intervals closely approximate those of the pion lifetime intervals, although they differ somewhat in detail. The fact that similar lifetime patterns occur in two systems whose actual lifetimes are shifted by two powers of α (four orders of magnitude) with respect to one another, and whose masses and quark flavors are different, suggests that this α-quantized lifetime pattern is a general result, and not specific to one hadron family.

The particle masses that correspond to the particle lifetimes of Fig. 3.22.1 are displayed in Fig. 3.22.2, where the pion and D-meson spin $J = 0$ "platform states" are placed at the origin of the relative-mass excitation scale shown along the abscissa. In each case there are first-excited-states at m_π, followed by second-excited states at $X \equiv 3m_\pi$. The interest in Figs. 3.22.1 and 3.22.2 is that they display two different "reciprocal-lifetime-and-mass" systems which both feature the same "α-leap" lifetime intervals and the same "α-mass" energy intervals. The low-mass PS pion system contains the α-quantized m_π mass generator in the form of the 137 MeV "α-leap" from the spin-0 leptonic e^-e^+ pair to the spin-0 hadronic pion. The higher-mass hadronic D meson system then uses this same $m_\pi \sim 137$ MeV

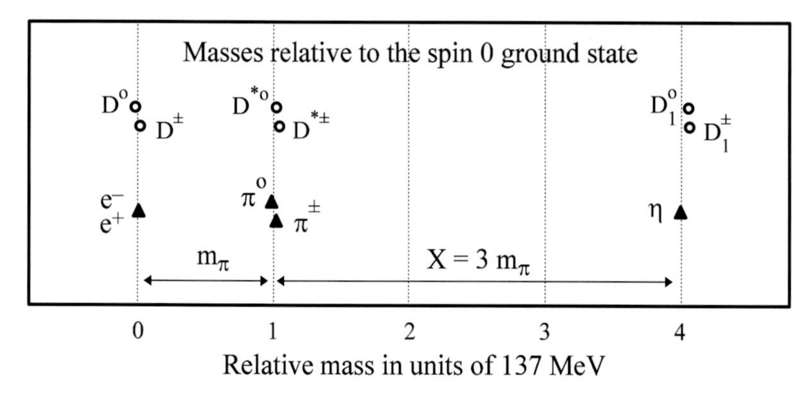

Fig. 3.22.2 Matching α^{-1}-quantized mass patterns for the low-mass non-strange pseudoscalar mesons (bottom) and the higher-mass charmed D mesons (top). In these spin 0 systems the PS meson "ground-state" mass and D meson "platform-state" mass are set equal to zero, and the higher excitations are shown plotted on a 137 MeV mass grid. As can be seen, the mass accuracies are comparable in these two quite different systems. These α^{-1}-quantized masses are reciprocal to the α-quantized particle lifetimes displayed in Fig. 3.22.1, which reinforces the reality of the observed α-dependence in both of these hadronic systems. The D^{*0} meson mass is displayed in Fig. 3.22.2, but the corresponding D^{*0} lifetime does not appear in Fig. 3.22.1 because it has not yet been accurately measured [10]. It will be interesting to find out how this reciprocal D^{*0} lifetime eventually fits into the α-quantized lifetime grid.

"α-mass" and $X \equiv 3m_\pi$ excitation quantum in its own excitation processes. The lifetime α-quantizations of these states logically follow from the mass α-quantizations, although the details remain to be worked out.

As the final result in this section, we show yet another spin-0 system with this same mass quantization — the D_s meson, whose excited-state lifetimes have not been measured. Figure 3.22.3 combines the mass diagrams for all three of these systems — the non-strange PS pions, the charmed D mesons, and the charmed-strange D_s mesons — all of which have spin-0 "platform states." As can be seen these three flavor-types each have a first excitation level about $m_\pi \sim 140$ MeV above the ground state, and then a second excitation level about $X = 420$ MeV above the first level. The excited-state lifetimes for the D_s mesons have not yet been accurately measured. The first excitation level for the PS mesons is an "α-leap" of $m_\pi = e^-e^+/\alpha$, and the first excitation levels for the D and D_s mesons are m_π "α-masses". The second levels for all of these systems are $X = 3m_\pi$ excitations. This m_π α-quantization of the masses first appears in the α-leap from a spin-0 electron–positron pair to a pion, but it is evidently a more general process

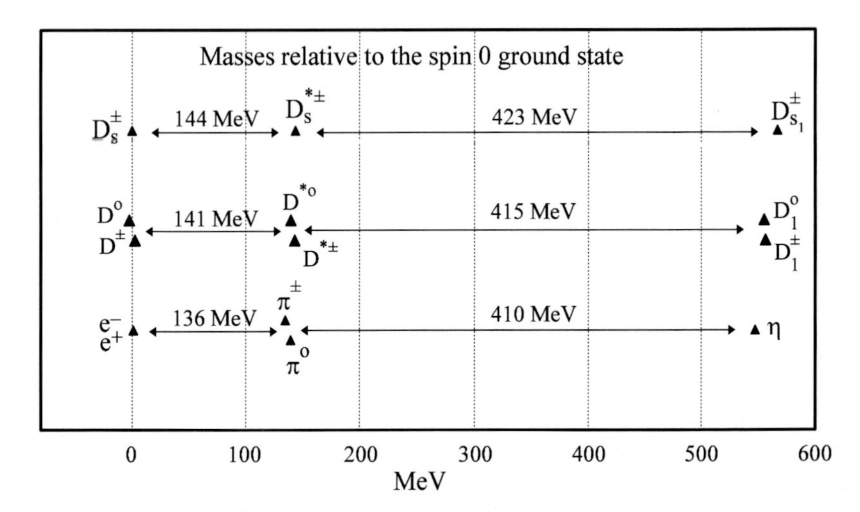

Fig. 3.22.3 The two α^{-1}-quantized mass systems of Fig. 3.22.2 shown plotted here in comparison with a third spin-0 system, the charmed-strange D_S mesons. The same excitation mass intervals are observed in all three of these quark-flavor families. The lifetimes of the $D_S^{*\pm}$ and $D_{S_1}^\pm$ excited states have not yet been determined [10]. The m_π excitation quantum first appears in the α-leap from a spin-0 electron pair to a pion (Sec. 3.3), but the mass excitations displayed in Fig. 3.22.3 suggest that it is a more general process which can occur throughout the spectrum of threshold-state particle excitations. Since the m_π mass unit is (from the present perspective) spinless, and since the Standard Model u, d, s, c, b quarks are spin-1/2 fermions, we have spin-1/2 plus spin-0 hybrid excitations occurring in the D and D_s mass systems. Figure 3.23.2 illustrates a hybrid excitation in the Ω hyperon. The lowest-mass hybrid state is the $\omega(783) = q\bar{q}m_\pi$ meson, where $q \equiv (u,d)$. An important feature of the excitations shown in Fig. 3.22.3 is that the $D_S - D$ mass difference, which we would expect on the basis of their constituent-quark masses to be about 210 MeV, is actually only 100 MeV (Table C1), and this same 100 MeV $D_S - D$ mass interval is maintained for the m_π and $3m_\pi$ excitation levels, which indicates that the basic structures and binding energies of these two systems are determined entirely by the D and D_s platform states, and are not appreciably affected by the higher excitations.

that occurs throughout the spectrum of threshold-state elementary particle excitations. Since the m_π mass unit is (from the present viewpoint) spinless, and since the SM u, d, s, c, and b quarks are spin-1/2 fermions, we have spin 1/2 plus spin 0 hybrid excitations occurring in the D and D_S mass systems. The lowest-mass hybrid state is the $\omega(783) = q\bar{q}\pi$ meson, where $q \equiv (u,d)$. We will see in Fig. 3.23.2 that the Ω^- hyperon ground state also appears as a hybrid excitation.

The main conclusions we would like to draw from Figs. 3.22.1–3.22.3 are their experimental validation of the concept of the *reciprocal α-quantization of elementary particle masses and lifetimes*, and of the universality of this concept in different families of particles.

3.23 The q, s, c, b Quark Benchmark Test: Calculate 16 Unpaired-Quark Ground States

In the quest for an explanation — a Mendeleev mapping — of elementary particle masses, the most important states to study are the long-lived ground states. The three lowest-mass particle states, and three of the most stable, are the "Rosetta stone" electron, muon, pion triad discussed in Sec. 3.3. Another important particle grouping is the pseudoscalar meson nonet summarized in Sec. 3.21. The mass systematics of these two particle groups leads to the identification of a new class of spinless substates composed of $m_b = 70$ MeV boson mass quanta, which we can denote as "pion quarks" The SM u, d, s, c, b quarks, when treated from the constituent-quark point of view, all have masses which are multiples of the spin-1/2 fermion mass quantum $m_f = 105$ MeV, and can be denoted as "muon quarks" (Sec. 3.25). The importance of the s, c and b quarks was brought out clearly in Chapter 2, where they were seen to dominate the lifetimes of the particles that contain them, particularly in the cases where they occur as unpaired quarks which are not matched up with their antiquark counterparts. These same unpaired-quark s, c and b particle states also play an important role in the mass systematics of these particles. They function as quark "ground states," upon which short-lived excited states are constructed. A total of 15 baryon and meson s, c and b quark ground states have been observed to date [10] (Eq. (3.23.4)), and the task of accurately reproducing their masses, plus the proton mass, is a benchmark test for any proposed set of s, c and b quark mass values. In Sec. 3.23 we demonstrate how the α-quantized constituent-quark masses deduced in previous sections of Chapter 3 fare when subjected to this new benchmark test.

The central theme in the present book has been to demonstrate the α-quantization of elementary particle lifetimes and the reciprocal α^{-1} quantization of elementary particle masses. The lifetime α-quantization occurs for the long-lived particles that have mean lives $\tau > 10^{-21}$ s, which is the empirical zeptosecond boundary between the long-lived *threshold-state* particles and the much-short-lived *excited-state* resonances. Experimentally, there are 38 elementary particles (including the electron and proton) with measured lifetimes $\tau > 1$ zs [10]. In addition, the $\phi(1020) = s\bar{s}$ meson is an important threshold state, even though it has a slightly shorter lifetime due to its $\phi \mapsto K\bar{K}$ strangeness conserving decay mode. These threshold-state particles are listed in Table 3.6.1, where they are arrayed according to spins and flavors. We also display them here, with the eight isotopic spin multiplets placed inside parentheses:

$$(\pi^\pm, \pi^0),\ \eta,\ \eta',\ e^\pm,\ \mu^\pm,\ (p^\pm, n^0),\ \tau^\pm,\ (K^\pm, K_L^0, K_S^0),\ \phi,$$
$$\Lambda^0,\ (\Sigma^+, \Sigma^0, \Sigma^-),\ (\Xi^0, \Xi^-),\ \Omega^-,\ (D^\pm, D^0),\ D_s^\pm,\ D^{*\pm},\ J/\psi_{1S},\ J/\psi_{2S},$$
$$\Lambda_c^+,\ (\Xi_c^+, \Xi_c^0),\ \Omega_c^0,\ (B^\pm, B^0),\ B_s^0,\ B_c^\pm,\ \Upsilon_{1S},\ \Upsilon_{2S},\ \Upsilon_{3S},\ \Lambda_b^0,\ \Xi_b\,. \tag{3.23.1}$$

The *lifetime* α-dependence that is displayed in Chapter 2 occurs in this set of particles. Thus it logically follows that these are the particles we should examine for a possible reciprocal *mass* α-dependence, as suggested by the Heisenberg relationship between conjugate lifetimes and mass widths (Eqs. (1.7.1) and (1.7.2)). If an α^{-1}-scaling can be demonstrated in the masses of these particles, it reinforces the significance of the observed lifetime α-scaling.

The particles that are contained in the eight isotopic spin multiplets of Eq. (3.23.1) have different lifetimes. Thus they contribute individually to the lifetime α-dependence. However, the mass values within each multiplet are quite similar. Their mass splitting is on the order of few tenths of a percent for all but the lowest-mass multiplets. More to the point, we presently have no established way of accounting for these mass splittings in a constituent-quark context. So the most reliable way of phenomenologically handling the multiplet masses is to let their average values represent the entire multiplet. The one exception we make to this rule is to have the proton mass represent the $(p^\pm, n^0)$ nucleon doublet, due to the special importance we attach here to the proton mass. The multiplet-averaged mass values used here are summarized in Appendix B. Many of the particle states listed in Eq. (3.23.1) have corresponding equal-mass antiparticle states, which do not have to be considered here.

In addition to the mass degeneracy in the isotopic spin multiplets, there are also other changes we make in the particle set of Eq. (3.23.1) for the purpose of mass calculations. The J/ψ_{2S}, Υ_{2S} and Υ_{3S} mesons are excited states of the J/ψ_{1S} and Υ_{1S} ground states, and thus are removed from the data set. The $D^{*\pm}(2010)$ meson (see Figs. 3.22.2 and 3.22.3) is similarly removed. And the somewhat composite Ξ_b baryon lifetime measurement has no corresponding quoted mass value. Along with these subtractions from the lifetime data set, there are some additions that are useful for the mass systematics. The Ξ_{cc} baryon has a measured mass but only an upper limit for its lifetime [10]. The Σ_c multiplet has lifetimes $\tau < 1$ zs, but its mass values are informative. Finally, for good measure we add in the shortest-lived and most massive particles of all — the $W^{\pm}$ and Z^0 isotopic spin doublet $\overline{WZ}$. When these changes are applied to the 39-particle lifetime data set of Eq. (3.23.1), we emerge with the following 27-particle mass data compilation:

$$[8+e]: \quad e,\ p,\ \mu,\ \tau,\ \pi,\ \eta,\ \eta',\ K,\ \phi; \tag{3.23.2}$$

(M^X octet leptons and low-mass hadrons of Tables 3.13.1 and 3.16.1)

$$[3+\phi]: \quad \phi,\ J/\psi_{1S},\ \Upsilon_{1S},\ \overline{WZ}; \tag{3.23.3}$$

(M^T mass-tripled vector meson states of Secs. 3.17 and 3.19)

$$[15+p]: \quad p,\ \Lambda,\ \Lambda_c,\ \Lambda_b,\ \Sigma,\ \Sigma_c,\ \Xi,\ \Xi_c,\ \Xi_{cc},\ \Omega,\ \Omega_c,\ D,\ D_s,\ B,\ B_s,\ B_c; \tag{3.23.4}$$

(unpaired q, s, c, b quark states of Sec. 3.23)

where ϕ and p are double-entered, as noted at the beginning of Eqs. (3.23.3) and (3.23.4). The 27 particles listed in Eqs. (3.23.2)–(3.23.4) are the electron ground state and threshold states that hold the key to the α-quantization of elementary particle masses. As is indicated below these equations, the pseudoscalar mesons, leptons, proton and ϕ meson are contained in the M^X platform excitations of Secs. 3.5–3.16. The $\phi = s\bar{s}$, $J/\psi = c\bar{c}$ and $\Upsilon = b\bar{b}$ flavor threshold states are contained in the M^T mass-tripling generation process of Secs. 3.17 and 3.18, and the M^T mass tripling process is extended upward in energy in Sec. 3.19 to accurately encompass the average mass $\overline{WZ}$ of the gauge boson isotopic-spin doublet (but see the later results in Sec. 3.20). The remaining 15 long-lived particle states (Eq. (3.23.4)) are all states which contain *unpaired* combinations of s, c, and b quarks, and are the ones, together with the proton, that we consider as the 16-particle constituent-quark test set in the present section. The α-quantized lifetime-to-mass data base conversion process discussed here is summarized in Table 3.23.1.

The 16 threshold-state particles listed in Eq. (3.23.4) constitute one of the most important tests of the present α-quantized mass systematics. There are five reasons for this:

Table 3.23.1 The conversion from a 36-member elementary particle α-quantized *lifetime* data set to a corresponding 27-member α-quantized *mass* data set, and then to a 16-member unpaired-quark test set of α-quantized masses (bottom row).

The 36-member α-quantized lifetime data set, which consists of the elementary particles that have lifetimes $\tau > 10^{-21}$ s:

$\pi^{\pm}$, π^0, η, η', $\mu^{\pm}$, n^0, $\tau^{\pm}$, $K^{\pm}$, K_L^0, K_S^0, Λ^0, Σ^+, Σ^0, Σ^-, Ξ^0, Ξ^-, Ω^-, $D^{\pm}$, D^0, $D_s^{\pm}$, $D^{*\pm}$, J/ψ_{1S}, J/ψ_{2S}, Λ_c^+, Ξ_c^+, Ξ_c^0, Ω_c^0, $B^{\pm}$, B^0, B_s^0, $B_c^{\pm}$, Υ_{1S}, Υ_{2S}, Υ_{3S}, Λ_b^0, Ξ_b.

Subtractions from the lifetime data set:

$D^{*\pm}$, J/ψ_{2S}, Υ_{2S}, Υ_{3S} (excited states), Ξ_b (no measured mass).

Additions to the data set that are relevant to α-quantized masses:

$e^{\mp}$, $p^{\pm}$, ϕ, Σ_c, Ξ_{cc}, $W^{\pm}$, Z^0.

Use of mass-averaged isotopic spin multiplets:

$(\pi^{\pm}, \pi^0) \rightarrow \pi$, $(p^{\pm}, n^0) \rightarrow p$, $(K^{\pm}, K_L^0, K_S^0) \rightarrow K$, $(\Sigma^+, \Sigma^0, \Sigma^-) \rightarrow \Sigma$, $(\Xi^0, \Xi^-) \rightarrow \Xi$, $(D^{\pm}, D^0) \rightarrow D$, $(\Xi_c^+, \Xi_c^0) \rightarrow \Xi_c$, $(B^{\pm}, B^0) \rightarrow B$.

The final 27-member α-quantized mass data set, shown divided into (a) M^X (Tables 3.13.1 and 3.16.1), (b) M^T (Secs. 3.17 and 3.19), and (c) unpaired-quark ground-state (Sec. 3.23) mass α-quantization groups, with the proton and ϕ meson each listed in two groups.

(a) M^X::$(e, p, \mu, \tau, \pi, \eta, \eta', K, \phi)$;

(b) M^T::$(\phi, J/\psi_{1S}, \Upsilon_{1S}, \overline{WZ})$,

(c) unpaired-quark test set::$(p, \Lambda, \Lambda_c, \Lambda_b, \Sigma, \Sigma_c, \Xi, \Xi_c, \Xi_{cc}, \Omega, \Omega_c, D, D_s, B, B_s, B_c)$, which is displayed in Fig. 3.23.1.

(1) these 16 particles are composed entirely of SM spin-1/2 u, d, s, c, b quarks;
(2) the 16 particles contain the s, c, and b quarks in all possible unpaired combinations, and also the $p = qqq$ baryon ground state, where $q \equiv (u, d)$.
(3) each quark combination represents the lowest-mass, and therefore simplest, particle state for that combination — these are the quark-combination "platform states;"
(4) the α-quantized constituent-quark mass values used here for the u, d, s, c, b quarks were deduced from the paired-quark M^X and M^T systematics of Eqs. (3.23.2) and (3.23.3), so the 16 unpaired-quark particle states of Eq. (3.23.4) represent 16 essentially independent tests of these paired-quark mass values.

(5) the mass values of the u, d, s, c, b quarks that we use here do not represent five independent adjustable parameters which were "fine-tuned" to meet the constraints posed by the particles of Eqs. (3.23.2) and (3.23.3), but rather are closely related to one another, and in fact are all calculated from theory via the mass generation operation of the coupling constant α on the electron ground state (Secs. 3.3 and 3.4), combined with the systematics of the relativistically spinning sphere (Chapter 4); what we have here in essence is a "no-adjustable-parameter" α-quantized mass formalism.

If the u, d, s, c, b "constituent-quark" basis set that was obtained from the mass systematics of the particles in Eqs. (3.23.2) and (3.23.3) can meet the mass challenge posed by the 16 particles in Eq. (3.23.4), this will raise the bar for the results that should be demanded from any "current-quark" approach to the masses of these particles.

In addition to the intrinsic masses of the constituent quarks, the other factors that affect the particle masses are the charge-splitting effects discussed in Sec. 3.14 and the hadronic binding energies considered in Sec. 3.15. These are small effects, on the order of about one or two percent, and will be ignored here since we are looking for broad general agreement over a wide range of mass values. Thus we calculate the mass of a particle by simply adding up the masses of its constituent quarks. Roughly speaking, these quark masses are multiples of the muon mass: $(q \equiv u, d)$, s, c, $b \cong 3$, 5, 15, 45 $m_\mu \cong 315$, 525, 1575, 4725 MeV. But since we have precise theoretical values available, we will use them as an illustration of the point that there are *no adjustable parameters* in these mass calculations. The ground-state mass is the electron mass m_e, and the α-scaled fermion mass quantum is $m_f \equiv 3m_e/2\alpha$ (Eq. (3.4.2)). The fermion constituent-quark masses are

$$m_f = 3m_e/2\alpha = 105.04 \text{ MeV},$$

$$q = m_e(1 + 3m_f) = 315.6 \text{ Mev}, \quad s = m_e(1 + 5m_f) = 525.7 \text{ MeV},$$

$$c = m_e(1 + 15m_f) = 1576.1 \text{ Mev}, \quad b = m_e(1 + 45m_f) = 4727.2 \text{ MeV}. \tag{3.23.5}$$

Figure 3.23.1 shows calculated (dashed lines) and experimental (solid lines) mass values for the 16 particles of Eq. (3.23.4). The quark configuration is shown below each set of mass values and the percent accuracy of the calculation (relative to the experimental mass) is shown above. The average mass accuracy for these 16 ground states is 3.5%. As can be seen, there are

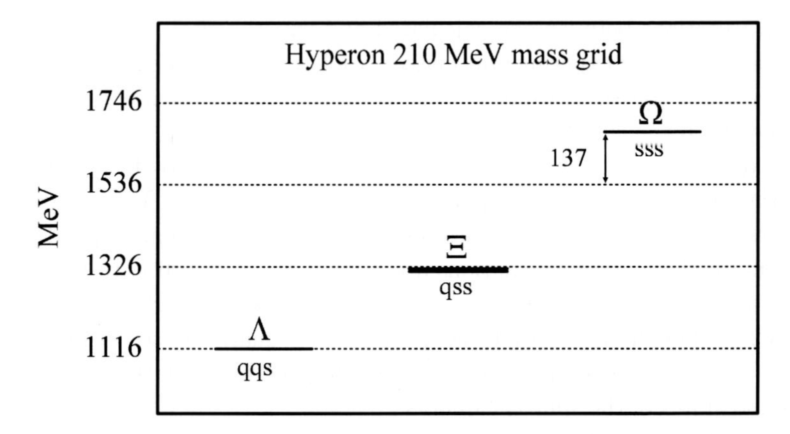

Fig. 3.23.2 Experimental mass values for the Λ, Ξ and Ω hyperons, plotted on a 210 MeV mass grid. Since the $s - q$ constituent-quark mass difference is about 210 MeV (Eq. (3.23.5)), we expect these masses to each step upward by 210 MeV. However, the Ω mass is 137 MeV higher than this projected value. This might be a manifestation of the Pauli exclusion principle, which prevents the three identical s quarks in the Ω from being in the lowest quantum state. It might also be related to the excitation systematics displayed in Table 3.23.2, where a dominant XXX excitation quantum would drive the final-state Ω to a mass value that is one pion mass higher than expected.

to their qss and qqs quark states. The Ω has the quark configuration sss, which has a calculated mass 210 MeV above the mass of the Ξ, at about 1535 MeV. However, the resonance that appears at 1535 MeV is a Ξ excitation, not the Ω. Quantum mechanically, three spin-1/2 fermion s quarks in a particle cannot all appear in an S state configuration, due to the operation of the Pauli exclusion principle. Thus a higher excitation seems required. What is interesting is that the Ω appears 137 MeV higher than expected (including expected binding energies), as we show in Fig. 3.23.2, where the Λ, Ξ and Ω hyperons are plotted on a 210 MeV mass grid which is centered on the Λ. The appearance of this 137 MeV excitation interval, when compared with the $\sim$ 70 MeV excitation interval for the Σ above the Λ, suggests that these pionic excitations fit into the α-quantized mass scheme.

We can gain further information about these hyperon excitations by studying their α-quantized principal production channels, which are summarized in Table 3.23.2. The $\pi + p \rightarrow \Lambda + \bar{K}$ channel requires 534 MeV of energy in the center-of-mass frame, which corresponds to the creation of a

Table 3.23.2 The excitation systematics of hyperon associated production. The required center-of-mass Q-values suggest the following α-quantized production channels, where the curly brackets represent creation operators. The production channels may in some cases determine the final-state particle configurations.

$$\pi + p + 537 \text{ MeV} = \pi\{\pi X\}p \Rightarrow \Lambda + \bar{K}$$
$$\pi + p + 614 \text{ MeV} = \pi\{\mu\bar{\mu}X\}p \Rightarrow \Sigma + \bar{K}$$
$$K + p + 380 \text{ MeV} = K\{X\}p \Rightarrow \Xi + \bar{K}$$
$$K + p + 1230 \text{ MeV} = K\{XXX\}p \Rightarrow \Omega + \bar{K} + \bar{K}$$

πX excitation unit that breaks apart in the formation of the $\Lambda + \bar{K}$. The $\pi + p \to \Sigma + \bar{K}$ channel has a Q value of -614 Mev, which requires a $\mu\mu X$ excitation quantum. The $K + p \to \Xi + \bar{K}$ channel Q value of -380 MeV requires an X quantum. And the $K + p \to \Omega + \bar{K} + \bar{K}$ channel Q value of -1230 MeV requires an XXX excitation. Thus the quantum X also plays a role, at least phenomenologically, in these associated production channels, and the form of the production excitation may in some cases dictate the mass of the final excited-state particle.

The final piece of information about these hyperon excitations comes from the charmed hyperon resonances. Figure 3.23.3 shows the $\Lambda_c = qqc$, $\Xi_c = qsc$ and $\Omega_c = ssc$ charm baryons plotted on a 210 MeV mass grid which is centered on the Λ_c. As can be seen, these three charm states exhibit the 210 MeV mass intervals that are expected from their constituent-quark mass values. Thus when freed from the constraint of the Pauli exclusion principle (or possibly of the Ω excitation channel shown in Table 3.23.1), the mass of the Ω_c occurs right where we would expect it to be relative to the masses of the Λ_c and Ξ_c. It should be noted here that in the present α-quantized constituent-quark model, the difference in mass between an s quark and a u or d quark is 210 MeV, and this is the characteristic mass difference we observe when dealing with Λ, Ξ, Λ_c, Ξ_c and Ω_c hyperons, whereas in the Standard Model, the mass difference between an s quark and a u or d quark is about 100 Mev, which bears no relationship to the observed hyperon mass differences.

On the global mass scale of Fig. 3.23.1, the charmed D mesons and bottom B mesons are reproduced fairly accurately by the zero-binding-energy quarks masses of Eq. (3.23.5). However, when studied in more detail, they exhibit complexities. The calculated mass of the $D = c\bar{q}$ meson is 1.3% too high, which is in rough agreement with the binding energies displayed in Table 3.15.1 and Figs. 3.15.1 and 3.15.2. However the $(D_s = c\bar{s}) - (D = c\bar{q})$

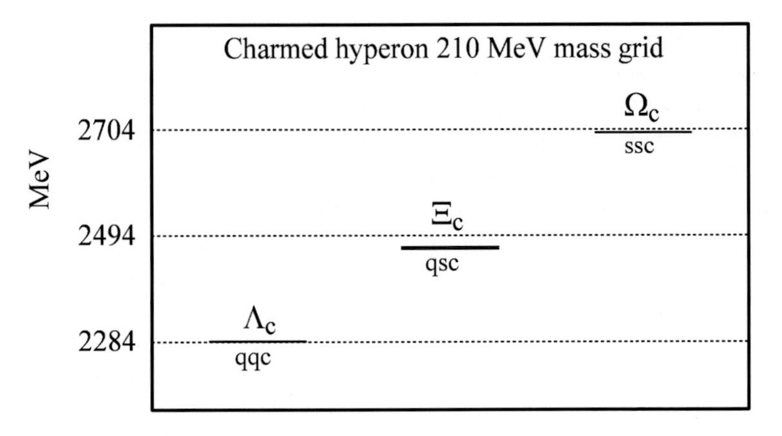

Fig. 3.23.3 Experimental mass values for the Λ_c, Ξ_c and Ω_c bayrons, plotted on a 210 MeV mass grid. These charmed baryons follow the expected 210 MeV interval rule. The 137 MeV mass anomaly in the Ω mass that was displayed in Fig. 3.23.2 has disappeared. The Pauli constraint imposed by having three identical particles in the Ω has also vanished.

experimental mass difference of 101 Mev does not correspond to the intrinsic $s - q$ quark mass difference of 210 MeV. The calculated masses of the $B = b\bar{q}$ and $B_s = b\bar{s}$ mesons are qualitatively correct, but are *below* the experimental masses, which is a difference that cannot be attributed to binding energy effects. On the other hand, the experimental mass value of the $B_c = b\bar{c}$ meson was recently lowered from 6400 to 6286 MeV [10]. It now has a calculated mass that is 17 MeV *above* the experimental mass, and its hadronic binding energy of 0.27% is in reasonable agreement with the trend exhibited by the $s\bar{s}$, $c\bar{c}$ and $b\bar{b}$ threshold states (Table 3.15.1 and Fig. 3.17.2). We can see from these results that using the SM quarks with the α-mass values assigned in Eq. (3.23.5) gives reasonable estimates for the experimental masses of these states, but not always to the 1% precision of the threshold states of Table 3.16.1. The low-mass M^X octet particles are the ones that exhibit a simple α-mass structure.

The pseudoscalar mesons, which were discussed in Secs. 3.7 and 3.10, are reproduced as combinations of spinless m_b boson mass quanta. The hyperons and charmed and bottom mesons displayed in Fig. 3.23.1 are reproduced as combinations of spin-1/2 m_f fermion mass quanta. There are also particles that are reproduced as mixed combinations of m_b and m_f quanta. The lowest-mass example is the $\omega(782)$ meson, as we now discuss. The thresholds for producing $s\bar{s}$, $c\bar{c}$ and $b\bar{b}$ quark pairs are the ϕ,

J/ψ and Υ vector mesons (Figs. 3.17.1 and 3.17.3), which are produced in the M_ϕ platform excitation tower discussed in Secs. 3.12 and 3.17. The generation of $u\bar{u}$ and $d\bar{d}$ quark pairs logically occurs in this same tower, but their required $\frac{1}{2}X$ quantization (Fig. 3.19.2) does not conserve initial-state particle–antiparticle symmetry in each channel of the tower, since each channel contains an odd number of m_b mass quanta (under the assumption that these spinless mass quanta represent the first step in the excitation). Thus no narrow 630 MeV $u\bar{u}$ or $d\bar{d}$ peak is observed. However, a "mixed" 770 MeV $u\bar{u}\pi$ or $d\bar{d}\pi$ excitation adds an m_b mass quantum to each channel, which makes it possible to conserve this symmetry, and the narrow-width $\omega(782) \to \pi\pi\pi$ meson appears at about this excitation energy. The D and D_s charmed quark mass excitations displayed in Figs. 3.22.2 and 3.22.3 represent other examples where α-quantized fermion and boson basis states are combined together.

The conclusion we draw from these analyses is that the various elementary particle quark ground states do have an α-quantized *mass* structure, but these particles have to be properly arrayed in order to bring out the quantitative aspects of this structure. In the same way, the α-quantized structure of threshold-state particle *lifetimes* only becomes apparent when these lifetimes are properly arrayed. The α-quantized lifetimes and masses are experimental regularities, whose reciprocal interpretation requires only quantum mechanics combined with the role played by the coupling constant α. The lifetime α-scaling and the mass α^{-1}-scaling when taken together form a very consistent dual phenomenology, and the explanation of these results probably resides in both sets of data.

3.24 The Short-Lived Excited-State Masses: Evidence from Excitation Clusters

In these Chapter 3 studies of elementary particle masses, we have focused on the lower-mass, long-lived, threshold-state particles that have lifetimes $\tau > 10^{-21}$ s. These particles logically have the simplest mass structures, which should facilitate their interpretation. The short-lived higher-mass particles occur as excited states of these threshold particles, and their excitations are more complex. This situation is roughly analogous to that of the lifetimes studied in Chapter 2. The longer-lived particles have lifetimes which are grouped into widely-spaced clusters, and their patterns can be singled out and identified. The shorter-lived particles, on the other hand, have essentially a continuum of closely-spaced lifetimes, and discerning their lifetime structure is a more difficult task. Determining their mass structure is also more difficult. In the present studies we do not attempt to reproduce the masses of these short-lived particles, but we make a few general observations as to the type of mass structure they seem to represent.

One of the seminal events in elementary particle physics was the 1974 "November revolution that turned the wheel" [45]. In November 1974 simultaneous announcements from Brookhaven (J) and SLAC (ψ) heralded the discovery of a sharp J/ψ resonance peak at 3100 MeV, which after some time was identified as a $c\bar{c}$ bound state. Prior to its discovery, all of the long-lived narrow-width particle states were at lower energies, and the high-energy regions were occupied by broad-width high-spin resonances. The J/ψ demonstrated that very stable particles ($\tau > 1$ zs) exist at 3 GeV, and the subsequent discovery of the Υ showed that even stabler particles exist at 9 GeV. These discoveries changed the way we think about these particles in a number of areas of particle theory. The area we are interested in here is the phenomenology of the excited-state masses. Prior to the November Revolution, it made a certain amount of sense to picture the high-mass, broad-width, high-spin resonances as nuclear-physics-type rotational levels that occur in bands above lower-mass, narrow-width, low-spin "ground states." These rotational bands featured an $\ell(\ell + 1)$ mass interval rule, where ℓ is the orbital angular momentum. Papers on this rotational-band approach are listed in Refs. [63–65]. Figure 3.24.1 shows a 1974 plot [66] of the N, Δ, Λ and Σ baryon and hyperon rotational bands.

After the November Revolution, with its high-energy narrow-width particle states, and after the completion of more accurate measurements at the higher energies, it became increasingly clear that the α-quantized excitation spectrum observed at the lower energies continues up to higher energies, although with more complexity in the excitations [67]. Rotational bands were not readily able to accommodate these new data, which seemed to indicate quark excitations of some kind all the way up the energy scale. This necessitated a fresh look at the spectrum of short-lived excited states.

In the present section we make a global examination of the mass values of the 157-particle lifetime data base of Appendix A, whose masses are listed in Appendix B. Figure 3.24.2 is a mass plot of the *nonstrange* and *strange* particles in Appendix B, and Fig. 3.24.3 is a similar plot of the *charmed* and *bottom* particles. The particle masses in Fig. 3.24.2 are spread out over the whole excitation range for each type of particle. However, they tend to be clustered together in groups rather than distributed uniformly.

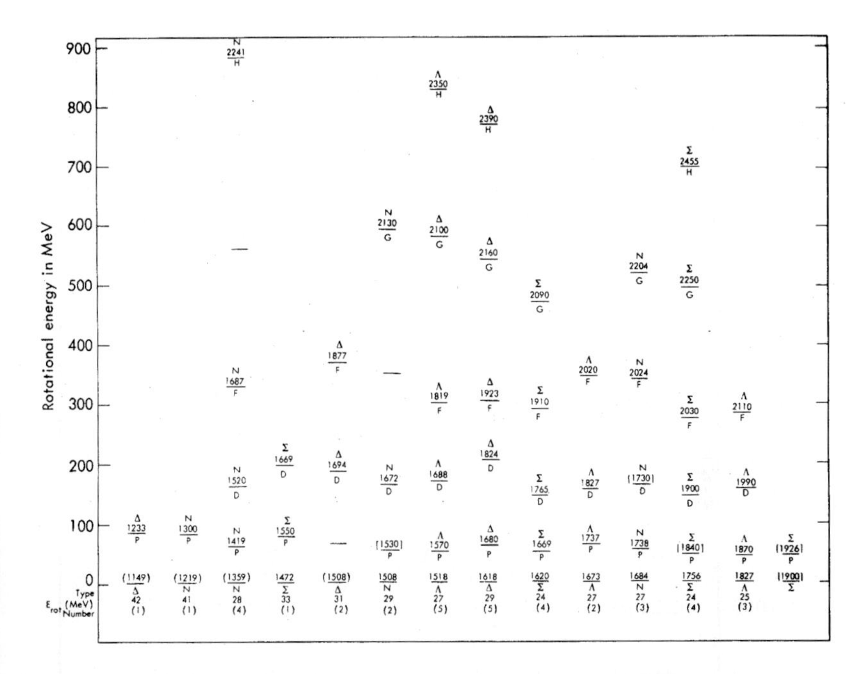

Fig. 3.24.1 A 1974 mass plot of N, Δ, Λ and Σ particles, arranged in $\ell(\ell+1)$-spaced rotational bands, where ℓ is the angular momentum of the state [66]. At that time, all of the higher-mass states were broad-width resonances with large values ℓ-values. When the very-narrow-width J/ψ meson subsequently appeared at a high mass value, it did not fit in with this nuclear-physics-type approach to hadron resonances. Hence the concept of particle rotational bands no longer seemed viable.

These mass clusters may provide significant information as to the nature of the excitations. In particular, the meson, kaon, N, Δ and Λ excitation bands each contain one dominant cluster. Table 3.24.1 lists the resonances contained in these clusters, and also a few lower-mass clusters that can be singled out. The dominant clusters in the N and Λ excitation bands each have an S-state resonance as the lowest-mass state, with a variety of randomly-occurring spin states at slightly higher masses. The important point about these clusters is that they contain a variety of states (six for N, five for Δ, and four for Λ) at essentially the same mass. This indicates that the cluster mass arises as an aggregation of smaller mass units, which can be arranged in different ways to produce various spin states and decay modes. These comments also apply to the dominant meson (seven states) and kaon (five states) clusters, thus indicating that the basic mass units

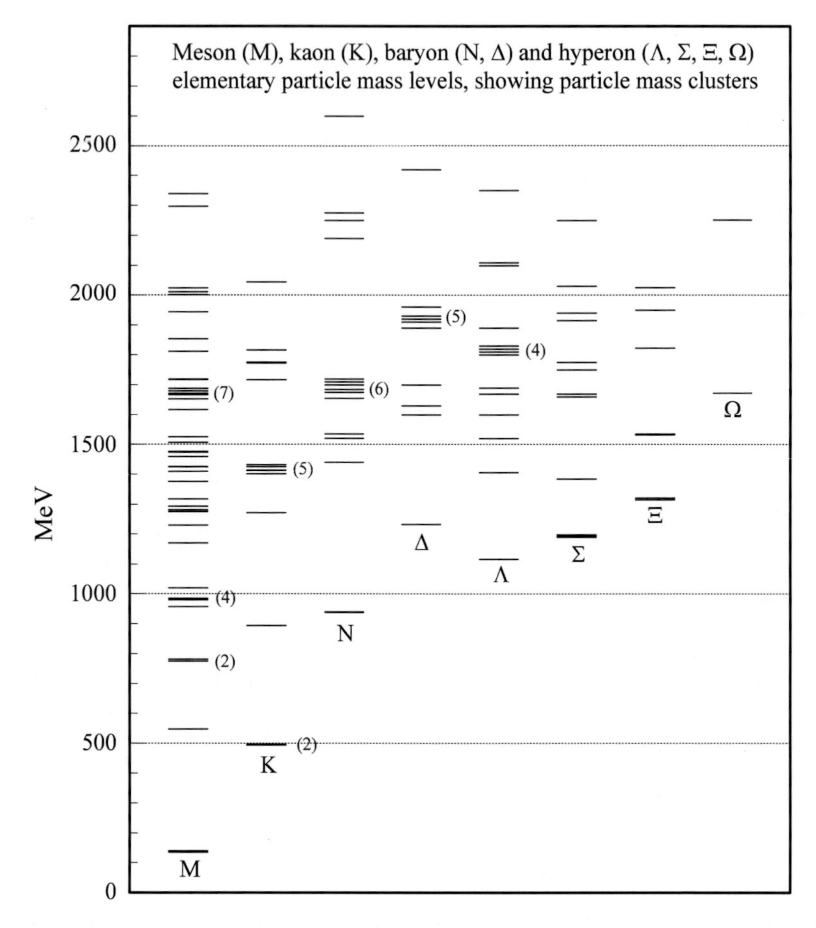

Fig. 3.24.2 A mass plot of all the well-measured meson, kaon, baryon and hyperon elementary particle states [10]. As can be seen, these states are spread out over a wide range of mass values. A clue as to the possible mass structure of these excitations is provided by the existence of relatively narrow-width clusters of masses, where several particles with different spin values occur at almost the same mass. Prominent mass clusters are indicated in Fig. 3.24.2 by numbers in parentheses that give the number of levels in a cluster. The particle states included in these clusters are listed in Table 3.24.1. As this table shows, a variety of spin states are contained in a single cluster. This demonstrates that the masses of the particles do not depend directly on the overall spin value, and it suggests that the cluster may be an aggregate of mass substates which can have different intrinsic spin values in different configurations. This indicates in turn that the mass excitation process at higher energies is a continuation of the mass excitations at lower energies, but with an increasing complexity of the aggregate mass.

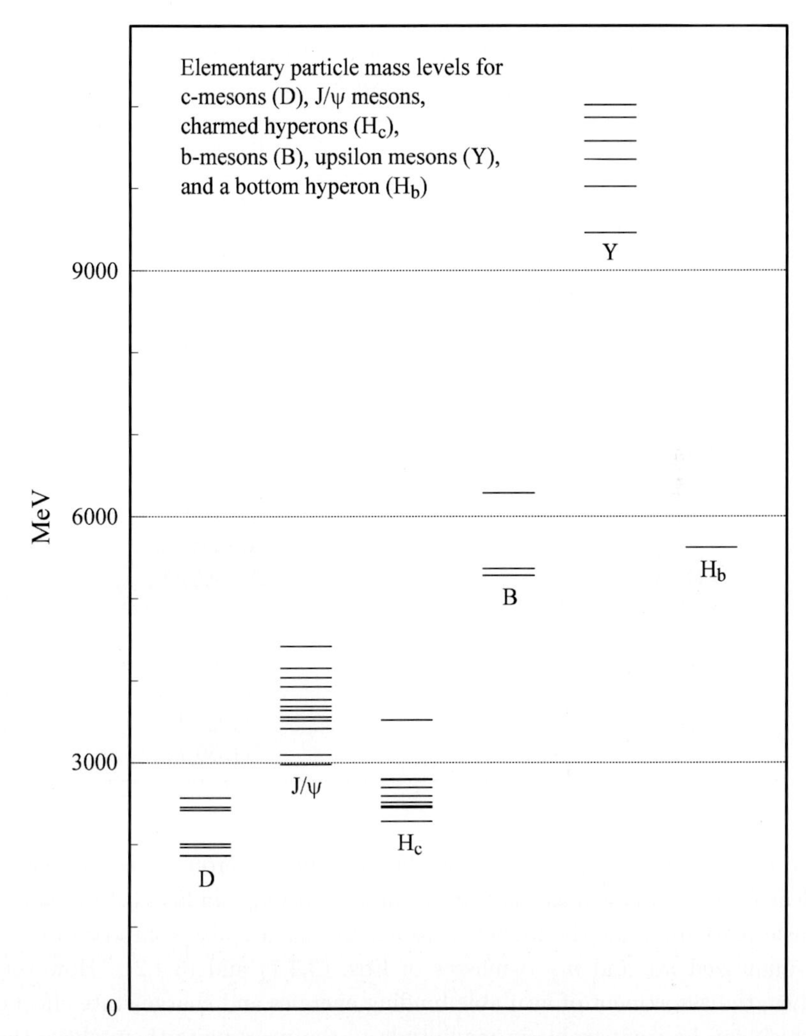

Fig. 3.24.3 A mass plot of all the well-measured charmed and bottom meson and baryon elementary particle states [10]. These states occur at higher masses than those shown in Fig. 3.24.2, but they spread out over a range of mass values in the same manner as the lower-mass states do, indicating that they represent the same basic types of mass excitations. An important result which follows from the masses shown in Fig. 3.24.3 is that the paired-quark $J/\psi = c\bar{c}$ and $\Upsilon = b\bar{b}$ particles appear at about twice the mass values that we deduce for the c and b quarks in the unpaired $D(c)$ and $B(b)$ particles. Hence these charmed and bottom quark states are clearly formed from constituent quarks, wherein the mass of the state is derived primarily from the intrinsic masses of the component quarks, and not from an accompanying gluon field.

Table 3.24.1 Dominant mass clusters in mesons, baryons and hyperons. These clusters are displayed graphically in Fig. 3.24.2.

Mesons M	J^P	Kaons K	J^P	Baryons N, Δ, Λ
$\rho(775.8)$	1^-	$K^\pm(493.7)$	0^-	$N(1650)S_{11}$
$\omega(782.7)$	1^-	$K^0(497.6)$	0^-	$N(1675)D_{15}$
				$N(1685)F_{15}$
				$N(1700)D_{13}$
$\eta'(957.8)$	0^-	$K_1(1402)$	1^+	$N(1710)P_{11}$
$f_0(980)$	0^+	$K_0^*(1414)$	0^+	$N(1720)P_{13}$
$a_0(984.7)$	0^+	$K^*(1414)$	1^-	
$\phi(1019.5)$	1^-	$K_2^{*\pm}(1425.6)$	2^+	
		$K_2^{*0}(1432.4)$	2^+	
				$\Delta(1905)F_{35}$
$\omega_3(1667)$	3^-			$\Delta(1910)P_{31}$
$\omega(1670)$	1^-			$\Delta(1920)P_{33}$
$\pi_2(1672.4)$	2^-			$\Delta(1930)F_{37}$
$\phi(1680)$	1^-			$\Delta(1950)D_{35}$
$\rho_3(1688.8)$	3^-			
$f_0(1718)$	0^+			
$\rho(1720)$	1^-			$\Lambda(1800)S_{01}$
				$\Lambda(1810)P_{01}$
				$\Lambda(1820)F_{05}$
				$\Lambda(1830)D_{05}$

are common to all of these meson, baryon and hyperon states. It seems clear from the lower-mass excitations described (e.g.) in Sec. 3.16 that the basic mass units for the higher-mass excitations are also composed of the α-quantized m_b and m_f α-masses of Eqs. (3.4.1) and (3.4.2). However, given the assortment of available binding energies and charge-state effects, which can be comparable in magnitude to the mass units themselves, the mass α-quantization does not stand out so clearly at the higher energies.

A slightly different way of portraying the nonstrange and strange particle masses is shown in Fig. 3.24.4, which is taken from a 1990 publication [67]. This is a plot of the well-established meson, kaon, N, Δ, Λ and Σ resonances listed in RPP1988 [68]. However, instead of plotting all of the masses on an absolute mass scale, as in Fig. 3.24.2, the N and Δ resonances are plotted as excitation energies above a 939 MeV "ground state," and the Λ and Σ resonances are similarly plotted above 1116 and 1193 MeV "ground states," respectively. The symbols represent the masses, and the

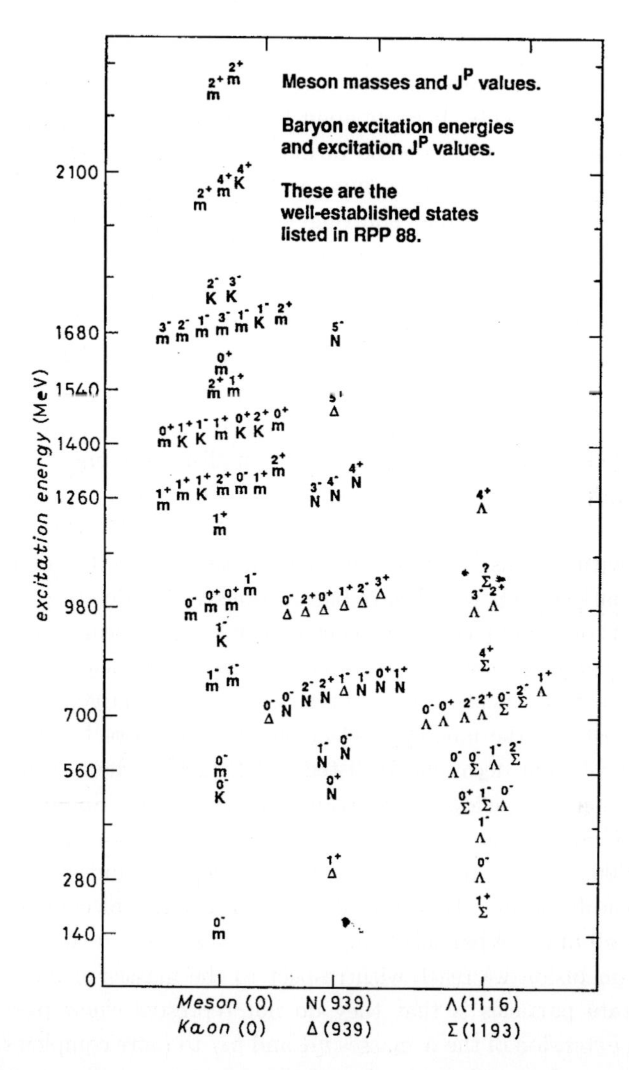

Fig. 3.24.4 A 1990 mass plot of meson, baryon and hyperon elementary particle states [67], with the ground-state excitation for each family type set at zero on the mass scale. Thus these mass levels represent excitation energies above the basic "ground states." As can be seen, these levels fall into mass clusters similar to those portrayed in Fig. 3.24.2, but with mesons and kaons, N and Δ baryons, and Λ and Σ hyperons intermingled within each of these three types. As can be seen, each cluster contains a variety of spins and parities. Also, the clusters follow a mass quantization pattern that is common to all of these ground-state excitations, and which seems correlated with the 140-MeV mass grid shown along the abscissa of the plot. This systematics suggests that α-quantized mass aggregates constitute the structural form of the excitations.

spin-parity assignments are shown above the symbols. As can be seen, these particle excitations fall into clusters that each contain a variety of spin-parity assignments. It is apparent from these clusters that the masses are only mildly dependent on the various spin values, so that the spin values really do represent intrinsic spins and not rotational excitations, which would give much larger mass splittings (Fig. 3.24.1).

The charm and bottom masses displayed in Fig. 3.24.3 are few in number, and do not present evidence of dominant clusters. However, they extend the mass systematics of Fig. 3.24.2 in two significant ways: (a) they extend the narrow-width masses to much higher energies, thus showing that large and stable "mass aggregates" are possible — which is a result that was not anticipated theoretically; (b) they make it clear that what we have here is indeed a constituent-quark mass formalism: the masses of both the paired-quark and unpaired-quark states are clearly dominated by the intrinsic masses of the c and b quarks. If the field of particle physics had been initiated with the discovery of these high-mass quarks, it is doubtful that current-quark models would have been seriously considered.

There is one final piece of evidence we can bring to bear on this problem of particle masses: namely, the decay modes of particles such as the τ lepton. The τ is logically a structureless, integrally charged, non-hadronic particle (just like the muon and electron, the other two leptons), and yet it has 203 different decay modes listed in RPP2006. The τ seems to decay into every conceivable lower-mass state that is available. Similarly, the J/ψ, $B^{\pm}$ and B^{0} mesons have 150, 341 and 306 decay modes, respectively, listed in RPP2006. Thus transformations between leptons and mesons of every variety readily occur. This indicates a commonality of combinable basis states for all of these particles.

The conclusion we reach with respect to the masses of the short-lived excited-state particles is that they do not represent "new physics," but rather an extension of the α-masses m_b and m_f to more complex structures at higher energies, with a blurring of the sharpness of the α-quantization due to binding energy and charge splitting effects.

3.25 Muon (Fermion) Masses and Quarks; Pion (Boson) Masses and Generic Quarks; Superheavy Muon α-Quark Masses

The α-quantization of the long-lived, low-mass, threshold-state elementary particles has led to the identification of two fundamental α-mass excitation quanta — the spin-0 boson mass $m_b = 70$ MeV and spin-1/2 fermion mass $m_f = 105$ MeV (Eqs. (3.4.1) and (3.4.2) — which serve as the structural elements for bosons and fermions up to 11 GeV. These two mass quanta are mathematically related to one another via the relativistically spinning sphere model described in Chapter 4. The spin-1/2 fermion mass quantum m_f is a familiar concept in particle physics, although not with the universality shown for it here, since its mass is almost identical to that of the muon. The muon itself is the lightest known fermion above the electron ground state. Thus it seems appropriate to denote the fermion entities that are composed of m_f mass quanta as "muon-mass states." In particular, we can place the SM $q \equiv (u, d)$, s, c and b quarks, which are composed of 3, 5, 15 and 45 m_f subquanta, respectively, in the category of "muon quarks." The muon-mass states also include the "muon triad" μ, p and τ, which are composed of 1, 9 and 17 m_f subquanta, respectively.

The spin-0 boson mass quantum m_b, on the other hand, represents a new type of elementary particle excitation that is not included in the paradigm of the Standard Model. Its most prominent manifestation is in the pseudoscalar meson nonet π, η, η', K, where it accurately reproduces the masses of these states. But it also appears in hybrid combinations with fermions, where it represents bosonic excitations of fermion "platform states." The lowest-mass boson-fermion hybrid is the $\omega = q\bar{q}\pi$ meson. Other hybrid examples are displayed in the D and D_s charmed-quark excitations shown in Fig. 3.22.3, where the $D = c\bar{q}$ and $D_s = c\bar{s}$ fermion quark excitations are the lowest-mass particles containing these quark combinations, and thus function as "quark platform states" for the excitation of higher-mass and shorter-lived particles. The spin-0 mass quantum m_b, in contrast to the spin-1/2 mass quantum m_f, is not manifested as an observed particle. (It can be observed on a license plate at the web site "70mev.org", and also in Figs. P1–P3 in the postscript to this book.) The lowest-mass boson state (above the spin-0 electron–positron boson ground state) is the $\pi = m_b\bar{m}_b$ pion. Thus it is descriptive to denote these boson entities as "pion-mass states," and to denote their quark-like substates as "pion quarks." In the present work we deal only with the particle–antiparticle masses of these quark-like substates, and not with their charge states, so they are more appropriately described as "generic pion quarks." With this understanding, we will simply refer to them as pion quarks. The relevance of the designations "muon quarks" and "pion quarks" will become more apparent in Sec. 3.26, where we introduce the concepts of "muon strangeness" and "pion strangeness."

A third fundamenta α-mass, the muon mass quantum $m_f^\alpha = m_f/\alpha =$ 14,394 MeV, makes its appearance in the massive W and Z bosons and the top quark t, as we discuss in Secs. 0.10 and 3.20. It represents a second α-leap upward from the "first-α-leap" m_f mass quantum, and it gives rise to the $q^\alpha \equiv (u^\alpha, d^\alpha) = (u/\alpha, d/\alpha) = 3m_f^\alpha =$ 43,182 MeV set of "α-quarks". The quark configuration $\overline{WZ} = q^\alpha \bar{q}^\alpha$ accurately reproduces the average mass of the $W^\pm$, Z^0 isotopic spin doublet (Eq. (3.20.6)) and also its three charge states, and the quark mass assignment $t = 4q^\alpha$ (Fig. 3.20.2) gives a precision fit to the mass of the top quark (Eq. (3.20.8)).

In the present section we formalize the notation for "muon masses" and "pion masses," and we recast the muon and pion platform excitations in a slightly different form than displayed in previous sections of Chapter 3. The fundamental "muon" and "pion" mass quanta m_f and m_b were defined in Eqs. (3.4.1) and (3.4.2), and are

$$m_f = 3m_e/2\alpha\,, \quad m_b = m_e/\alpha \quad (m_e = \text{electron mass})\,, \tag{3.25.1}$$

with matching $\bar{m}_f$ and $\bar{m}_b$ antimuon and antipion quanta. The composite "muon masses" (particles and muon quarks) and "pion masses" (pion quarks) are

$$\mu_n \equiv m_e + n m_f\,, \quad \pi_n \equiv m_e + n m_b\,. \tag{3.25.2}$$

The observed M_μ platform-excitation "muon-mass" states (which are produced in particle–antiparticle pairs) are

$$\begin{aligned} &\mu_1 = \mu\,, \quad \mu_9 = p\,, \quad \mu_{17} = \tau \quad (\text{particles})\,, \\ &\mu_3 = (q \equiv u, d)\,, \quad \mu_5 = s \quad (\text{quarks})\,, \end{aligned} \tag{3.25.3}$$

where the muon itself represents the muon platform state M_μ (Secs. 3.8 and 3.11). The M_μ platform excitation unit that we employ here for these directly-excited fermion states is the excitation quantum Q_f, which we characterized notationally as

$$Q_f = 2m_f \simeq \mu\bar{\mu} \simeq 210 \text{ MeV}\,, \tag{3.25.4}$$

where the particle–antiparticle composition of this mass quantum is not clearly specified. The quantum Q_f is one-half the size of the excitation quantum $X \simeq 420$ MeV that is featured in the M^X generation process. The muon-mass states of Eq. (3.25.3) in units of Q_f are

$$(\mu_1, \mu_3, \mu_5, \mu_9, \mu_{17}) = \mu + (0, 1, 2, 4, 8)Q_f\,, \tag{3.25.5}$$

as displayed in Figs. 0.2.4 and 3.25.1. These figures show how the $Q_f \simeq \mu\bar{\mu}$ excitation quanta successively double for each occupied level of the $M_\mu \simeq \mu$ excitation column in the $M_{\mu\mu}$ muon-mass excitation tower.

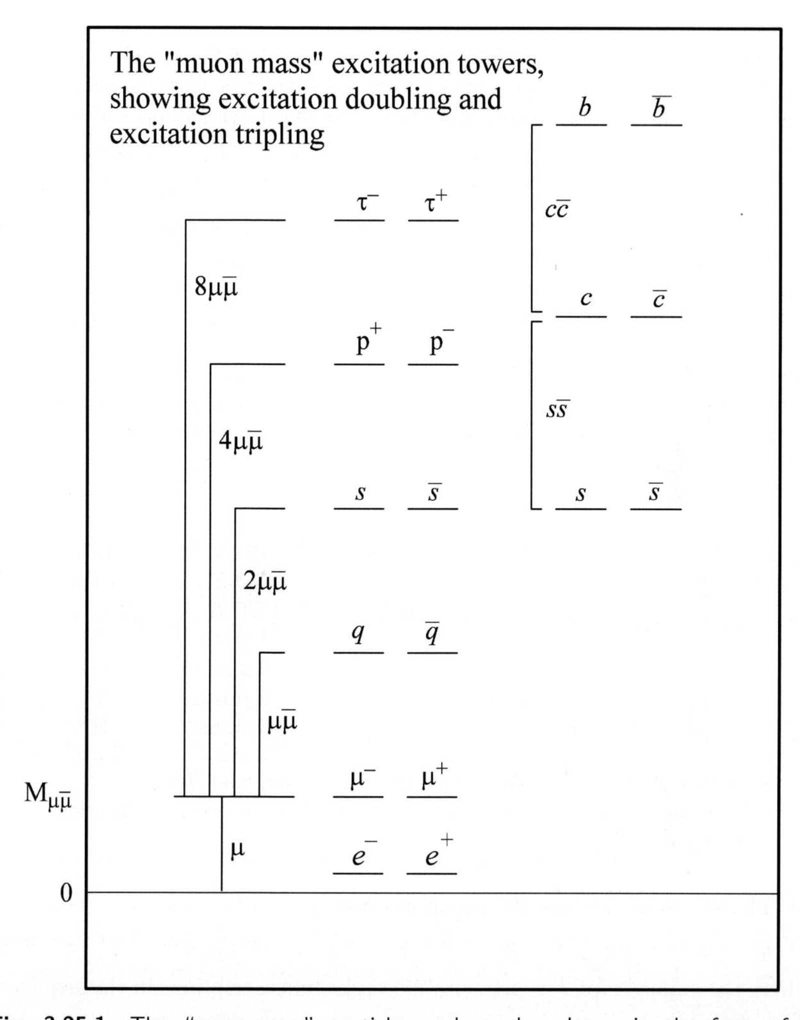

Fig. 3.25.1 The "muon-mass" particles and quarks, shown in the form of a schematic "excitation-doubling" tower built on the $M_{\mu\mu}$ platform state of Sec. 3.8 and featuring the excitation quantum $Q_f \simeq \mu\bar{\mu}$ of Eq. (3.25.4), plus an "excitation-tripling" tower constructed on the $\phi = s\bar{s}$ vector meson platform state of Sec. 3.12. The mass values for these muon-mass excitations are displayed in Table 3.25.1. This muon-state generation process encompasses the three leptons, the proton, and the SM quarks $q \equiv (u, d)$, s, c and b. The SM quarks in turn generate the 19 quark platform-state particles listed in Table 3.25.2. These muon-mass basis states accurately reproduce ($\sim$3%) the masses of the 21 particles listed in Tables 3.25.1 and 3.25.2 without using any adjustable parameters or binding energy corrections. They cannot reproduce the pseudoscalar meson nonet particles, which require the "pion-mass" basis states shown in Fig. 3.25.2.

Table 3.25.1 Muon-mass calculated (Eqs. (3.25.1)–(3.25.7)) and experimental [10] values in MeV for platform-state elementary particles, using no adjustable parameters or binding energy corrections. The muon-mass states with superscripts T represent mass triplings of the μ_5 quark state. The average calculated mass accuracy for the μ, p, τ muon-mass triad is 0.48%. The quoted mass accuracies for the q, s, c, b muon quarks were obtained by averaging over the accuracies of the quark platform-states calculations in which they appear in Table 3.25.2. The completeness and accuracy of this quark platform-state mapping (see Fig. 3.23.1) verifies the usefulness of α-quantized muon constituent-quark masses in reproducing these threshold-state particles. The improved accuracy of the quark calculations with increasing mass reveals their constituent-quark nature. The large errors shown for the s-quark states arise from the fact that there probably are two s quarks, muon and pion, with different mass values, as discussed in Sec. 3.26.

Muon mass	Particle state	Calc. mass	Exp. mass	Error
μ_1	μ	105.549	105.658	0.10%
μ_9	p	945.85*	938.27	0.81%*
μ_{17}	τ	1786.16	1776.99	0.52%
μ_3	$q \equiv (u, d)$	315.63	11 states**	3.73%**
μ_5	s	525.70	9 states**	3.66%**
μ_5^T	c	1576.08	9 states**	3.22%**
μ_5^{TT}	b	4727.22	5 states**	2.34%**

*Only one electron mass included ($p = 3\mu_3$ would have $3m_e$).
**Average of the listed errors for the quark platform-state combinations displayed in Table 3.25.2.

The *muon quarks* are the muon masses

$$q \equiv (u, d) = \mu_3\,, \quad s = \mu_5\,, \quad c = \mu_{15} \cong s^T\,, \quad b = \mu_{45} \cong c^T\,, \qquad (3.25.6)$$

where the superscript T represents successive mass-triplings in the sequence (Fig. 3.19.1)

$$c = s^T\,, \quad b = s^{TT}\,, \quad \overline{WZ} = (b\bar{b})^{TT}\,. \qquad (3.25.7)$$

Table 3.25.1 lists the calculated muon mass values for the particle and muon quark states of Eqs. (3.25.5) and (3.25.6), together with the experimental mass values for the particle states. The μ, p and τ particles are reproduced to an average mass accuracy of 0.48%, in a calculation that features no adjustable parameters and no binding energy corrections. The quoted errors in the q, s, c, b quark-mass calculations in Table 3.25.1 were obtained from the mass fits displayed in Table 3.25.2, averaged over the number of particles that contain each particular quark.

Table 3.25.2 Muon-mass calculated (Table 3.25.1) and experimental [10] values in MeV for the 16 measured q, s, c and b unpaired-muon-quark platform-state combinations of Fig. 3.22.1, plus the paired $s\bar{s}$, $c\bar{c}$, and $b\bar{b}$, quark threshold states, using no adjustable parameters or binding energy corrections. The average mass accuracy (relative to the experimental value) for these 19 fundamental muon-quark configurations is 3.24%. The significant result here is not the numerical accuracy of these uncorrected mass calculations, but rather the fact that the comprehensiveness of this quark mapping of the various quark platform states, when combined with the overall level of accuracy, furnishes a powerful argument in favor of the use of constituent-quark masses in elementary particle theories.

Quark state	Particle state	Calc. mass	Exp. mass	Error
$s\bar{s}$	ϕ	1051.40	1019.46	3.13%
$c\bar{c}$	J/ψ	3152.16	3096.92	1.78%
$b\bar{b}$	Υ	9454.45	9460.30	0.06%
qc	D	1891.71	1866.9	1.33%
qb	B	5042.85	5279.2	4.48%
sc	D_s	2101.78	1968.2	6.79%
sb	B_s	5252.92	5367.5	2.14%
cb	B_c	6303.30	6286.0	0.28%
qqq	p	946.88*	938.27	0.92%*
qqs	Λ	1155.93	1115.68	3.61%
qqs	Σ	1155.93	1193.15	3.12%
qss	Ξ	1366.01	1318.07	3.64%
sss	Ω	1576.08	1672.45	5.76%
qqc	Λ_c	2206.31	2286.46	3.51%
qqc	Σ_c	2208.36	2453.89	10.01%
qsc	Ξ_c	2416.39	2469.45	2.14%
qcc	Ξ_{cc}	3466.77	3518.9	1.48%
ssc	Ω_c	2626.46	2697.5	2.63%
qqb	Λ_b	5357.45	5624.0	4.74%

*Three electron masses included.

Table 3.25.2 lists the calculated and experimental mass values for the paired-quark $s\bar{s}$ $c\bar{c}$ and $b\bar{b}$ threshold-states plus the 16 platform states formed from q, s, c, and b unpaired-quark combinations. This is a total of 19 quark-combination platform states, which are reproduced using the four calculated quark masses shown in Table 3.25.1. No adjustable parameters or binding energy corrections are applied. These four constituent-quark masses reproduce the 19 quark platform-state masses to an average accuracy of 3.24%. This level of accuracy, when combined with the completeness of the mapping (all observed quark platform-state combinations are included), makes a compelling argument in favor of the use of constituent-quark masses for this set of muon quarks.

The *pion quarks* are the pion masses (Eq. (3.25.2))

$$\pi_1 , \quad \pi_4 , \quad \pi_7 . \tag{3.25.8}$$

(In Sec. 0.7 these three generic pion quarks were denoted as q_π, q_η and q_k, respectively.) The pion platform state M_π (Sec. 3.7) is the hadronically bound pion-quark pair

$$\pi \equiv \pi_1 \bar{\pi}_1 . \tag{3.25.9}$$

The M_π platform excitation quantum that reproduces the pion quarks of Eq. (3.25.8) is

$$Q_b = 3m_b \simeq 210 \text{ MeV} . \tag{3.25.10}$$

This Q_b pion excitation unit (like the Q_f muon excitation unit for the platform M_μ) is one-half the size of the X quantum used in M^X particle generation. The Q_b excitation quanta occur in an excitation-doubling sequence above the M_K platform (Fig. 0.2.12) that closely mirrors the Q_f excitation-doubling sequence above the M_μ platform (Fig. 0.2.04). The M_π pion-mass generic quarks in units of Q_b are

$$(\pi_1, \pi_4, \pi_7) = m_e + m_b + (0, 1, 2)Q_b , \tag{3.25.11}$$

as displayed in Fig. 3.25.2. The observed pion-mass particles are

$$\pi = \pi_1 \bar{\pi}_1 , \quad \eta = \pi_4 \bar{\pi}_4 , \quad \eta' = \pi_7 \bar{\pi}_7 , \quad K = \pi_7 , \quad \bar{K} = \bar{\pi}_7 , \tag{3.25.12}$$

which are the (isotopic-spin-averaged) members of the pseudoscalar meson nonet. The calculated and experimental mass values for these particles are listed in Table 3.25.3.

There are two main objectives to be attained from the discussion in the present section. The first is to establish (again) the fact that the experimental elementary particle masses exhibit a constituent-quark substructure. To accomplish this result, we calculated two sets of α-quantized constituent quark masses — fermion masses and boson masses — using no adjustable parameters, and then combined each of these sets in various combinations so as to reproduce the observed spectrum of long-lived threshold-state particles. For clarity, we did not employ binding energy corrections. As demonstrated in Tables 3.25.1–3.25.3, the masses of 25

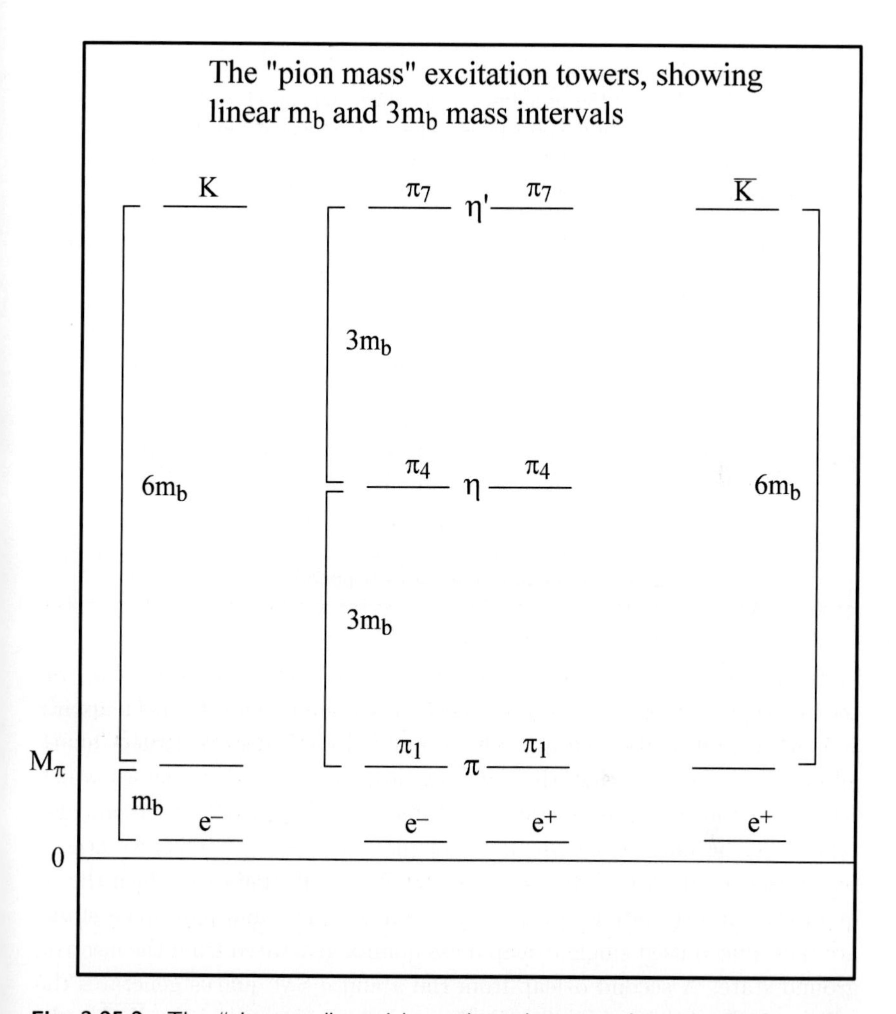

Fig. 3.25.2 The "pion-mass" particles and quark states, shown in the form of a linear excitation tower built on the M_π tower of Sec. 3.7. The mass values for these states are displayed in Table 3.25.3. The π_1, π_4 and π_7 states shown in Fig. 3.25.2 (Eq. (3.25.8)) are "generic pion quarks," in the sense that we deal here only with their mass values and do not assign charges to them. (In Figs. 0.2.12 and 0.2.12 they are labeled as q_π, q_η and q_K quarks, respectively.) The SM quark assignments for the PS mesons do an accurate job of reproducing the isotopic spin states, but have difficulties with the masses. And the experimental PS *nonet* does not fit comfortably into the SU(3) assignment of an *octet* plus an essentially unrelated *singlet*. Experimentally, the η and η' mesons appear in the PS nonet on an equal footing with respect to masses, lifetimes and decay modes (Sec. 3.21), which seems to belie the asymmetric SM assignments for them.

Table 3.25.3 Pion-mass calculated (Eqs. (3.25.1)–(3.25.12)) and experimental [10] values in MeV for the pseudoscalar meson nonet, using no adjustable parameters or binding energy corrections. The average calculated mass accuracy for these four states is 2.19%. Most of this error can be attributed to the fact that we do not apply hadronic binding energies for the paired π, η and η' pion-mass states, as indicated in Table 3.15.1. The main point we want to bring out here is that these α-quantized pion masses do a good job of reproducing the pseudoscalar meson nonet, which cannot be accomplished with the muon masses of Tables 3.25.1 and 3.25.2.

Pion mass	Particle state	Calc. mass	Exp. mass	Error
$\pi_1 + \bar{\pi}_1$	π	141.07	137.27*	2.77%
$\pi_4 + \bar{\pi}_4$	η	561.23	547.51	2.51%
$\pi_7 + \bar{\pi}_7$	η'	981.38	957.78	2.46%
π_7	K	490.69	495.66*	1.00%

*Mass averaged over isotopic spin states (Appendix B).

individual particles and isopic spin multiplets were reproduced by this proceedure to an average accuracy of 2.84%. The completeness of this mapping is as important as its accuracy. One ramification of this constituent-quark approach is the fact that the constituent-quark q, s, c, b masses which reproduce the muon-mass states of Tables 3.25.1 and 3.25.2 cannot also reproduce the lower-mass pseudoscalar pion-mass states. Thus we need a second set of quark-like states to account for the PS mesons, which the α-quantization fortunately provides. These muon-mass and pion-mass states are reproduced used single-α-leap mass quanta generated from the electron ground state. A second α-leap from the u and d SM quarks generates the α-enhanced supermassive u^α and d^α quarks that reproduce the $\overline{WZ}$ and t quark masses. Thus the α-quantized constituent-quark formalism extends over the entire range of elementary particle masses, with no need for employing adjustable parameters.

The second objective in this section is largely one of semantics. We use the Rosetta-stone spin-0 pion and spin-1/2 muon particle states as the basis for dividing the threshold-state elementary particles into the dichotomy of *fermion* "muon states" and *boson* "pion states," with "muon quarks" to reproduce the former and "pion quarks" to reproduce the latter. As soon as we leave the lowest-mass threshold-state or platform-state masses and move on to higher excitations, we encounter "hybrid" states that have both fermion and boson components, but at the lowest energies this dichotomy is quite accurate. One point which follows from this analysis is that in the Standard Model the attempt is made to use just the muon quarks to reproduce all of the particle states, including the pseudoscalar pions. This approach in fact works remarkably well for isotopic spins, but it leads to an impasse when applied to masses.

The concept of dividing the elementary particle ground states into muon states and pion states is seemingly just a labeling convention. However, it acquires more significance when we confront the problems raised by the strangeness quantum number s, wherein two strangeness modes — muon and pion — seem to be required, each with a different constituent-quark mass value. This is the topic of Sec. 3.26.

3.26 Evidence for the $s^* = 595$ MeV Strange Quark Excited State

In the last section we formally labeled the spin-1/2 fermion-mass states and spin-0 boson-mass states as constituting two separate types of particle excitations — *muon* masses and *pion* masses. The fundamental muon platform-state masses are multiples of the fermion mass $m_f \cong m_\mu$, and the fundamental pion states (the PS pions) are multiples of the (unobserved) boson mass m_b, which first appears as $\pi = m_b \bar{m}_b$. The SM quarks u, d, s, c, b are muon quarks. In the present α-quantized mass formalism, these quarks occur with constituent-quark mass values that accurately reproduce the 21 basic muon platform states listed in Tables 3.25.1 and 3.25.2. Two usefuls aspect of the SM quarks are that they provide the correct isotopic spin quantum numbers for the particles they reproduce, and they carry quark "flavor" quantum numbers that are conserved in strong interactions. These SM quarks, on the other hand, do not reproduce the mass values of the PS mesons, which require the "generic pion quarks" of Eq. (3.25.8), as displayed in Table 3.25.3. The pion quarks are beyond the paradigm of SM physics.

The identification of two classes of elementary particle quarks — muon quarks and pion quarks — raises an interesting question with regard to these quarks:

Are the muon quarks and pion quarks related to one another?

The long-lived low-mass particle states are composed either of all muon quarks or all pion quarks, with hybrid combinations generally occurring in the short-lived excited states. Let us focus on the long-lived states in answering this question. The u, d, s, c, b constituent-quark masses are not appropriate for the PS mesons, so the muon quarks and pion quarks are not strongly-linked by their mass values. The masses of the c and b quarks are so large that the *charm* and *bottom* quark flavors do not play a role in the PS meson system. However, the strange quark mass s does fall inside of the PS meson domain, and the *strange* quark flavor carried by the quark s plays an important PS role: it is also carried by the K mesons. Furthermore, the strangeness carried by the kaons is the same strangeness that is carried (e.g.) in the Λ hyperon. In the Associated Production reaction $\pi^- + p \rightarrow \Lambda + K^0$ [54], the π^- and proton, each with strangeness $S = 0$, interact to form a Λ that has $S = -1$ and a kaon that has $S = +1$, thus conserving overall strangeness. Hence strangeness is a common attribute of both the muon-mass and pion-mass systems. Phenomenologically, the strangeness quantum number S seems to be clearly linked to the particle–antiparticle symmetry or asymmetry of the mass substates that compose the particle. Thus if strangeness is common to both the muon masses and pion masses, it indicates that the muon-mass and pion-mass particle–antiparticle symmetries must be related, which in turn indicates that the muon and pion masses m_f and m_b are related. This relationship is established in Chapter 4 via the relativistically spinning sphere model (Sec. 4.3), which reveals that m_f and m_b are spin-1/2 and spin-0 modes of the same fundamental mass quantum.

One of the most interesting aspects of phenomenology is to establish relationships between two types of phenomena which at first glance seem to be unrelated. One famous example occurred with Maxwell's unification of light and electromagnetism. We have just established another, but much more modest, example here in the elementary particle domain:

> *The process of Associated Production, when combined with the systematics of mass α-quantization, demonstrates that the strangeness quantum number S applies to both the spin-1/2 mass quanta in the hyperons and the spin-0 mass quanta in the kaon, and thus mandates the existence of a link (the relativistically spinning sphere) between these quanta.*

The occurrence of "strangeness" in both the spin-1/2 muon masses and spin-0 pion masses enables us to phenomenologically resolve a long-standing problem in elementary particle physics. This problem appears most clearly

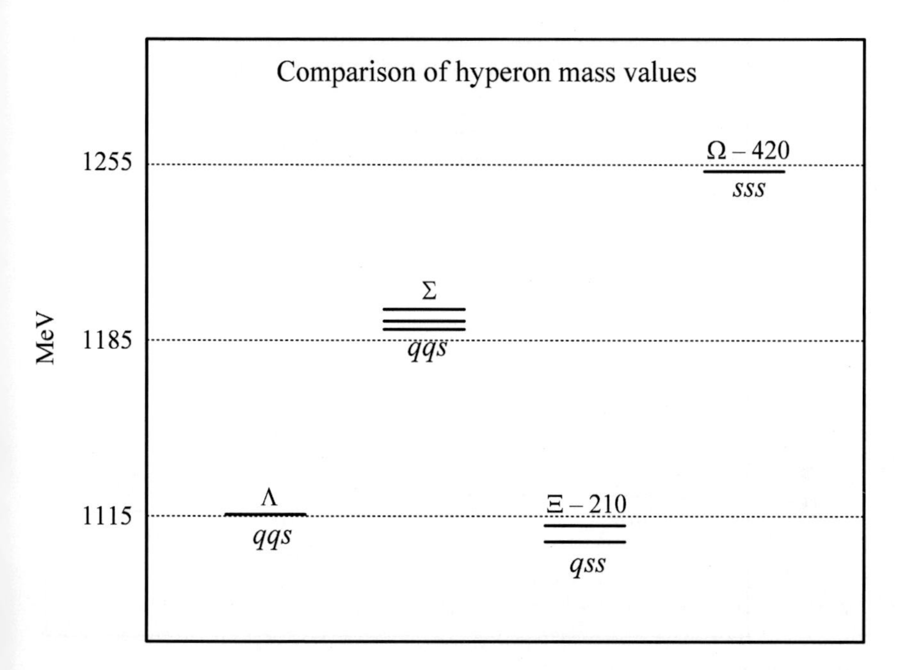

Fig. 3.26.1 The platform-state Λ, Σ, Ξ and Ω hyperon masses, shown with a 70 MeV mass grid anchored on the Λ mass. The Λ and Σ levels are at the experimental mass values, the Ξ levels are shifted downward by 210 MeV, and the Ω level is shifted down by 420 MeV. If these hyperon masses all reflect the basic $q = 315$ and $s = 525$ MeV constituent-quark mass values, then the unshifted and down-shifted mass levels should appear at the same mass value. As can be seen, the Σ masses actually appear 70 MeV higher, and the Ω mass appears 140 MeV higher. The $\Lambda - \Sigma$ mass splitting suggests the existence of two different strange quarks s and s^* with different masses, as discussed in the text. The high value for the Ω may also follow for this same reason, with constraints on the s quarks in the Ω imposed by the exclusion principle (which does not allow three fermions in an S-state), and/or by the nature of the Ω production process (Table 3.23.2).

in the hyperon ground states $\Lambda = qqs$, $\Sigma = qqs$, $\Xi = qss$ and $\Omega = sss$. The constituent-quark masses for these states are $q \equiv (u, d) = 315$ MeV and $s = 525$ MeV. Thus as the number of s quarks is increased in the hyperons, we expect the mass to increase in steps of 210 MeV. This relationship is obeyed by the Λ and Ξ hyperons, but not by the Σ and Ω hyperons, as is demonstrated in Fig. 3.26.1. The Σ masses are too large by 70 MeV, and the Ω mass is too large by 140 MeV. In Fig. 3.26.2 we move on to the charmed hyperon ground states $\Lambda_c = qqc$, $\Sigma_c = qqc$, $\Sigma_c = qsc$ and $\Omega_c = ssc$, where we discover that this mass interval rule is obeyed by the

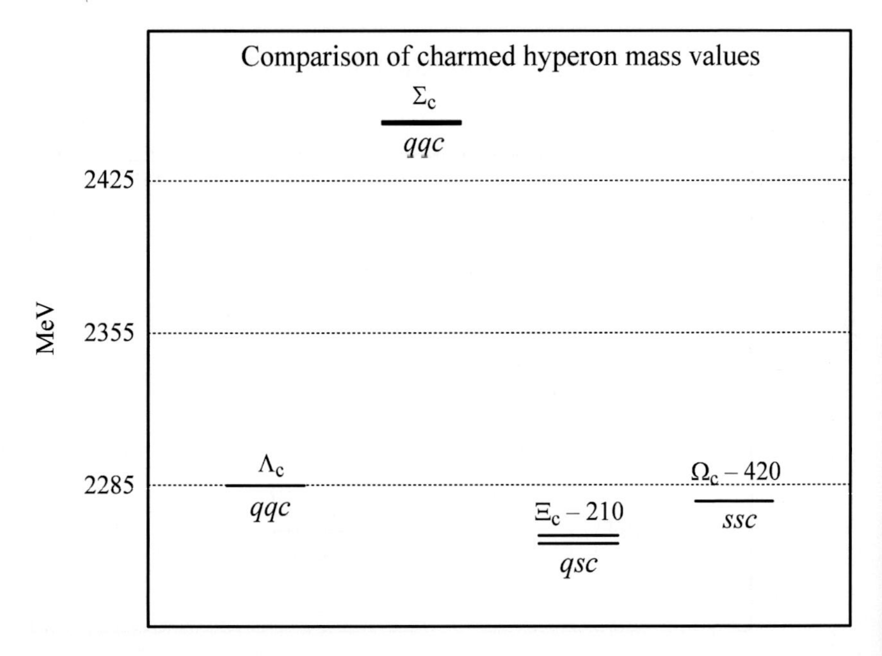

Fig. 3.26.2 The platform-state Λ_c, Σ_c, Ξ_c and Ω_c hyperon masses, shown with a 70 MeV mass grid anchored on the Λ_c mass. The Λ_c and Σ_c levels are at the experimental mass values, the Ξ_c levels are shifted downward by 210 MeV, and the Ω_c level is shifted down by 420 MeV. The Λ_c, Ξ_c and Ω_c levels all appear at the same mass value, showing that they exhibit the same s quark mass values (and same q quark and c quark mass values). The anomalous Ω mass interval displayed in Fig. 3.26.1 has disappeared, possibly reflecting the fact that the exclusion principle limitation has been removed by the s quark to c quark transfer. However the Σ_c level remains high by a little more than 140 MeV, showing that the Σ hyperons have a more complex quark structure than the basic Λ hyperons.

Λ_c, Ξ_c and Ω_c charmed hyperons, but not by the Σ_c charmed hyperons. The fact that the Σ and Ω masses deviate from the 210 MeV strangeness interval rule by mass units of 70 MeV suggests that these deviations fall within the α-quantized mass formalism. The occurrence of the strangeness quantum number S in both the muon and pion mass systems, which have different mass structures, opens up the possibility that the mass composites which carry these strangeness quantum numbers may have different masses. The strange quark s is created by adding a "strange" 210 MeV "muon pair" $Q_f = m_f m_f$ (or possibly a "strange" triad $Q_b = m_b m_b m_b$) to a q quark, as illustrated for example in Fig. 3.25.1. This produces the strange muon quark $s_\mu = 525$ MeV. The strange kaon contains contains seven m_b subquanta, as shown in Fig. 3.25.2, and the manner in which the particle–antiparticle asymmetry of these seven substates is reflected in the strangeness quantum number that it carries is not immediately apparent. But we can assume for the sake of argument that it is carried in the π_4 quark state of Fig. 3.25.2, and that the higher-mass s quark in the Σ hyperon is created as a hybrid excitation by adding a 280 MeV "strange" π_4 quark to a q quark, thus creating a strange quark hybrid excited-state $s^* = 595$ MeV. Since we are dealing here with the phenomenology of mass values, our best clue to as to what is actually going on resides in the mass fits that we can obtain with this kind of reasoning. If we accept the existence of the strange quarks $s = 525$ MeV and $s^* = 595$ MeV, we can construct the hyperon ground states as the quark configurations $\Lambda = qqs$, $\Sigma = qqs^*$, $\Xi = qss$ and $\Omega = ss^*s^*$. These give the hyperon masses displayed in Fig. 3.26.3, which obey an extended 210 MeV and 280 MeV "two-strange-quark" interval rule. The α-quantized constituent-quark approach that forms the basis for the present work does not seem consistent with "nuclear-physics-type" spin-orbit or other mass splitting mechanisms, so the s and s^* "two-mass" strange quark emerges as the most logical resolution of the Λ–Σ mass-splitting problem.

There is one other particle — the $K^*(892)$ meson — that seems to clearly require the $s^* = 595$ MeV hybrid strange quark. The K^* has an isotopic-spin-averaged mass of 894 MeV [Appendix B]. From its spin value $J = 1$, we know that the K^* is necessarily composed of a pair of spin-1/2 quarks. The quark assignment $K^* = q(315) + s(525) = K^*(840)$ is too low in mass by 54 MeV. The alternate assignment $K^* = q(315) + s^*(595) = K^*(910)$ gives a K^* meson with a hadronic binding energy of 16 MeV, or 1.8%, which is comparable to the hadronic binding energies displayed in Table 3.15.1. The reason for the anomalously high value (by 70 MeV) of the K^* may lie in its M_K platform production mechanism, which is displayed in Fig. 0.2.19 and discussed in Sec. 0.9. The two M_K platform excitations shown there are $K = m_bX$ and $K^* = m_bXX$. The spin-0 K meson can be constructed of spinless pion quarks, but the spin-1 K^* meson requires spin-1/2 muon quarks. The XX platform excitation can be written in a variety of isoergic forms, but the K^* requires it to be written as $XX = q\bar{s}$ or $\bar{q}s$. The K^* itself is then $q\bar{s}m_b \equiv q\bar{s}^*$ or $\bar{q}sm_b \equiv \bar{q}s^*$. Since the K^* is the lowest-mass *strange* spin-1 particle, its mass value is of particular phenomenological significance. The K^* isotopic-spin mass splitting of 4.44 MeV closely resembles that of the pions and kaons in magnitude, and the K^{*0}, like the K^0, has the heavier mass. So the K

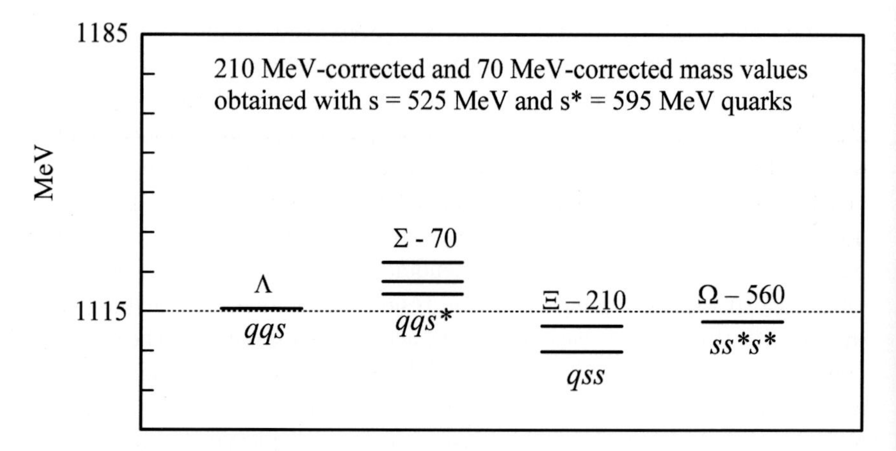

Fig. 3.26.3 The platform-state Λ, Σ, Ξ and Ω hyperon masses, shown with a 70 MeV mass grid anchored on the Λ mass, with downshifted Σ, Ξ and Ω masses, and with a postulated s and s^* two-mass quark structure. The s quark is the familiar muon quark, with a mass of 525 MeV. The s^* quark is an excited-state quark, with a mass of 595 MeV. As can be seen in Fig. 3.26.3, the s and s^* strange quark formalism properly reproduces the Λ and Σ mass values, and can be used to properly reproduce the Ω mass value. The necessity for the s^* strange quark arises from the $\Lambda - \Sigma$ mass dichotomy, and perhaps even more compellingly (as discussed in the text) from the necessity of reproducing the $K^*(892)$ mass.

and K^* occur as closely-related excitations that share the M_K excitation platform.

Additional evidence for the s^* excited quark state has recently appeared in a high-mass baryon excitation experiment (see Ref. [147] in Sec. A of the Postscript). A charm baryon Ω_c^* excited state was produced that decays radiatively to the Ω_c^0 ground state. The measured energy difference between these two states is 70.8 ± 1.0 (stat.) $\pm$ 1.1 (syst.) MeV. This may be the most accurate "direct" measurement to date of the α-quantized boson mass quantum $m_b = 70.025$ MeV. The Ω_c^0 has the quark flavor configuration $\Omega_c^0 = ssc$. If we postulate that $\Omega_c^* = ss^*c$, then we have accounted for the observed $\Omega_c^* - \Omega_c^0$ mass difference as an $s^* - s$ quark mass difference, and we have also helped to confirm the existence of the s^* quark state. Alternately, we can assume that $\Omega_c^0 = ss^*c$ (see Tables C1 and C2 in Appendix C), and then postulate that $\Omega_c^* = s^*s^*c$, which gives the same mass difference and flavor content. An important point here is that these higher-mass "ground state" quark structures are quite clear-cut, and their mass values seem to be particularly straightforward (as in the B_c meson, for example).

The lowest-mass *nonstrange* spin-1 particles are the $\rho(776)$ and $\omega(783)$ mesons, which are reproduced here as the hybrid excitations $q\bar{q}(630)\pi(140) = 770$ MeV. We can denote these as the excited quark pair $q^*\bar{q}^*$, but they do not carry the strangeness quantum number. The reason for their existence can be inferred from the X-quantized nature of the quark-producing M_ϕ platform state (Figs. 3.17.1 and 3.19.2), which does not permit asymmetric $\frac{1}{2}X = m_b\bar{m}_b m_b$ excitation units in each channel, but may permit symmetric $m_b\bar{m}_b m_b\bar{m}_b$ excitation units.

I would like to close this section on the strangeness quantum number with a personal anecdote. In the early 1950's I attended a physics colloquium by Murray Gell-Mann at UC Berkeley in which he introduced a new (to me at least) quantum number S. I can still picture him as he took a piece of chalk and wrote a large capital S on the blackboard. This was my first experience with the wonders of pions and kaons and hyperons as they rapidly appeared in a succession of experiments in the 1950's. To be able to witness these developments in "real time" was a privileged experience, and to be still writing about the wonders of "strangeness" (and possibly even adding something to the concept) a half century later is an opportunity to be treasured.

3.27 The Universal 35 MeV Mass Grid

The production of α-quantized elementary particles from an electron or electron–positron ground state yields two basic mass quanta — the boson mass $m_b = m_e/\alpha \cong 70$ MeV (Eq. (3.4.1)) and the fermion mass $m_f = (3/2)m_e/\alpha \cong 105$ MeV (Eq. (3.4.2)). These can appear in pure boson (spin-0) or pure fermion (spin-1/2) substate combinations, unpaired or paired, and also in mixed boson–fermion combinations. The largest common mass factor for these excitations is the mass element $m^U = 35$ MeV, so if a universal quantized mass grid is desired that will accommodate all α-quantized particles, it should be in units of m^U.

Mass intervals of 35, 70, 140 and 210 MeV occur throughout the elementary particle mass spectrum [69], and this fact is documented in the literature every so often. A brief review of early papers is given in the Postscript to the present book. A recent discussion is that of David Akers [70]. An elegant modern approach was implemented by Paolo Palazzi [71], who starts with a computerized version of the RPP particle database [10], divides the particles into PDG spin-parity-flavor families, and analyzes each family separately, using integer multiples P of the 35 MeV mass quantum

m^U to reproduce the masses in each family, where P = even for mesons (bosons) and odd for baryons and leptons (fermions). Detailed statistical fits are made to obtain an optimized m^U mass value for each family, and at the same time select the best values of P for the states of the family. The meson analysis [71] demonstrates that the masses of all of the meson families can be reproduced with mass units m^U which are distinct but close to 35 MeV. The analysis also shows that the mass quantum m^U, as obtained for the various families, is itself discreet, with a grid spaced by 0.25 MeV mass intervals that is centered at the value $m^U = 35.4$ MeV. In addition, the positions of the m^U values for the various meson families on the m^U mass grid show interesting correlations with the quantum numbers of the families. Palazzi's results for the low-mass threshold states are similar to those of the present paper, but he extended them beyond the present work by demonstrating that this 35 MeV mass quantization also applies to the higher-mass and shorter-lived particles. This result suggests in particular that the higher-mass states are not rotational excitations, which would have a different type of mass quantization.

The Palazzi universal mass grid phenomenologically establishes the fact that elementary particles exhibit a quantized mass structure in mass units of $m^U \cong 35$ MeV. This fact is commonly agreed upon by workers in this area of phenomenology, as documented in papers that extend back for half a century (see Postscript). However, these workers tend to diverge in various directions when seeking explanations for this mass structure. One approach, that of α-quantized particle lifetimes and reciprocal α^{-1}-quantized particle masses, is contained in the present book. In this approach, the two basic low-energy mass quanta are the spin-0 boson mass quantum $m_b = 70$ MeV and the spin-1/2 fermion mass quantum $m_f = 105$ MeV, which appear mathematically as the non-spinning and relativistically-spinning configurations of a uniform spherical mass quantum (Fig. 0.2.14). This relativistically spinning sphere (RSS) representation of a muon, $\mu \cong m_f$, gives calculated values for its spin and magnetic moment that correctly reproduce its gyromagnetic ratio g (Sec. 4.4). The $q \equiv (u, d)$ constituent-quark masses are reproduced as 315 MeV mass units $m_q = 3m_f$. We can tell from the magnetic moments attributed to these quarks [11] that they are RSS's with Compton radii $R_c = \hbar/(3m_f c)$. This assignment gives the magnetic moments correctly (Sec. 4.4), and it yields a three-quark uud cluster that has the measured size of the proton [21]. This tells us two things: (1) when a single spherical fermion mass quantum m_f is tripled in magnitude to form a u or d quark, the resultant quark is still a single spherical mass quantum (with a radius 1/3 that of the muon); (2) when three quarks *uud* are combined to form the mass-tripled proton, the resultant proton is *not* a single mass quantum, but is a cluster of three individual mass quanta. We know this from (a) the measured size of the proton, (b) the measured magnetic moment of the proton, and (c) the three-center Bjorken scattering of electrons off protons at very high energies.

In the present book we have not directly addressed the question of the shapes of large-mass elementary particle quark clusters. The b quark, for example, contains 45 m_f quanta. It most probably is a spin-1/2 RSS with a very small Compton radius, as is the c quark. A $b\bar{b} = \Upsilon$ or $b\bar{c} = B_c$ quark pair thus has fractional charges that are very close together, with "rubber band" hadronic binding energies that are very small (asymptotic freedom), as displayed in Fig. 3.15.1. The $c\bar{q} = D$ meson is somewhat larger and has a larger HBE that fits in with the HBE systematics shown in Fig. 3.15.1. The $c\bar{s} = D_s$ meson, on the other hand, has a much larger than expected HBE (Table 3.25.2), which may signal a more extended structure. The D and D_s charmed mesons both serve as platform states for π and 4π excitation levels (Fig. 3.22.3), and it is not clear what geometric forms should be assigned to these hybrid excitations. The $\Lambda = qqs$ and $\Sigma = qqs \cdot m_b$ mass configurations present a similar problem, as do the meson and baryon mass clusters displayed in Table 3.24.1 and Fig. 3.24.2. Spectroscopic data would provide some answers to these representational problems, but it is difficult to obtain for particle states which last for only very brief periods of time. The bottom line here is that whereas we can make some reasonable inferences about the geometries of the stablest particles, the geometries of the complex higher-mass excitations have not been pinned down. One exception is the highest-mass state of all, the top quark t, which from its zero binding energy (Eq. (3.20.8)) is probably a very small spin-1/2 relativistically spinning sphere.

A somewhat different approach to elementary particle geometries is offered by Palazzi, who made mass-lifetime correlation analyses which suggest that particles might be shell-structured. In particular, the mesons show a shell population sequence that is similar to the nuclear shell-model sequence, including indications at energies below 2 GeV of "doubly-magic" states and sub-shell structure [71].

Palazzi's mass analysis also gives (from the viewpoint of the present paper) some information about meson binding energies. The various spin-parity-flavor families were statistically analyzed in order to obtain the best average mass fits as functions of m^U and P (where P is an even integer).

The most strongly-bound meson family has an m^U value of 33.86 MeV, which in terms of the mass values of Eqs. (3.4.1) and (3.4.2) in the present paper corresponds to a (negative) binding energy of 3.3%. This is in close agreement with the hadronic binding energies displayed in Table 3.15.1 and Figs. 3.15.1 and 3.15.2. Other spin-parity-flavor families have smaller binding energies that seem to be characteristic for each of the various members within the family, and which plausibly reflect (again from the present viewpoint) a blend of particle–antiparticle symmetries, isotopic spin effects, and lifetime constraints within the families.

The important point about the work of Akers, Palazzi and others (including the studies outlined in the present book) is the emphasis on experimental data for guiding theoretical developments. Theories live or die on the basis of experimental confirmations or refutations, and theorists ignore experimental results at their peril.

3.28 Mass Freedom in Quantum Chromodynamics (QCD)

In this, the concluding section of our Chapter 3 discussion of elementary particle masses, we address a fundamental question with respect to the Standard Model:

> *Can the Standard Model be formulated, and its very striking successes be maintained, while using different mass representations for the u, d, s, c, b, t quark states?*

This question is a generalization of another question that emerges from the present phemonenological studies of elementary particle lifetimes and masses:

> *The α-quantization of the long-lived elementary particle lifetimes reveals the manner in which these particles are cleanly sorted into single-flavor quark lifetime groups, so that this observed α-quantization is a confirmation of the SM quarks; and yet the observed α-quantization of the masses of these particles points to a mass structure for the quarks that differs from the SM values: how can this be?*

Can we change the mass formulation of the Standard Model while still retaining its isotopic spin and interaction systematics, which have served as a guide for the development of a comprehensive elementary particle theory? We argue here that the answer to this question is *yes*.

Historically, the quark mass problem initially posed no difficulties. The quark model *circa* 1969 was developed under the assumption that quarks are very massive constituent-quark states ($q_m > 5$ GeV, the energy limit at that time), which are held together by their enormous binding energies [36], so that precise quark masses were of no particular consequence. Some of the guiding phenomenology for the quark model and SU(3) symmetries involved accurate *mass intervals* between related resonances, but these mass intervals do not bear directly on the absolute masses for these states. In the context of the present α quantized masses, it should be noted that some of these were 140 MeV mass intervals, which are now seen to abound in these states [69]. When high-energy accelerators were subsequently developed that pushed the required quark binding energy up past 5 GeV to much higher values, and free quarks were still not observed, it was realized that conventional binding energies did not apply to the quark bonding, and the mass problem was then opened up to the twin tracks of *current* quarks (which have small intrinsic masses and large gluon masses), and *constituent* quarks (which have large intrinsic masses and modest 3% gluon masses). This ambiguity was made possible in part because the striking isotopic spin successes of the quark model have to do with the quark charges, which are not directly tied to the quark masses. The Standard Model is presently centered on the use of current quarks [10], although as first the c quarks and then the b quarks were discovered, their experimental properties were creeping closer and closer to the systematics of constituent quarks. After the concept of very large binding energies was abandoned in favor of asymptotic freedom, then the fact that the Standard Model uses the same u and d quarks for the low-mass pseudoscalar mesons as it does for the higher-mass baryons and hyperons meant that a current-quark approach was essentially mandated. These quark assignments give the correct isotopic spin rules for the PS mesons, but they have led to continuing problems with the masses. Also, leptons and hadrons belong to completely different mass formalisms in the Standard Model, which has precluded any ability to calculate, *e.g.*, the proton-to-electron mass ratio.

In the context of this discussion, let us recapitulate the present work. We started by examining the spectrum of elementary particle *lifetimes*. These are simple to work with, and they are conjugate to elementary particle *mass widths*, which in turn are logically related to the *masses* themselves. This analysis demonstrated that the long-lived threshold-state particle lifetimes occur singly or in groups, with separation intervals in powers of $\alpha = e^2/\hbar c$. These lifetime groupings include all of the observed long-lived leptons and hadrons. When we then examined the spectrum of elementary particle masses from this same α-quantized viewpoint, we were able to identify an $m_b = 70$ MeV boson mass and an $m_f = 105$ MeV fermion mass that serve as basis states for constructing the threshold-state particles, including both leptons and hadrons. These are single-α-leap mass enhancements of the electron ground state. We also identified an $m_f^\alpha = 14{,}394$ MeV fermion mass that is a double-leap mass α-enhancement which serves as the basis state for precisely constructing the $\overline{WZ}$ average mass and the mass of the top quark t. Finally, looking ahead to Chapter 4, we can invoke the relativistically spinning sphere (RSS) model to tie together the m_b and m_f masses. The mass values that we calculate in Tables 3.16.1, 3.25.1–3.25.3, C1 and C2 are constituent-quark masses, so we have gone full circle from the original constituent-quark masses to current-quark masses and back to (α-quantized) constituent-quark masses, but now with moderate binding energies. The non-observability of individual quarks arises in this α-quantized formulation from the tenacious binding of fractional quark charge clusters, which cannot be broken apart and separated into non-integer fragments. Charges can be separated into fractions within a charge cluster that encloses a group of quarks, but they reunite if the quarks in the group are separated or fragmented. This α-quantized lifetime and mass formalism exhibits the QCD flavor structure, as for example in the lifetime groupings of Fig. 2.11.10, and it must necessarily incorporate or allow for the plethora of striking results that have been achieved in the Standard Model [10]. But it differs in details of the treatment of masses, and in the handling of leptons and hadrons. If we examine the spectroscopy of the electron, using our standard formalisms of mechanics and electrodynamics, it becomes apparent [1] that the "mechanical mass" of the electron is non-interacting (neutrino-like), so that the interactions of the electron are purely electromagnetic. This result carries over by association to the intrinsic quark masses, so that the hadronic quark interactions are due to their fractional charges. QCD is a theory of charges, not masses.

In Sec. 1.2, where we first discussed the particle mass mystery, we cited Richard Feynman's comment at the very end of his insightful book *QED*, where he noted that "*Throughout this entire story there remains one especially unsatisfactory feature: the observed masses of the particles, m.*" [7]. In this same book, Feynman pointed out an important contribution to the theory of quantum electrodynamics made by Hans Bethe and Victor

Weisskopf [72], two outstanding pioneers in this field, who noted that if cutoffs are systematically applied to divergent integrals, then the results can be carried over from one calculation to another. In their classic two-volume *Concepts of Particle Physics* [2], which summarizes the status of elementary particle physics and quantum chromodynamics, Gottfried and Weisskopf addressed the quark mass problem as follows:

> *Unfortunately, QCD has nothing whatsoever to say about the quark mass spectrum, nor, for that matter, does any other existing theory.* [3]

This quotation, which we also cited in Sec. 1.1, delineates the boundaries of QCD.

In order to move beyond this impasse, we have in the present work turned directly to experiment, where the mass values shown in Table 3.16.1, 3.25.1–3.25.3, C1 and C2 emerge in a manner that is essentially independent of theory. It should be possible to retain the truly impressive achievements of quantum chromodynamics, and to reformulate it within the context of a set of constituent rather than current quarks — a set that includes both spinning and non-spinning basis states, and both leptons and hadrons.

The need for incorporating leptons and hadrons in a common mass formalism is brought out in a recent American Institute of Physics Bulletin of Physics News item [73] with the heading "Indications of a change in the proton-to-electron mass ratio." In the center of this news item the following sentence appears:

> *There is at present [April 19, 2006] no explanation why the proton's mass should be 1836 times that of the electron's.*

Chapter 4 The Mathology of the Elementary Particle: The Relativistically Spinning Sphere

4.1 Introduction to Mathology

Phenomenology is the study of a collection of phenomena, in order to see what kind of organization exists. In the case of physics, the phenomena are the experimental data — the measurements of the features of the world around us. Thus *phenomenology*, to a physicist, is usually *experimental phenomenology*. A prime example of phenomenology, but in the field of chemistry instead of physics, was the work of Dmitri Mendeleev, who discovered the atomic table of elements. He decided that the mass of an element was its salient feature, and he wrote out a card for each element which contained its mass and other important characteristics. He also measured some of these masses more accurately than had been done before. Arranging the cards according to mass, he was able to observe a periodicity in the order in which the basic characteristics occurred. The gaps in his layout were subsequently filled in by the discovery of new elements with the properties he had predicted. As Lederman, for example, points out [74], Mendeleev never got the recognition during his lifetime that he merited for this accomplishment.

In elementary particle physics, *experimental* phenomenology — the patterning of the experimental data — involves a study of particle lifetimes, masses, decay modes, interaction cross-sections, and so forth. There is also *theoretical* phenomenology — a mathematical phenomenology — that can be used to reproduce the spectroscopic properties of the particles, and hence the particles themselves. We can denote this mathematical phenomenology as *mathology*. The way it works is to look at the data, create a mathematical formula, and fit the formula to the data. If the formula works, then maybe it means something, even if we presently have no explanation for it. Atomic physics offers two prime examples of mathology. The first is the Ritz combination principle, in which it was demonstrated that the wave numbers of the atomic spectral lines could be expressed as differences between terms of the general form $1/n^2$, which led to the equation $1/n^2 - 1/m^2$, where n and m are integers. This formalism worked so well that it had to have some physical significance, although it, like the Mendeleev chart, had to await the Bohr atom for its explanation. The second example of mathology in atomic physics was the insertion of half-integer angular momentum quantum numbers into the Landé interval rule for the anomalous Zeeman effect. This procedure solved a problem that had baffled physicists for years. According to one story, Wolfgang Pauli was walking down the street with a dour expression and met a friend. When the friend questioned him about his woeful visage, he answered: "Who can be happy when he is thinking about the anomalous Zeeman effect?" The Landé rule worked very well in reproducing the anomalous spectral lines, but there was no physical justification at that time for using half-integers. This justification was later supplied by Goudsmit and Uhlenbeck in their concept of the spinning electron with its $\frac{1}{2}\hbar$ spin value.

Mathology is not a widely-employed tool at the present time for exploring elementary particle spectroscopy. Abstract theoretical calculations abound, but attempts to reproduce the properties of particles with specific mathematical models are frowned upon, partly because of a belief that these particles do not admit this type of "classical" representation. In Chapter 4 we invoke one mathematical model, the relativistically spinning sphere (RSS), mainly because it correctly accounts for the empirically determined factor of 3/2 ratio between the masses of spinning and nonspinning quanta, and thus relates the 211 MeV muon pair to the 140 MeV pion. The RSS model gives the correct answer, so it may in fact mean something. It also gives some other correct answers that were not anticipated. In Chapter 5 we will go a little farther afield and consider the mathological properties of a particle-hole pair, which is an essential ingredient for "zero mass" models that reproduce the spectroscopy of the photon. We begin this sojourn into the mathology of the elementary particle with a brief discussion of the mathological aspects of QED, which is surely the "crown jewel" of all mathologies.

4.2 The Most Accurate Example of Mathology: Quantum Electrodynamics (QED)

The theory of quantum electrodynamics (QED) is calculationally one of the most impressive achievements in modern theoretical physics, and the matching experimental measurements are equally impressive. The calculation of the anomalous magnetic moment of the electron stands as a representative example. The original Dirac theory predicted that the magnetic moment of the electron is $\mu = e\hbar/2mc$. Accurate experiments starting around 1948 showed that the experimental value is about 0.1% larger than this value, and this anomaly was ascribed to the effect of electron–photon interactions, which were not included in Dirac's formalism. When QED was first used to calculate the effect of the electron–photon interaction on the magnetic moment μ, the calculations yielded infinite integrals. However, renormalization techniques were developed, primarily by Schwinger, Feynman and Tomonaga, which demonstrated how to systematically eliminate the infinities in the QED calculation, and the results they obtained were in close agreement with experiment. Since that time, both the experiments and the theoretical calculations have been greatly improved. The values for these quantities as quoted in Feynman's 1985 book *QED* are [75]

$$\mu_{\text{exper}} = 1.00115965221 \times e\hbar/2mc\,;$$

$$\mu_{\text{theory}} = 1.00115965246 \times e\hbar/2mc\,.$$

The experimental uncertainty is about 4 in the last digit, and the theoretical uncertainty is five times as much. The experimental value listed in RPP2006 [10] is even more precise:

$$\mu_{\text{exper}} = (1.0011596521859 \pm 0.0000000000038) \times e\hbar/2mc\,.$$

The extreme accuracy of the agreement between theory and experiment indicates that the theory is mathematically sound, and that the infinities are being properly dealt with.

The mathematical details of the QED calculation are displayed, for example, in Jauch and Rohrlich's monograph *The Theory of Photons and Electrons* [76]. Feynman's *QED* [6] discusses this topic without the use of any mathematics, but with a careful description of the Feynman diagrams that show the various ways electrons and photons can interact in going from the initial to the final electron state. The more diagrams that are added in, the more accurate the answer becomes. Although the final result obtained by adding up all of the possible Feynman diagrams is stunningly accurate, it leaves one mathological question unanswered:

> *What physical attribute of the electron reflects this anomalous magnetic moment?*

The QED calculation does not provide a clue as to the correct answer. As Feynman pointed out long ago [77]:

> *We don't even know why the sign is positive (other than by computing it).*

Thus we have a mathematics that works, but we do not know just what it is calculating. This in a sense is pure mathology, with no interpretive underpinning. The various infinities, which are equally mysterious, are evidently being handled properly, but their origin is left up in the air. As we will see, elementary particle mathology may be able to provide some insight into these enigmas.

QED supplies the mathology of the interactions between an electron and the photons in its coulomb field. We can add in a mathology for the electron itself by examining its structural properties when pictured as a relativistically spinning sphere (Secs. 4.3–4.7). This electron mathology provides an answer (Sec. 4.6) to the above question about the physical effect behind the magnetic moment anomaly. We can also add in a mathology for the photon by picturing it as a combination of particle–hole states (Sec. 5.2). This photon mathology suggests a possible explanation (Sec. 5.3) for at least some of the infinities that plague the QED calculation.

We now move on to the relativistically spinning sphere model, which supplies mathological answers to questions that were not even being asked while searching for it.

4.3 The Mechanical Mathology of Relativistically Spinning Spherical Masses

The subject of relativistically spinning masses has been one of controversy among theorists for the past century. Spinning spheres can be subdivided into spinning discs, which are less complex to deal with mathematically, and are the objects usually discussed. The questions center around two main points: (1) is the geometry on the rotating disc non-Euclidean? (2) what is the nature of the purely relativistic stresses, if they exist? The rotating disc has been extensively analyzed by Henri Arzelies in his book *Relativistic Kinematics* [78]. He comments as follows:

> *Despite the divergences of opinion which appear in the literature, agreement seems to have been reached on the first of these problems, the nature of the spatial geometry on the disc* [it is non-Euclidean]. *The second problem concerns the stresses of purely relativistic origin which may be created within the disc. It is still the object of numerous arguments among specialists.* [79]

A collection of papers and books on this topic is referred to in Ref. [80]. In the spirit of mathology, we can address the problem of the rotating sphere without having to consider these controversies. Our goal here is to reproduce the properties of a particle such as the electron. The electron does after all exist, and it does have a well-measured spin angular momentum J and magnetic moment μ. If a relativistically spinning sphere (RSS) model can accurately reproduce these properties, then this result may tell us as much about the RSS as it does about the electron.

The relativistically spinning sphere is one of the most intriguing, and most unrecognized, mathological concepts in elementary particle physics. Mathematically, it is the essence of simplicity. Take a uniform sphere of matter that is at rest in the laboratory frame of reference. Give it a rest mass M_0. Now spin it. The spin energy adds to the rest-mass energy, so the total mass (or energy) increases. As it spins faster and faster, the mass continues to increase, and eventually requires a relativistic equation to properly describe the motion. How much can the mass increase? When the instantaneous velocity at the equator reaches the velocity of light, the sphere can rotate no faster. A sphere that is spinning at this full relativistic limit (or infinitesimally below the limit) is denoted here as a *relativistically spinning sphere* — RSS. What is the calculated spinning mass M_s at the RSS limit? It turns out that the spinning mass is $M_s = 3/2\ M_0$:

> *A fully relativistic spinning sphere is half again as massive as it was at rest.*

This result holds for any radius, and requires only that the (invariant) rest mass be in the shape of a uniform sphere of matter.

We can extend the above result by assigning a specific RSS radius to the sphere: namely, the Compton radius $R_s = \hbar/M_s c$, which is mandated spectroscopically (Sec. 4.4). This fixes the angular momentum of the RSS at the calculated value $J = \frac{1}{2}\hbar$. Thus we arrive at a result which is of considerable importance for elementary particle physics:

> *Spin-1/2 particles are half again as massive as their spinless counterparts.*

This is where we come into contact with the α-quantized mass systematics of Chapter 3. The lowest massive particle–antiparticle-symmetric state is the electron–positron pair, which appears at 1 MeV. The spinless pion (which is its own antiparticle, and hence also possesses particle–antiparticle symmetry) appears at about 137 MeV, which is an α^{-1}-leap above the mass of the electron pair. A spin-1/2 muon–antimuon pair (which likewise possesses particle–antiparticle symmetry) appears at 211 MeV — a factor of 3/2 higher in energy. The muons are essentially unbound particles, and they each have only a single charge state, so their threshold energy is easy to interpret (Figs. 0.2.3 and 3.3.2). The pion, on the other hand, is a composite particle which has matching hadronically bound particle and antiparticle substates, and also mass-dependent charge states (Fig. 3.4.1). Thus the "intrinsic" mass of the pion that should be compared to the "intrinsic" mass of a muon pair in order to determine their precise experimental mass ratio is not easy to ascertain. The analysis of these "Rosetta stone" masses (Sec. 3.3) reveals the existence of two basic *α-masses* — the spinless boson mass quantum $m_b = m_e/\alpha \cong 70$ MeV of Eq. (3.4.1) and the spin-1/2 fermion mass quantum $m_f = (3/2)m_e/\alpha \cong 105$ MeV of Eq. (3.4.2), which have an assigned mass ratio of precisely 3/2. (A third *α-mass*, the spin-1/2 fermion supermass $m_f^\alpha = (3/2) \cdot m_e/\alpha^2$, generates the super-massive $W^\pm$, Z^0 and top quark t particle states, as described in Secs. 0.10 and 3.20.) Calculational linkage between the m_b and m_f mass quanta is provided by the RSS model, which ties them together as spinless and spinning configurations of the same fundamental mass quantum. The physical relevance of this model to elementary particle physics, apart from its spectroscopic properties, is that particles which are composed of m_b quanta (the pseudoscalar mesons) decay into particles which are composed of m_f quanta (e.g. the muon), and vice versa. These decays are easier to account for if the same mass quantum is involved on both sides of the decay process. And if strangeness is conserved for a mixture of m_b and m_f quanta (Sec. 3.26), this commonality of masses seems essential.

The formal definition of the relativistically spinning sphere is straightforward, and has been well documented in the literature [81, 82]. The case of interest here is when the sphere is rotating at the *relativistic limit*, which we formally define as follows:

> *A relativistically spinning sphere (RSS) is fully relativistic when its equator is moving at, or infinitesimally below, the velocity of light, c.* (4.3.1)

Assuming this limiting rotation, we set forth the four mechanical RSS lemmas that are displayed in Table 4.3.1, together with three related corollaries. The proofs for these lemmas are in the literature [81, 82], and follow analytically from the equations of classical mechanics. We sketch the main results here. The basic mass quantum M_0 is represented as a uniform sphere of matter. When the sphere is set into rotation at an angular velocity ω, its observed mass increases relativistically in accordance with the equation

$$m(r) = m_0(r)/\sqrt{1-\omega^2 r^2/c^2}\,, \tag{4.3.2}$$

where $m(r)$ is a mass element at a distance r from the axis of rotation. In the context of special relativity, this mass increase arises from the instantaneous velocity $v = \omega r$ of the mass element. It can also be viewed in the context of general relativity as arising from the gravitational potential of the rotational motion [83]. The volume becomes non-Euclidean [84–86] and increases as

$$V(r) = V_0(r)/\sqrt{1-\omega^2 r^2/c^2}\,. \tag{4.3.3}$$

To get the total mass and volume increase, we integrate over the volume of the sphere.

The mass equation in cylindrical polar coordinates (r, θ, z) is

$$M_s = \frac{3M_0}{R_s^3}\int_0^{R_s}\sqrt{\frac{R_s^2 - r^2}{1-\omega^2 r^2/c^2}}\,r\,dr\,, \tag{4.3.4}$$

where M_s and M_0 denote spinning and spinless masses, and ω is the angular velocity. The corresponding moment of inertia equation is

$$I_s = \frac{3M_0}{R_s^3}\int_0^{R_s}\sqrt{\frac{R_s^2 - r^2}{1-\omega^2 r^2/c^2}}\,r^3\,dr\,, \tag{4.3.5}$$

As the $m(r)$ mass element approaches the equator, its mass diverges, but the volume element it represents vanishes so as to exactly compensate for the diverging mass. Hence these integrals are convergent. The integration in Eq. (4.3.4) gives $M_s = \frac{3}{2}M_0$, which is Lemma 1 in Table 4.3.1. A similar integration for the volume gives $V_s = \frac{3}{2}V_0$, which is Lemma 2. The

Table 4.3.1 Mechanical properties of the fully relativistic spinning sphere (RSS), whose equator is moving at, or infinitesimally below, the velocity of light, c (Eq. (4.3.1)). Assuming this limiting rotation, we set forth the following four lemmas.

Lemma 1: The mass of an RSS is half again as large as that of its nonspinning counterpart: $M_s = \frac{3}{2}M_0$.

Lemma 2: The volume of an RSS is half again as large as that of its nonspinning counterpart: $V_s = \frac{3}{2}V_0$.

Corollary A: The density distribution $\rho(r) = m(r)/v(r)$ of an RSS is an invariant, independent of its rotational velocity.

Corollary B: Since the density distribution $\rho(r)$ of a massive object is a measure of its applied stresses, the invariance of $\rho(r)$ can be taken as an indication of the absence of relativistic stresses.

Corollary C: The relativistic radius of curvature of an RSS vanishes at the equator of the sphere, so that a mass element placed there is effectively in linear motion, and the centrifugal forces vanish.

Lemma 3: The calculated moment of inertia of an RSS is $I = \frac{1}{2}M_sR_s^2$, where R_s is the radius of the sphere.

Lemma 4: If the RSS radius is the Compton radius $R_s = \hbar/M_sc$, then its calculated spin angular momentum is $J_s = \frac{1}{2}\hbar$.

integration in Eq. (4.3.5) gives $I_s = \frac{3}{4}M_0R_s^2 = \frac{1}{2}M_sR_s^2$, Lemma 3. Finally, inserting the Compton radius $R_s = \hbar/M_sc$ and setting $\omega R_s = c$ gives $J = I_s\omega = \frac{1}{2}\hbar$, Lemma 4. If the equatorial velocity ωR_s of the spinning sphere is infinitesimally less than c, then the corrections to the relativistic equations are also infinitesimal.

Corollaries A, B, and C suggest that the relativistic stresses which are customarily assumed to apply to a relativistically spinning sphere [87–89] may not be a factor. For example, the relativistic stretching of the length of a string has been pointed out [88], but the corresponding increase of the mass of the string so that its mass per unit length remains invariant [86] was apparently overlooked.

The relevance of this RSS model to the α-quantization of particle masses is that it ties together the $J = 0$ and $J = 1/2$ α-mass excitation quanta m_b and m_f as rotationless and rotational states of the same basic spherical mass quantum, $m_b = m_e/\alpha = 70.025$ MeV. The supersymmetric "crossover" excitation quantum $X = 6m_b = 4m_f = 420.15$ MeV (Sec. 3.9) is the lowest-mass excitation unit that can produce $m_b \leftrightarrow m_f$ transformations which are both isoergic and particle–antiparticle-symmetric. These transformations are accomplished by the process $m_b[m_b]\bar{m}_b + m_b[\bar{m}_b]\bar{m}_b \Leftrightarrow 2m_f\bar{m}_f$, where the brackets denote mass annihilations. Since spinless pions decay into spinning muons, and spinning tau leptons decay into spinless pions, these two-way transformations between spinless and spinning mass quanta are required experimentally.

In demonstrating the validity of a theory in physics, it helps to establish *cross links* between the different elements of the theory. These cross links serve to strengthen the entire theoretical framework. The main theoretical idea we are attempting to establish in the present work is the relevance of the fine structure constant $\alpha = e^2/\hbar c$ as a coupling constant in quantum chromodynamics (QCD). Its relevance as a coupling constant in quantum electrodynamics (QED) has already been clearly established [6]. The observed α-quantization of the long-lived threshold-state elementary particle *lifetimes* (Chapter 2) was the initial evidence presented here, and cross linkage is supplied by the pattern of α-quantization that is observed in the corresponding threshold-state elementary particle masses (Chapter 3). When this cross-linkage is invoked, the combined α-quantized lifetime and mass data provide a more compelling argument for the relevance of the fine structure constant $\alpha = e^2/\hbar c$ in QCD than either data set does separately.

Another example of cross-linking is provided by the RSS model when it accounts for the empirical 3/2 mass ratio that occurs between the $J = 1/2$ and $J = 0$ electron-generated mass units m_f and m_b in the α-mass systematics of the threshold-state elementary particles. This additional cross-linkage strengthens the case for the α-quantized lifetime and mass formalism. Furthermore, the cross-linkage not only confirms the significance of the experimentally-indicated 3/2 mass ratio, but the experimental mass results serve in turn to verify the relevance of the mathematical RSS model.

The RSS model is also of mathological interest. The original impetus for investigating relativistically spinning spheres [89] was supplied by the desire to reproduce the gyromagnetic ratio $g = 2$ of the electron (Sec. 4.4). The fact that the mass increase turned out to be 3/2 was an unexpected bonus. Also, the vanishing of the electric quadrupole moment (Sec. 4.5) and the identification of the physical basis of the anomalous magnetic moment (Sec. 4.6) are results that were not anticipated. The point here is that if a theory or model is both mathologically correct and physically relevant, it may produce results that are more far-reaching than originally envisioned.

4.4 The Spectroscopic Mathology of the Electron: A Classical Representation Does Exist

It may well be that, in the future, we will have to give up the notion of a point-like mass. It might, after all, be that the masses of leptons and quarks are a consequence of a hitherto unobserved substructure of these particles, just as the proton mass is due to its substructure.

Harald Fritzsch [90]

It has been frequently asserted in the literature, *e.g.* in *The Feynman Lectures on Physics* [91], that "there is no classical explanation" for the gyromagnetic ratio $g = 2$ of the electron. However, the relativistically spinning sphere (RSS) stands as a counterexample to this assertion, and thus is of pedagogical interest over and above any relevance it may have to the "real world" of physics.

The search for the RSS was based on the following line of reasoning. (1) The proton mass is quite accurately given as nine muon masses, which can be divided into three quarks that are each equivalent to three muon masses. Thus the 105 MeV muon mass serves in some sense as a fundamental mass unit. (2) The proton has a measured size of about one fermi (10^{-13} cm). The muon, on the other hand, has no measurable size: it appears as a point-like entity. So how can we logically reproduce a finite-size proton as a collection of nine muon points? (3) As an answer to (2), we must somehow devise a finite-size model that has the spectroscopic properties of the muon? (4) As a start, we classically reproduce the muon magnetic moment μ as arising from a current loop of radius r, where $\omega r = c$ is the angular velocity of the rotating charge e, which gives $\mu = \pi r^2 \cdot i = \pi r^2 \cdot e/c \cdot \omega/2\pi = er/2$. If we now set r equal to the Compton radius of the muon, $r = \hbar/mc$, we obtain $\mu = e\hbar/2mc$, which is the correct first-order value for the muon magnetic moment. (5) We then observe that if we had $I = \frac{1}{2}mr^2$ as the moment of inertia of the muon mass, setting $J = I\omega$ would give $J = \frac{1}{2}mrc = \frac{1}{2}\hbar$, where $\omega r = c$ and $r = \hbar/mc$. This is the correct value for the spin of the muon. Thus this Compton-sized model correctly reproduces the gyromagnetic ratio $g = (\mu/J) \cdot (2mc/e) = 2$ of the muon, and it also applies to the electron if we let m be the electron mass instead of the muon mass.

The task of obtaining a finite-size model for the muon has thus devolved into the task of finding a mass distribution which has a moment of inertia

$I = \frac{1}{2}mr^2$. Classically, the moment of inertia of a uniform sphere of matter m is $I = \frac{2}{5}mr^2$, which is smaller than the required value of $I = \frac{1}{2}mr^2$. In order to obtain a larger value for I, we need to have relatively more mass at large radii than a uniform distribution allows. This is the result that we in fact obtain if the mass of the sphere is increased relativistically by the rapid rotation of the sphere. The question then becomes one of determining just how much this relativistic increase in I would be if the sphere is spinning at the relativistic limit. The answer to this problem was worked out on a 3" $\times$ 5" card at 5:30 in the morning in the kitchen of our house in Livermore, California, on a sunny Sunday in the fall of 1969, and it turned out to be $I_s = \frac{3}{4}M_0R_s^2$. This in turn raised the question as to how much the mass M_0 was increased by this relativistic rotation. The calculated answer to this question was $M_s = \frac{3}{2}M_0$. Plugging this value for M_0 into the expression for I_s gives $I_s = \frac{1}{2}M_sR_s^2$, which is the sought-after value! Thus a Compton-sized relativistically spinning sphere with an equatorial charge distribution e does in fact reproduce the gyromagnetic ratio $g = 2$ of the muon or electron. The mass relationship $M_s = \frac{3}{2}M_0$ also serves to relate the 105 MeV spinning muon mass to an unobserved 70 MeV nonspinning mass that logically represents half of the 140 MeV particle–antiparticle-symmetric pion. Hence the RSS mathology, in addition to accounting for the gyromagnetic ratio g of the muon and electron, provides a bonus in the form of the 70 MeV mass quantum needed for the pion.

On the basis of the above discussion, we can add an electromagnetic RSS lemma and an additional corollary to those listed in Table 4.3.1. These are Lemma 5 and Corollary D at the top of Table 4.4.1.

As Lemma 5 states, an equatorial *point* charge on the equator correctly reproduces the magnetic moment μ of an electron or muon. An equatorial *distributed* charge would also reproduce the magnetic moment, but it would not reproduce the point-like scattering that is observed in electron–electron (Møller) and electron–positron (Bhabha) scattering [92]. Also, a ring of charge would interact with itself and produce an energy term. The mass calculation of Lemma 1 and the moment-of-inertia calculation of Lemma 3 are both based on the assumption that there is no electromagnetic mass sitting on the equator of the RSS. If we assume that the charge e on the electron or muon is truly a point, as required experimentally, there are only two choices we can make for its self-energy W_E by using classical considerations: zero or infinity. Infinity does not work, so the answer has to be that the point charge e has zero self-energy.

Table 4.4.1 Electromagnetic properties of the relativistically spinning sphere.

Lemma 5: An equatorial massless point charge e on the RSS corresponds to a current loop with a calculated magnetic moment $\mu = e\hbar/2mc$.

Corollary D: The classical RSS model correctly reproduces the gyromagnetic ratios of the spin-1/2 electron and spin-1/2 muon.

Lemma 6: An equatorial point charge on an RSS gives rise to an electric quadrupole moment, but if the axis of rotation of the RSS is at the quantum-mechanically prescribed angle of 54.7° to the z quantization axis, the electric quadrupole moment vanishes identically in the z direction, and it also vanishes in the x and y directions when averaged over a cycle of precessional motion.

Lemma 7: The energy of the magnetic field that corresponds to the magnetic moment of the electron can be estimated by integrating over and extrapolating inward from the asymptotic equations for the magnetic field, and it is on the order of 0.1% of the total electron energy. It can also be estimated from the self-energy of a thin current loop, and it shows the expected functional dependence. Since the magnetic energy, in contrast to the other energy components of the electron, is irrotational, it affects the magnetic moment-to-spin ratio of the RSS, increasing the observed gyromagnetic ratio by about 0.1%.

One time at a cocktail party long ago at a physicist's apartment in Europe, in a conversation with the present author about the representation of the electron as a spinning sphere, Eugene Wigner remarked: "That used to be my model."

4.5 The Vanishing Electric Quadrupole Moment of the Electron

There is a problem that needs to be addressed with respect to the concept of a Compton-sized electron. A rotating equatorial electric charge e represents a current loop, and it gives rise to an electric quadrupole moment that should be detectable in atomic orbitals. But these orbitals show no such effect. Why not? This is the question we now consider.

In Sec. 4.3 we displayed the equations for the mass distribution and moment of inertia of a relativistically spinning sphere. If the Compton radius $r = \hbar/mc$ is assigned to the RSS, its calculated spin value is $J = \frac{1}{2}\hbar$. Then in Sec. 4.4 we placed an electric charge e on the equator of the spinning sphere and demonstrated that the current loop it represents generates a magnetic moment $\mu = e\hbar/2mc$. Thus this mathological model reproduces the gyromagnetic ratio $g = 2$ of the electron or muon. In order to reproduce

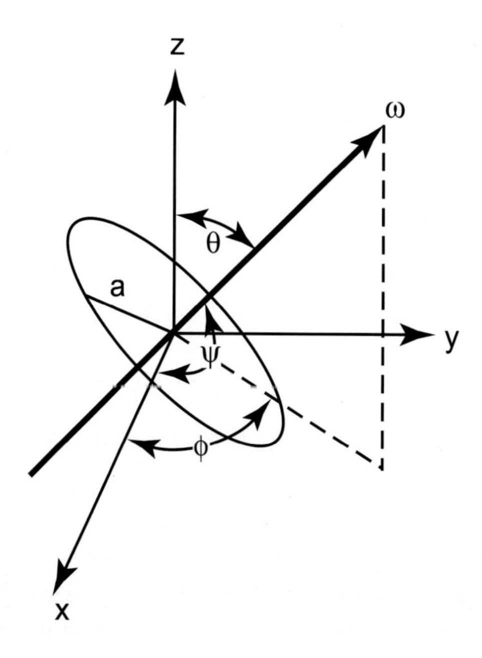

Fig. 4.5.1 A current loop of radius a, shown projected on x, y, z axes. The loop axis is tilted at an angle θ with respect to the quantization axis z, and its precessional motion around this axis is defined by the angle ϕ. If θ has the quantum-mechanically prescribed value of 54.7° (Eq. (4.5.4)), then the electric quadrupole moment of the current loop vanished identically along the z axis, and its average over a cycle of precession vanishes along the x and y axes.

point-like Møller or Bhabha scattering, the electric charge e has to be point-like ($< 10^{-17}$ cm), and to preserve the correct inertial equations for the RSS it has to have zero self-energy. But this Compton-sized current loop has an electric quadrupole moment that should be observable in atomic orbitals. Since no such effect has been reported, we need to look for conditions under which the quadrupole moment can effectively vanish.

Figure 4.5.1 shows a current loop of charge e and radius a whose spin vector ω is tilted at an angle θ with respect to the z axis of quantization. The projection of ω on the x, y plane makes an angle ϕ with respect to the x axis. The z axis is the quantum mechanical spin quantization axis, and the angle ϕ specifies the precession of ω around the z axis. The equation for the electrostatic potential V at points r along the z axis is [93]

$$V(r) = e/a \sum_{n=0}^{\infty} (-1)^n \frac{1 \cdot 3 \cdots (2n-1)}{2 \cdot 4 \cdots 2n} (a/r)^{2n+1} P_{2n})(\cos\theta) \,, \qquad (4.5.1)$$

where e is the absolute value of the electric charge, r is the distance from the center of the loop to the point on the z axis, and $r > a$. The first term in this series is

$$V_0 = e/r\,, \tag{4.5.2}$$

which is the Coulomb potential of a point charge e. The second term is

$$V_2 = -\frac{1}{4}(e/r)(a/r)^2(3\cos^2\theta - 1)\,, \tag{4.5.3}$$

which is the electric quadrupole term. Thus the quadrupole moment vanishes at the angle

$$\theta_{\mathrm{QM}} = \arccos(1/\sqrt{3})\,. \tag{4.5.4}$$

The interesting thing about the angle θ_{QM} is that this is the angle which projects a particle spin vector $J = \sqrt{\frac{1}{2}(\frac{1}{2}+1)}\hbar$ onto the z axis with the value $J_z = \frac{1}{2}\hbar$. Hence if the RSS rotates with its spin axis at the quantum mechanically prescribed angle θ_{QM}, the electric quadrupole moment vanishes identically along the z axis. Furthermore, when the projection of V_2 along the x and y axes is averaged over a cycle of precessional motion, it also vanishes [94]. The third term in this expansion, V_4, is of order α^4 smaller than the Coulomb term, and can be neglected. We summarize these results as another electromagnetic RSS lemma — Lemma 6 in Table 4.4.1.

From the standpoint of mathology, a significant point about the relativistically spinning sphere model, which features a Compton-sized electron, is that it not only reproduces the spin and magnetic moment of the electron, but also furnishes a plausible reason for the non-observability of its most obvious "size effect," the electric quadrupole moment of the electron (or muon).

The concept of the electron as a "point particle" is the result that seems to be mandated experimentally from scattering measurements. And yet there is no way that we can visualize its spin and magnetic moment values in terms of known physical laws if we do not ascribe to it a much larger size. In this connection, it is interesting to ponder a quotation by Asim Barut [95]:

> *If a spinning particle is not quite a point particle, nor a solid three dimensional top, what can it be? What is the structure which can appear under probing with electromagnetic fields as a point charge, yet as far as spin and wave properties are concerned exhibits a size of the order of the Compton wave length?*

One way to try and observe a size effect in a Compton-sized electron is to scatter it off a charged object that is more localized than it is, such as an atomic nucleus, and see if the spiraling motion of the point charge on the electron is detectable. If the electron energy is too low, it will not get near the nucleus, and if the energy is too high there will be no time for the charge to spiral during the interaction. But there might be a KeV region where the spiraling effect would be apparent. Extensive computer calculations, which involved thousands of hours of computer time, have been carried out to define conditions under which this effect might be observable [96, 97].

4.6 The Physical Basis for the Anomalous Magnetic Moment of the Electron: The Answer to Richard Feynman's Challenge for a First-Order Model

The calculation of the anomalous magnetic moment of the electron is extremely accurate, as we discussed in Sec. 4.2, and yet the physical understanding of the phenomenon that is being calculated is still a mystery. This situation has been vividly described by Richard Feynman [77]:

> *It seems that very little physical intuition has yet been developed in this subject. In nearly every case we are reduced to computing exactly the coefficient of some specific term. We have no way to get a general idea of the result to be expected. To make my view clear, consider, for example, the anomalous electron moment .... We have no physical picture by which we can easily see that the correction is roughly $\alpha/2\pi$; in fact, we do not even know why the sign is positive (other than by computing it). ... We have been computing terms like a blind man exploring a new room, but soon we must develop some concept of this room as a whole, and to have some general idea of what is contained in it. As a specific challenge, is there any method of computing the anomalous moment of the electron which, on first rough approximation, gives a fair approximation to the α term ...?*

The relativistically spinning sphere gives us a mathological model with which we can respond to Feynman's challenge. With the aid of the RSS, we will demonstrate that the key to answering Feynman's challenge was in essence provided by Rasetti and Fermi in 1926 (the year the present author was born). However, both the RSS and the Rasetti–Fermi calculation are required in order to establish this result. We start with the Dirac equation for the magnetic moment:

$$\mu_{\mathrm{D}} = e\hbar/2mc\,. \tag{4.6.1}$$

This equation gives the magnetic moment μ of the electron as a function of just its mass m and charge e. This same basic equation also applies to the muon. Taking the ratio of these two equations, we obtain $\mu_{\mathrm{muon}}/\mu_{\mathrm{electron}} = m_{\mathrm{electron}}/m_{\mathrm{muon}}$, which has an experimental accuracy of about 7 parts in a million [98]. This shows that the relationship between μ and m is the same for each of these particles, even though they have quite different masses.

The anomaly in the magnetic moment arises from the fact that very accurate measurements revealed the experimental value

$$\mu_{\mathrm{exp}} \cong (e\hbar/2mc)(1+\alpha/2\pi) \tag{4.6.2}$$

for both the electron and the muon. Since the magnetic moment depends on just the mass m, the anomaly logically also depends on the mass. Thus we can write

$$\mu_{\mathrm{exp}} = e\hbar/2(m-\Delta m)c\,, \tag{4.6.3}$$

which gives

$$\Delta m \cong m\cdot\alpha/2\pi\,. \tag{4.6.4}$$

The task now becomes one of identifying a mass component within the electron or muon that constitutes about 0.1% of the total mass of the particle. There are four kinds of mass (energy) that we can define for an electron or muon:

(a) *electrostatic self-energy* W_E;
(b) *magnetic self-energy* W_H;
(c) *mechanical mass* W_M;
(d) *gravitational mass* W_G.

A clue to selecting the relevant mass is provided by the fact that the magnetic moment anomaly also occurs in the gyromagnetic ratio. Hence this is an anomaly in the magnetic moment, and not in the spin. Viewing the anomaly in the context of Eq. (4.6.3), the anomaly logically corresponds to an *irrotational* mass component that affects μ but not J. The only irrotational mass component in the above list is the magnetic self-energy W_H, which is the magnetic field energy associated with the current loop that generates μ. The mass W_E of the point charge e is zero (Sec. 4.4), and the gravitational mass W_G is negligible at the length scales considered here. This leaves just W_H and W_M as masses that contribute to the gyromagnetic ratio. The mass W_M is responsible for almost all of the inertial properties of the particle, including the spin J and energy mc^2. Hence it is not irrotational. Thus W_H is singled out as the culprit that is causing the anomaly in g. Since the anomaly is a 0.1% effect, we expect to find $W_H/W_M \cong 0.1\%$. Does the RSS give us this result? Using the RSS, we need a calculation of W_H.

An early estimate of the energy in the magnetic field of the electron was supplied by Rasetti and Fermi [99], which we reproduce here. Using polar cylindrical coordinates $(r,\ \theta,\ z)$, consider the magnetic moment μ as arising from a current loop whose axis is along the z axis. The asymptotic magnetic field components are [100]

$$H_r = \frac{2\mu\cos\theta}{r^2}, \qquad H_\theta = \frac{\mu\sin\theta}{r^2}. \tag{4.6.5}$$

Assuming that μ is spread uniformly throughout a spherical volume of radius R_H, and that these equations apply all the way back to R_H, they obtained

$$\begin{aligned} W_H^{\text{ext}} &= \left(\frac{\mu^2}{8\pi}\right)\int_{R_H}^{\infty}\int_0^{\pi}\left(\frac{1}{r^6}\right)(3\cos^2\theta+1)2\pi r^2\sin\theta\,d\theta dr \\ &= \frac{\mu^2}{3R_H^3}, \quad (r > R_H). \end{aligned} \tag{4.6.6}$$

Several years later, Born and Schrödinger [101] made the same calculation, but this time assuming that the magnetic moment was distributed on the surface of the sphere. This gave $W_H^{\text{ext}} = \mu^2/2R_H^3$, in reasonable agreement with the Rasetti–Fermi calculation.

We can extend the Rasetti–Fermi result by making an estimate of the magnetic energy W_H^{int} that is contained inside of R_H. To do this, we evaluate the field equations (4.6.5) at the magnetic radius R_H, and then assume that these values hold constant all the way in to the origin. This gives

$$\begin{aligned} W_H^{\text{int}} &= \left(\frac{\mu^2}{8\pi R_H^6}\right)\int_0^{R_H}\int_0^{\pi}(3\cos^2\theta+1)2\pi r^2\sin\theta\,d\theta dr \\ &= \frac{\mu^2}{3R_H^3}, \quad (r < R_H). \end{aligned} \tag{4.6.7}$$

Thus the total magnetic field energy is

$$W_H^{\text{tot}} \cong \frac{2\mu^2}{3R_H^3}. \tag{4.6.8}$$

In the RSS model, with its equatorial charge distribution (Sec. 4.4), the magnetic radius R_H is equal to the Compton radius R_C of the particle:

$$R_H = R_C = \hbar/mc\,. \tag{4.6.9}$$

Substituting Eqs. (4.6.1) and (4.6.9) into Eq. (4.6.8) gives

$$W_H = \frac{\alpha}{6} \cdot mc^2\,. \tag{4.6.10}$$

Comparing this equation with Eq. (4.6.4) we see that

$$\frac{\alpha}{6} \simeq \frac{\alpha}{2\pi}\,, \tag{4.6.11}$$

so that we have satisfied Feynman's challenge! Of course, the close agreement shown in Eq. (4.6.11) must be regarded as somewhat fortuitous, given the approximations used for the magnetic field. But the key points here are that (1) we can unequivocally determine the *sign* of the anomaly (one of Feynman's requirements), and (2) we have supplied the *physical picture* that underlies the anomaly (another Feynman requirement).

The magnetic energy of the current loop can also be estimated by calculating the self-indictance of a thin classical current loop [102]. For a more complete discussion of these results see Refs. [103] and [104]. They are summarized as Lemma 7 in Table 4.4.1.

The present results not only satisfy a need requested by Richard Feynman, but they also constitute an argument in favor of spectroscopic mathology as applied to elementary particles. When physicists discovered that electrons interact in a point-like manner, they also came to the general conclusion that electrons *are* points. Once this conclusion was adopted, it became apparent that any effort to account for their spectroscopic properties in terms of the well-established theories of mechanics and electrodynamics was impossible. This position was clearly enunciated, for example, by Henry Margenau [105]:

> *An electron is an abstract thing, no longer intuitable in terms of the familiar aspects of everyday experience, but determinable through formal procedures such as the assignment of mathematical operators, observables, states, and so forth. In sum, the physicist, while still fond of mechanical models wherever they are available and useful, no longer regards them as the ultimate goal and the quintessence of all scientific description: he recognizes situations where the assignment of a simple model, especially a mechanical one, no longer works and where he feels called upon to proceed directly under the*

guidance of logical and mathematical considerations and at times with the renunciation of the visual aspects which classical physics would carry into the problem.

Margenau [105] and Feynman [77] were of course unaware of the relativistically spinning sphere, which did not appear in the literature until 1970 [89]. The importance of the RSS is that its mathology *does* work. It correlates the electron spin, magnetic moment, and anomalous magnetic moment; it provides a rationale for the vanishing of the electric quadrupole moment; and it accounts for the 3/2 mass ratio between muon pairs and pions that is revealed in the α-quantization of these masses. The price we have to pay for these RSS successes is to admit the electron as a Compton-sized particle into the company of the hadrons, all of whose measured sizes are Compton-like [106]. This "price" may in fact be the greatest achievement of the relativistically spinning sphere: it may finally convince physicists that whereas the size of an *electric charge* seems to be point-like, and hence also its interactions, the overall size of an *electron* that carries a charge with a large associated magnetic moment, and which has a small mass with a large associated spin angular momentum, must in fact be very unpoint-like.

The renunciation of spectroscopic mathology is one reason why QED theorists have been, as Feynman stated, "*computing terms like a blind man*" [77], and their rescue by string theorists seems to be nowhere in sight.

4.7 The Relativistic Transformation Properties of the Mathological Electron: Correct Transformations Occur only at the Rotational Relativistic Limit

In Sec. 4.3 we demonstrated that a relativistically spinning sphere correctly reproduces the experimental mass relationship $M_s/M_0 = 3/2$ between spinning ($J = 1/2$) and non-spinning ($J = 0$) mass quanta. Also, if the RSS is assigned the Compton wavelength $R_s = \hbar/M_s c$, the spin angular momentum $J_s = 1/2\hbar$ emerges as a calculated quantity. In Sec. 4.4 we showed that a point charge e placed on the equator of the RSS correctly reproduces the first-order magnetic moment $\mu = e\hbar/2M_s c$. A question which then arises is whether these properties are properly preserved under a Lorentz transformation from the rest frame of the electron to the laboratory frame of the moving electron. The Lorentz transformation is characterized by the parameter

$$\gamma = \frac{1}{\sqrt{1 - v^2/c^2}}, \tag{4.7.1}$$

where $\mathbf{v}$ is the velocity of the electron in the laboratory frame of reference. The correct transformation properties for the electron mass, spin and magnetic moment are [107, 108]

$$m_{\text{lab}} = \gamma m_{\text{cm}}\,; \quad J_{\text{lab}} = J_{\text{cm}}\,; \quad \mu_{\text{lab}} = \mu_{\text{cm}}/\gamma\,, \tag{4.7.2}$$

where the subscripts denote the laboratory and center of mass frames of reference.

A problem that arises in making a Lorentz transformation of the RSS is that the Lorentz transformation equations for a particle are formulated in such a way that they apply to the entire particle, whose internal structure is ignored. It is assumed that the particle is small enough that its motion can be categorized by the motion of its center of mass. Thus it is not clear how to obtain an analytical solution of the Lorentz transformation equations in the case of the relativistically spinning sphere with its relativistic inner structure. However, a numerical solution can be obtained by dividing the RSS into enough small components that each one of them is effectively a point, transforming the elements individually, and recombining them at the end of the calculation. This calculation was carried out as the following sequence of events [107]. An x, y, z Cartesian coordinate system was established in the center-of-mass frame of the electron, with the spin axis directed along the x axis. The RSS was divided into 20,000 mass elements [107], with a later calculation using 36,864 mass elements [108], and the x, y, z coordinates and $\dot{x}$, $\dot{y}$, $\dot{z}$ velocities were calculated for each element. Computer calculations showed that using a larger number of mass elements did not substantially change the results. Then a rotated x', y', z' coordinate system was established in the center-of-mass frame with the x' axis rotated at an angle θ with respect to the x axis, and the x', y', z' coordinates and velocities were calculated in this rotated frame. Finally, a transformation was made to the laboratory frame, with the electron moving at velocity $\mathbf{v}$ along the x' axis. The laboratory frame coordinates are

$$x_{\text{lab}} = x'/\gamma\,; \quad y_{\text{lab}} = y'\,; \quad z_{\text{lab}} = z'\,, \tag{4.7.3}$$

and the velocities are

$$\dot{x}_{\text{lab}} = \frac{\dot{x}' + v}{1 + \dot{x}v/c^2}\,; \quad \dot{y}_{\text{lab}} = \frac{\dot{y}'/\gamma}{1 + \dot{x}v/c^2}\,; \quad \dot{z}_{\text{lab}} = \frac{\dot{z}'/\gamma}{1 + \dot{x}v/c^2}\,. \tag{4.7.4}$$

Table 4.7.1 Calculated RSS Lorentz transformation values of the electron mass, spin, and magnetic moment as functions of the spin angle θ relative to the velocity vector, using linear electron velocities $v/c = 0.3$ (top and middle) and 0.7 (bottom).

Calculated RSS values as functions of θ.

Angle θ	Mass	Spin	Mag. Mom.
0°	+4.80%	+0.00%	−4.61%
30°	+4.81%	+0.04%	−4.68%
60°	+4.81%	+0.10%	−4.78%
Calculated angle-independent values from Eq. (4.7.2).			
	+4.83%	+0.00%	−4.61%

Separated coordinate and velocity components for the above calculations.

Angle θ	Component	Mass	Spin	Mag. Mom.
0°	coordinate	+0.00%	+0.00%	+0.00%
0°	velocity	+4.80%	+0.00%	−4.61%
0°	both	+4.80%	+0.00%	−4.61%
60°	coordinate	+0.00%	−1.73%	−1.73%
60°	velocity	+4.81%	+1.83%	−3.11%
60°	both	+4.81%	+0.10%	−4.78%
Calculated angle-independent values from Eq. (4.7.2).				
		+4.83%	+0.00%	−4.61%

Separated coordinate and velocity components, using the velocity $v/c = 0.7$.

Angle θ	Component	Mass	Spin	Mag. Mom.
60°	coordinate	+0.00%	−10.72%	−10.72%
60°	velocity	+40.01%	+15.03%	−21.88%
60°	both	+40.01%	+4.30%	−30.36%
Calculated angle-independent values from Eq. (4.7.2).				
		+40.03%	+0.00%	−28.59%

These quantities were then used to recalculate the RSS mass and spin values as observed in the laboratory frame. The magnetic moment calculation was made by dividing the equatorial charge e into a collection of fractional charges that were distributed around the equator so as to mimic the current distribution of the rotating charge e. The results of these calculations are displayed in Table 4.7.1.

The theoretical Lorentz transformation equations for the mass, spin and magnetic moment, which are independent of the angle of orientation θ of the spin axis to the forward velocity of the electron, are given in Eq. (4.7.2). Thus it is of interest to see how accurately this θ independence is maintained in the numerical calculations of the RSS. The top results in Table 4.7.1, which are based on a relativistic velocity $v/c = 0.3$, demonstrate that these three spectroscopic quantities are transformed, to a good approximation, in an angle-independent manner. Since these transformations involve both a relativistic contraction of length along the electron velocity vector and a relativistic velocity addition, we can separate out these two effects to see how they each affect the various transformed quantities. These results are shown in the middle of Table 4.7.1. As can be seen, the mass transformation depends only on the velocity components, as we expect. Interestingly, the spin transformation for angles $\theta \neq 0$ gets offsetting contributions from the length contraction and relativistic velocity addition, and thus remains constant. The magnetic moment transformation also gets transformations from both components (if $\theta \neq 0$), but these are both of the same sign, and they combine together to yield a $1/\gamma$ dependence [103, 104]. The final results at the bottom of Table 4.7.1 show the relativistic transformations for an electron velocity $v/c = 0.7$. The increased value for γ (Eq. 4.7.1) leads to larger relativistic spectroscopic changes in the mass, spin and magnetic moment, but the independence of θ is still maintained.

As a final relativistic examination of the relativistically spinning sphere — a result which is not displayed in Table 4.7.1 — we repeat these calculations for the case where the RSS is spinning at only 1/2 of the full relativistic limit; that is, with an equatorial velocity $v = 0.5c$. If we keep the radius fixed at the Compton radius (Lemma 4 in Table 4.3.1), then the calculated spin value is too low. This can be compensated for by increasing the RSS radius, but then the gyromagnetic ratio becomes too large, and there is no way to correct it. If we now apply the above Lorentz transformations to this slower-spinning sphere, we find that the mass and spin transform correctly for all angles θ. The magnetic moment transforms correctly when $\theta = 0$, but it becomes too large when $\theta \neq 0$. Thus the correct spectroscopic and Lorentz transformation equations are obtained for the relativistically spinning sphere only when it is rotating at the full relativistic limit of Eq. (4.3.1). The results discussed here are summarized in RSS Lemmas 8 and 9, which are displayed in Table 4.7.2.

This section concludes our treatment of the relativistically spinning sphere, which is an essential concept for mathological investigations of

Table 4.7.2 Lorentz transformation properties of the relativistically spinning sphere.

Lemma 8: A spinning sphere that is given a translational velocity w specified by the relativistic parameter $\gamma = \sqrt{1 - w^2/c^2}$ has the correct $M_\gamma = \gamma M_s$ (mass), $J_\gamma = J_s$ (spin) and $\mu_\gamma = \mu/\gamma$ (magnetic moment) relativistic transformation properties if, and only if, it is spinning at the full RSS relativistic limit $\omega = c/R$.

Lemma 9: The invariance of the spin angular momentum J_s of an RSS under translational motion is due to offsetting relativistic length contraction and mass increase effects, whereas the relativistic variation of the magnetic moment μ is due to effects of the same sign. The relativistic mass increase M_γ is independent of the length contraction.

the spectroscopic properties of Compton-sized elementary particles. In Chapter 5 we briefly consider another great mystery about elementary particles, the particle–wave dichotomy, where we must employ another essential concept — the particle–hole pair. The question of interpreting the electron as a particle or a wave continues down to the present time. David Hestenes summarized this problem very succintly [109]:

> *Is the electron a particle always, sometimes, or never? Theorists have come down on every side of this question.*

Physicists have recently devoted little attention to the mathological aspects of material *particles*, and they have probably devoted even less attention to the mathological aspects of immaterial *waves* and massless *photons*. But particle waves do exist that cause things to happen — or not happen; and circularly polarized spin-1 photons do exist whose angular momentum can rotate a quarter-wave plate. These entities should also admit a mathological explanation. As we will see, they do.

Chapter 5 The Mathology of Particle Waves: The Particle–Hole Pair

5.1 The Mathology of the Electron Phase Wave

All moving particles (e.g. electrons, neutrons and helium atoms) are accompanied by quantum mechanical de Broglie phase waves as they move along. The phase waves themselves travel at superluminal velocities, but a wave packet of superimposed phase waves can be constructed which has a group velocity that matches the velocity $\mathbf{v}$ of the particle. However, the wavelength λ_{group} of the wave packet plays no role in particle–wave interference experiments. It is the de Broglie wavelength $\lambda_{\text{phase}} = h/mv$ that enters into the spacings of the interference patterns.

The interpretation of the quantum mechanical wave function that accompanies the moving particle is a controversial topic. The conventional interpretation, first set forth by Max Born, is that the absolute square of the wave function amplitude in coordinate space gives the probability of finding the particle at a certain point in space, and the wave itself is a purely mathematical construct which is devoid of any measurable physical attributes. However, it would seem logical that a wave which produces physical effects should itself *be* physical. Regardless of interpretation, it is of mathological interest to see if a physical model can be devised that reproduces the properties of the de Broglie phase wave. That is the topic we address here.

The mathological *ansatz* we make is that a moving electron perturbs the vacuum state as it moves along and thereby excites "phase wave elements" that constitute the de Broglie phase wave. Since these wave elements seemingly have no directly measurable properties, we denote them as "zerons." The mathematical model we use is an *initial* state composed of just the

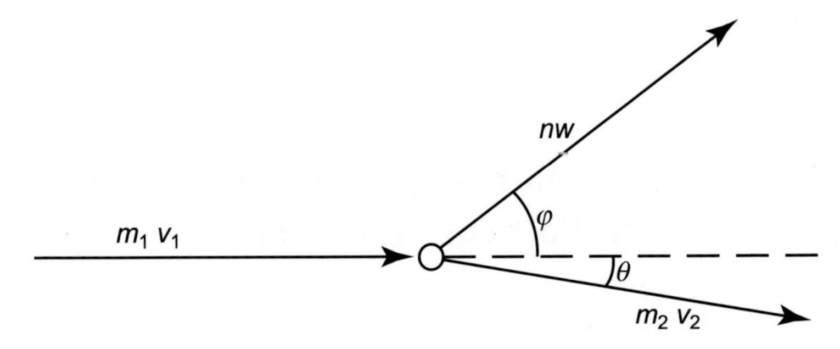

Fig. 5.1.1 A schematic diagram of a moving electron creating a spatial excitation in a process that conserves relativistic 4-momentum. The initial state is an electron that has relativistic mass m_1 and velocity v_1. The final state is the electron with mass m_2 and velocity v_2, plus an excitation quantum, denoted as a "zeron," that has mass n and velocity w. The final state electron and zeron are moving at angles θ and ϕ with respect to the axis defined by the incident electron. This problem is well defined relativistically. In order to conserve 4-momentum, the zeron has to be moving at a superluminal velocity, so it is a "tachyon" with a formally imaginary rest mass but a real energy value. The solutions to this problem in the limit where the zeron has much smaller energy than the electron are shown in Fig. 5.1.2.

electron, and a *final* state that has the slightly perturbed electron plus a zeron. This is a mathematically well-defined problem within the context of special relativity. Figure 5.1.1 shows the scattering diagram for this process [110–112].

The most direct way to solve this scattering problem is by using four-vectors to represent the electron and zeron [110]. The de Broglie wave equation is customarily written in the three-vector form

$$(1/\lambda)\boldsymbol{n} = \mathbf{p}/h\,, \tag{5.1.1}$$

where $\boldsymbol{p} = m\mathbf{v}$ is the relativistic momentum of the electron and λ and $\boldsymbol{n}$ are the wavelength and directional vector of the planar de Broglie phase wave. However, it can also be written in the covariant four-vector form [113]

$$\hat{P} = h\hat{L}\,, \tag{5.1.2}$$

where

$$\hat{P} = (E/c, \mathbf{p}) = m(c, \mathbf{v}) \tag{5.1.3}$$

is the energy–momentum four vector of a massive particle, and

$$\hat{L} \equiv (1/\lambda)(\mathrm{w}/c, \boldsymbol{n}) = \nu(1/c, \boldsymbol{n}/\mathrm{w}) \tag{5.1.4}$$

is the frequency four-vector, with w the velocity of the phase wave.

The energy–momentum four-vector equation for the scattering process displayed in Fig. 5.1.1 is

$$\hat{P}_1 = \hat{P}_2 + \hat{Q}\,, \tag{5.1.5}$$

where $\hat{P}$ and $\hat{Q}$ represent the electron and zeron states, which have masses m and n, respectively. It is useful to write the magnitude of the electron three-momentum in the form

$$p = c(m^2 - m_0^2)^{1/2}\,. \tag{5.1.6}$$

The magnitude of the zeron three-momentum is

$$q = n\mathrm{w}\,, \tag{5.1.7}$$

where the rest mass of the zeron is not yet specified. The four-vectors for Fig. 5.1.1 are

$$\hat{P}_1 = (cm_1, p_1, 0, 0)\,, \tag{5.1.8}$$

$$\hat{P}_2 = (cm_2, p_2\cos\theta, p_2\sin\theta, 0) = (cm_2, p_2^f, p_2^t, 0)\,, \tag{5.1.9}$$

$$\hat{Q} = (cn, q\cos\phi, q\sin\phi, 0) = (cn, q^f, q^t, 0)\,, \tag{5.1.10}$$

where the superscripts f and t denote forward and transverse momentum components, respectively. Inserting these three equations into Eq. (5.1.5) gives

$$p_1 = p_2^f + q^f\,, \tag{5.1.11a}$$

$$0 = p_2^t + q^t\,, \tag{5.1.11b}$$

$$\tan\phi = q^t/q^f\,, \tag{5.1.11c}$$

$$p_2^2 = (p_2^f)^2 + (p_2^t)^2\,. \tag{5.1.11d}$$

We now specify the input parameters m_0, m_1, n and ϕ, which are sufficient to determine the left-hand quantities in Eqs. (5.1.11). Hence we have four equations for the four unknowns p_2^f, p_2^t, q^f and q^t, and we can solve for all of them. The solutions to these equations in the perturbative limit

$$n \ll m_1 \tag{5.1.12}$$

are denoted [110] as the equations of *Perturbative Special Relativity* (PSR). We can also write the PSR limit of Eq. (5.1.12) in the form

$$\Delta p \equiv p_1 - p_2 \ll p_1\,. \tag{5.1.13}$$

The PSR equation we need here is the one for q^f, which is

$$q^f = [p_1 - \{p_1^2 - (p_1^2 - p_2^2)\sec^2\phi\}^{1/2}]/\sec^2\phi\,. \tag{5.1.14}$$

This equation is exact. Applying the PSR limit of Eq. (5.1.13), we set $p_2 = p_1 - \Delta p$ in Eq. (5.1.14), expand the square root, and discard terms in $(\Delta p)^2$. The $\sec^2\phi$ term cancels out, and Eq. (5.1.14) is reduced to the approximate PSR quantity

$$q^f \cong \Delta p\,, \tag{5.1.15}$$

which shows that the zeron forward velocity is independent of the scattering angle ϕ. As the final step, we use Eqs. (5.1.6) and (5.1.13) and the energy equation $m_1c^2 = m_2c^2 + nc^2$ to obtain the PSR equation (keeping terms of order n)

$$\begin{aligned}\Delta p &= c(m_1^2 - m_0^2)^{1/2} - c[(m_1 - n)^2 - m_0^2]^{1/2}\\ &\simeq cm_1n/(m_1^2 - m_0^2)^{1/2} = c^2n/\mathrm{v}_1\,.\end{aligned} \tag{5.1.16}$$

Since $q^f \equiv n\mathrm{w}^f$, Eqs. (5.1.15) and (5.1.16) lead to the equation

$$\mathrm{w}^f \simeq c^2/\mathrm{v}_1\,, \tag{5.1.17}$$

which is independent of both the zeron mass and the scattering angle ϕ. In the PSR limit, we have $\mathrm{v}_1 \simeq \mathrm{v}_2 \simeq \mathrm{v}$, so that we can write Eq. (5.1.17) in the vector form

$$\mathbf{v}\cdot\mathbf{w} \simeq c^2\,, \tag{5.1.18}$$

which is a generalization of the de Broglie equation $\mathrm{vw} = c^2$.

Figure 5.1.2 is a plot of the zeron forward velocity w^f as a function of the perturbation parameter n/m_1 for various values of the zeron scattering angle ϕ, and also a plot of the de Broglie velocity product $\mathrm{v}_1\mathrm{w}^f$ as a function of the angle ϕ, where $\mathrm{v}_1$is the incident electron velocity. The electron energy is 100 KeV, and the velocity of light c is set equal to unity. The left ordinate in Fig. 5.1.2 gives the scale for the forward velocity w^f. As can be seen, the velocities w^f are all superluminal, so the zeron is a *tachyon*. When the value of the perturbation parameter n/m_1 is less that 10^{-5}, the zerons at all scattering angles ϕ move with the same velocity w^f, and thus create an accurately planar wave. The de Broglie phase wave is the only physical phenomenon clearly identified to date that involves velocities greater than the velocity of light. The right ordinate in Fig. 5.1.2 gives the product $\mathrm{v}_1\mathrm{w}^f$ as a function of n/m_1 (in units where $c = 1$). This product accurately approaches the de Broglie value of unity in the PSR limit.

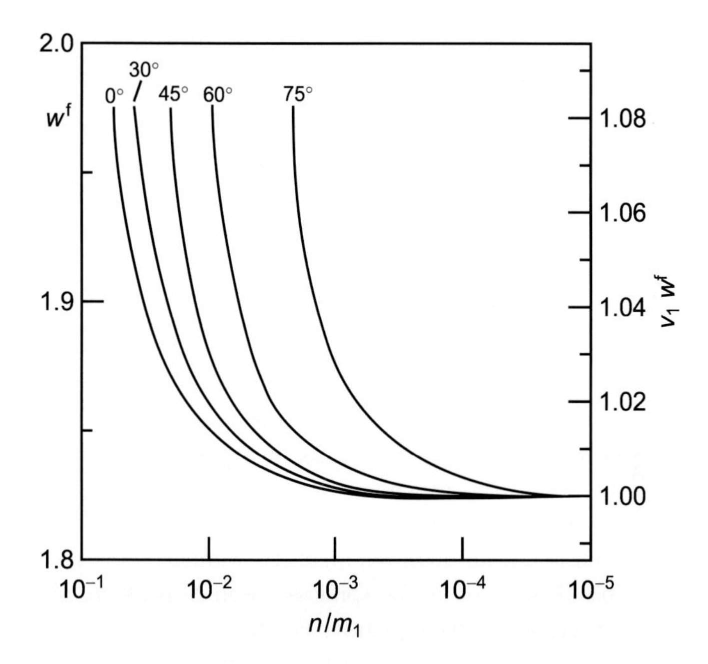

Fig. 5.1.2 Solutions to the electron spatial excitation process that is portrayed in Fig. 5.1.1. The velocity of light c is set equal to unity. With these units, the energy of the incident electron is m_1, the energy of the emitted zeron is n, and the "perturbative limit" is reached when the mass ratio n/m_1 is much less than unity ($n/m_1 < 10^{-5}$). The zeron moves outward with total velocity w at the scattering angle ϕ. The zeron forward velocity w^f is the projection of w along the line of the incident electron velocity v_1. In Fig. 5.1.2, the mass ratio n/m_1 is plotted along the abscissa. Curves are drawn for various values of the zeron scattering angle ϕ. The forward velocity w^f is plotted along the left ordinate, and the velocity product $v_1 \cdot w^f$ is plotted along the right ordinate. As can be seen, the velocity product $v_1 \cdot w^f$ approaches unity for all values of the scattering angle ϕ as the mass ratio n/m_1 decreases to 10^{-5} or less. Thus in the $n/m_1 < 10^{-5}$ perturbative limit, the forward velocity w^f of the scattered zeron matches the de Broglie phase velocity for all scattering angles ϕ, so that the de Broglie phase wave (a collection of zerons) is accurately planar.

As the final calculation in this section, we evaluate the rest mass m_0 of the zeron, which is obtained from the equation

$$\hat{Q} \cdot \hat{Q} = (\hat{P}_1 - \hat{P}_2) \cdot (\hat{P}_1 - \hat{P}_2) = n_0^2 \,. \tag{5.1.19}$$

Using $\hat{P}_1 \cdot \hat{P}_1 = \hat{P}_2 \cdot \hat{P}_2 = m_0^2 c^2$, we have

$$\hat{Q} \cdot \hat{Q} = 2c^2[m_0^2 - m_1 m_2 + (m_1^2 - m_0^2)^{1/2}(m_2^2 - m_0^2)^{1/2} \cos\theta] \,. \tag{5.1.20}$$

Setting $m_2 = m_1 - n$, going to the PSR limit $n \ll m_1$, using a square root expansion, inserting the PSR angle relationship

$$\sin\theta \simeq -m_1 n \tan\phi/(m_1^2 - m_0^2)\,, \tag{5.1.21}$$

and keeping terms through order n^2, we obtain

$$n_0^2 \simeq -n^2(m_0^2 + m_1^2 \tan^2\phi)/(m_1^2 - m_0^2)\,. \tag{5.1.22}$$

Since n^2 is a positive energy term, we see that n_0^2 is negative, so that n_0 is imaginary, as it must be for superluminal zerons in this relativistic formalism. We can also write Eq. (5.1.22) in the equivalent form

$$n_0 \cong in[(c^2 \sec^2\phi/\mathrm{v}_1^2) - 1]^{1/2}\,, \tag{5.1.23}$$

where v_1 is the velocity of the incident electron.

It should be noted that the rest mass of a particle is not always an observable state. From the viewpoint of the RSS model in Sec. 4.3, the rest mass of the spinning electron is the spinless mass $m_{e_0} = (2/3)m_e = 0.341$ MeV, which is not an observed quantity. Similarly, the rest mass of the spinning 105 MeV muon is the spinless boson α-mass $m_b = 70$ MeV, which is not a directly observed quantity either.

If we think only in terms of positive-mass particles, then objects such as the zeron which move at velocities greater than c are not allowed. And yet the zeron carries positive energy, as we have calculated here. How can this be? In seeking to address this problem, we should keep in mind the photon, which carries positive energy *at* the velocity c. The mathematical solutions to the zeron and the photon spectroscopic equations are obtained by using mathological models which contain both positive masses and hole states. The hole states, which are well known in solid state physics, function *mechanically* like "negative" masses, and thus cancel out the divergent positive masses at highly relativistic velocities; but they move under the action of *electromagnetic* forces in the same manner as positive masses. This is the subject of Sec. 5.2.

5.2 The Mathology of Particle–Hole Pairs: Zerons and Photons

The concept of the wave function in quantum mechanics remains a mystery. The mathematical formulations of particle motions under the influence of wave functions work so well that they are certainly correct. Quantum mechanics has undergone countless tests and emerged unscathed each time. And yet the particle wave itself seems to have no measurable attributes like energy or momentum. When a photon traveling from a distant star enters a telescope, it is contained in a wave packet that is larger than the reflecting mirror of the telescope, so that all portions of the mirror, which can be several meters in diameter, contribute to the focusing process. And yet as soon as the photon is detected, which is at a discrete point, the wave packet collapses and vanishes without a trace, and without having apparently removed any energy from the photon.

The wave mystery becomes more acute when we consider the electron double-slit experiment. A single electron is emitted from a source and travels to a detecting surface, passing through one of two closely-spaced slits on the way. It is detected as a localized object. After enough electrons have traveled separately through the apparatus, a diffraction pattern becomes apparent, which can be understood only under the assumption that both slits are contributing to the motion of the electron. Close one slit and the interference pattern is gone. Disturb the electron by measuring through which slit it went and the interference pattern is gone. But if the electron is a discrete particle, it cannot go through both slits. Thus something else is going through both slits — the electron wave. This means that we must ascribe physical reality to the wave. But it is a physical reality that is beyond the range of our most accurate measuring devices.

The wave mystery deepens even more when correlation experiments are made in which two simultaneously emitted photons travel in opposite directions, are scattered, and then exhibit correlations that reflect the scatterings of both of them, even though the scatterings take place outside of their mutual light cone. We can conceptually picture that the correlations are due to the fact that the photons share a common entangled wave function, but the wave must be transmitting information from one to the other at superluminal velocities, which is in conflict with special relativity.

These quantum mechanical particle–wave mysteries have been summarized by Feynman in a well-known quotation [114]:

> *I think it is safe to say that no one understands quantum mechanics .... Do not keep saying to yourself, if you can possibly avoid it, 'But how can it be like that?' because you will get 'down the drain' into a blind alley from which nobody has yet escaped. Nobody knows how it can be like that.*

In the face of these conceptual challenges, it seems like the last thing we need is a representational model based on essentially classical concepts to "explain" the nature of the quantum mechanical wave. And yet it is the duty of mathologists to see if models do exist that have the properties we ascribe to particle waves and to the photon itself. Whether the models correspond to physical reality is a different question, of course, but if the models do not even exist, then this is clearly the wrong track to pursue. And if the models do exist, then they deserve at least some additional follow-up attention to see just where, if anywhere, they are in conflict with experimental data.

The subject of particle waves is really beyond the scope of the present studies, which are centered on the phenomenological lifetime and mass systematics of massive particles. But our investigations into the mathology of the relativistically spinning sphere and its relevance to the spectroscopy of the electron have produced enough results that it seems reasonable to at least peek into the area of particle waves and see if we can find any related concepts. In particular, we were able to obtain the spin of the electron as the rotation of a real physical mass, and not just as a representation of the two-dimensional rotation group. Is there a model that reproduces the spin-1 $\hbar$ of the photon and the spin-0 $\hbar$ of the zeron, while at the same time paying attention to the other attributes of these entities? It turns out that the answer is in the affirmative *if* we are willing to admit "hole states" into our set of "particle" basis states. Of course, the question then arises as to what the hole states are holes *in*; they must be holes in the spatial fabric we denote as the "vacuum state." In Newton's time this concept would have raised no problems: they are holes in the "ether." But after Michelson and Morley, the ether disappeared. However, recently the ether has possibly reappeared in a somewhat different guise — as the unknown "dark matter" and/or "dark energy" that astrophysicists think must constitute most of the matter of the universe. In any event, we present the "particle–hole" mathological models of the zeron and photon here to complete the picture of particles and waves that we have developed in the preceding sections of this chapter. We cannot fully understand elementary particles until we understand particle waves, and conversely. They go together. The properties of particle–hole pairs have been discussed elsewhere [115, 116], and the present results are drawn from that work.

Hole states are well-known in semiconductor physics. They exist as the unfilled last-electron states that are required to close an atomic electron shell. The density of hole states is customarily determined in "Hall effect" measurements, where the holes are singled out from electrons by the reversal of the sign of the lateral Hall voltage. The holes themselves have no intrinsic physical properties. They are vacancies in distributed systems whose filled elements do have physical properties. Holes are zero occupation numbers, and they represent available degrees of freedom — not for themselves, but for their adjacent neighbors. The energy of a hole state is "negative" with respect to the average energy density in the region surrounding it, and the hole moves mechanically as if it represents a "negative mass" [117]. The electric charge of the hole is "positive" with respect to the average electron charge density of the region surrounding it, but it does *not* move electromagnetically as a negative mass, as we know for example from the Hall effect systematics. It is easy to picture just why this is so. If a region containing a hole is subjected to a *mechanical* force from the left, a particle to the left of the hole moves into the hole, thus carrying momentum to the right. The hole itself is moved to the left in this process, so it moves in the opposite direction of the momentum shift. We can represent this by assigning the hole a negative mass in the Newtonian equations of motion. If the region containing a hole is subjected to an *electrostatic* force that moves a positive charge to the right, it does not act on the hole, since the hole actually has no charge, but it pulls a negatively charged neighbor from the right into the hole, thus moving the hole to the right, in the direction of the applied force. Hence the *position* of the hole moves in the same direction that a positively charged positive-mass particle would move. We have

$$\mathbf{P} = +m\mathbf{v}, \quad d\mathbf{P}/dt = \mathbf{F} = +m\mathbf{a}\,, \quad \text{positive} - \text{mass state}\,, \tag{5.2.1}$$

$$\mathbf{P} = -m\mathbf{v}, \quad d\mathbf{P}/dt = \mathbf{F} = -m\mathbf{a}\,, \quad \text{negative} - \text{mass state}\,, \tag{5.2.2}$$

$$\mathbf{P} = -m\mathbf{v}, \quad d\mathbf{P}/dt = \mathbf{F}_{\text{mech}} = -m\mathbf{a}\,, \quad \text{hole or antihole state}\,, \tag{5.2.3}$$

$$\mathbf{P} = -m\mathbf{v}, \quad -d\mathbf{P}/dt = \mathbf{F}_{\text{elec}} = +m\mathbf{a}\,, \quad \text{hole or antihole state}\,, \tag{5.2.4}$$

where $\mathbf{v} = d\mathbf{x}/dt$, $\mathbf{a} = d^2\mathbf{x}/dt^2$, and $\mathbf{x}$ is the position vector of the particle or hole.

In Sec. 5.1 we presented the mathological equations for a vacuum-state excitation denoted as a "zeron" that was produced by a moving electron in a process that conserves relativistic energy and momentum (Fig. 5.1.1). The solution of the scattering equation showed that in the limit where the total energy of the zeron is much smaller than that of the electron (by a factor of at least 10^{-5}), the zeron moves with a forward velocity w^f equal to the de Broglie phase wave velocity, $\mathrm{w}^f = c^2/\mathrm{v}_{\text{electron}}$. We will now see if the spectroscopic properties of the zeron can be mathologically reproduced as a particle–hole pair.

Consider a particle–hole excitation that is created as a spatial perturbation, with the particle state carrying a negative charge (P^-) and the associated hole carrying a positive (effective) charge (H^+) with respect to the substrate. After the initial separation in the creation process, the particle and hole will be electrostatically attracted towards one another and will start to reunite. The one thing that can prevent them from quickly reuniting is if they are in rotational motion, so that the centrifugal force counterbalances the electrostatic attraction. Let us denote the electrostatic potential energy of the P^-–H^+ pair as

$$\Delta \equiv e^2/2r\,, \tag{5.2.5}$$

where $2r$ is their separation distance. The inward electrostatic force is $F = e^2/4r^2$. Consider first the dynamical equations for the particle state P^-, which has mass m. The inward electrical attraction must be matched by the outward centrifugal force, so that

$$m\omega^2 r = e^2/4r^2\,, \tag{5.2.6}$$

where ω is the angular velocity of the rotating P^-–H^+ pair. We now attribute a quantized angular momentum of $\hbar/2$ to P^-, which forms half of the pair. When we do this, we are letting a single P–H pair represent the zeron, which in actuality is probably a synchronized ensemble of P–H pairs. But this makes a useful limiting case for the subsequent characterization of the zeron as a multiplicitly of pairs. In this single-pair limit we have the angular momentum equation

$$\hbar/2 = mr^2\omega\,. \tag{5.2.7}$$

Eliminating the mass m from Eqs. (5.2.6) and (5.2.7) gives the energy equation

$$\hbar\omega = e^2/2r \equiv \Delta\,, \tag{5.2.8}$$

where Δ is the electrostatic potential energy defined in Eq. (5.2.5). Since $\hbar\omega$ is the energy quantum that Planck associated with electromagnetic radiation of frequency ω, we can think of Δ as the "Planck energy" of the rotating zeron. The above equations yield the following mc^2/Δ energy ratio:

$$mc^2/\Delta = 2/\alpha^2 = 37{,}558\,, \tag{5.2.9}$$

where $\alpha = e^2/\hbar c$. (The factor $2/\alpha^2$ is also the ratio of the electron mass to its binding energy in a hydrogen atom.) As can be seen, the rotating mass m in this single-mass limit is much larger than the electrostatic energy of the P–H pair, but the required mass rapidly gets smaller as the number of pairs in the wave element is increased.

We can see from Eq. (5.2.8) that the azimuthal velocity of the rotating P–H pair is

$$\mathrm{v} = \alpha c/2 = 1.1 \times 10^8 \text{ cm/s} \tag{5.2.10}$$

for all values of ω. (This is equal to the velocity of an electron or positron in the positronium ground state.) Since we have $\mathrm{v}/c \ll 1$, the rotational motion is non-relativistic. The corresponding equation for the radius r is

$$r = \mathrm{v}/\omega = \alpha c/2\omega\,. \tag{5.2.11}$$

A key mathological property of the rotating P^-–H^+ pair is that solutions exist for every value of the rotational frequency ω. The rotational motion is defined by the radial force equation (5.2.6) and the angular momentum quantization equation (5.2.7). These two equations contain the three variables m, r, and ω, and they serve as two constraints on these variables, leaving the third variable free to take any value. Thus if we want this zeron excitation to serve as a wave element in the electron phase wave (which is our goal here), we must allow for the fact that electron waves come with a full range of wavelengths and associated frequencies, as do photons and electromagnetic waves (another eventual goal).

The equations we have displayed here for the *mass* state P^- are standard results that apply for example to a classically rotating electron–positron pair. Now we must extend them to apply to the rotating *hole* state H^+. We can see from Eq. (5.2.4) that the acceleration $\mathbf{a}$ of the position of the hole H^+ under the action of the electrostatic field is the same as that of the mass P^-, but the *linear momentum* $\mathbf{P}$ is reversed. To picture the effect of this on the rotating system, let us first view it in a non-rotating frame of reference, as shown in Fig. 5.2.1, where the P^-–H^+ pair is rotating in the counterclockwise direction. The particle P^- at the top has its momentum vector $\mathbf{P}$ directed to the left, and the momentum change $d\mathbf{P}$ is directed inward. Thus the P^- momentum vector is rotating in the counterclockwise direction, and the particle motion follows the circular path shown in Fig. 5.2.1. The hole H^+ at the bottom has its momentum vector $\mathbf{P}$ directed to the left (opposite to the linear velocity $\mathbf{v}$), and the momentum change $d\mathbf{P}$ is directed outward (opposite to the acceleration $\mathbf{a}$). Thus the H^+ momentum vector is also rotating in the counterclockwise direction, and it follows the same circular path as P^-.

It is instructive to also view the P^-–H^+ rotation in a frame of reference that is rotating with the same angular frequency ω as the P^-–H^+ pair. In

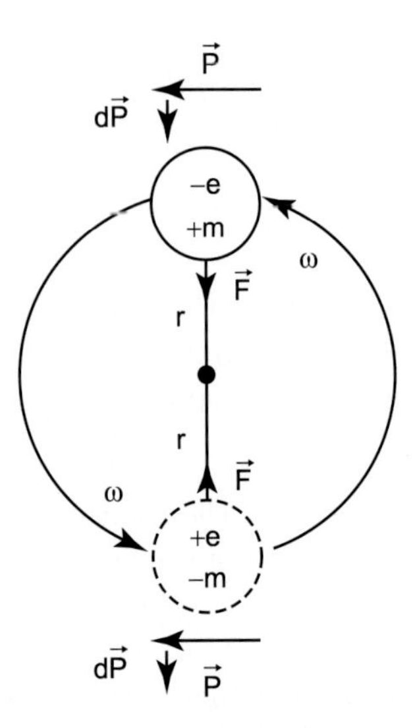

Fig. 5.2.1 The "zeron," a rotating P–H particle–hole pair, as observed in the laboratory frame of reference. The outward centrifugal force is balanced out by the inward electrostatic attraction between the − charge on the particle P and the (effective) + charge on the hole H. The momentum vectors for the hole H are in the direction opposite to the motion of the hole, but the angle they specify for the hole motion is the same as that for the particle motion, so both P and H follow the same circular path. The angular momentum and linear momentum of the P–H pair are formally equal to zero in first approximation (hence the name "zeron"). The angular frequency ω required for stable P–H rotation is the same as the frequency ω_d that is ascribed to the de Broglie phase wave.

this frame, which is displayed in Fig. 5.2.2, the positions of P and H are stationary. The inward electrostatic force on the particle P^- is counterbalanced by the outward centrifugal force $m\omega^2 r$. The forces operating on the hole H^+ are more complex. The inward electrostatic force on H^+ causes an inward acceleration of its position vector, which is associated with an outward acceleration of mass, so that $d\mathbf{P}/dt$ is outward, as shown in Fig. 5.2.2. In addition to this electrostatic force, there are two rotational forces operating: (1) the motion of H^+ to the right (as in Fig. 5.2.1) corresponds to a streaming of mass to the left. The centrifugal force associated with this mass motion is

$$F_{\text{centr}} = -m\boldsymbol{\omega} \times (\boldsymbol{\omega} \times \boldsymbol{r})\,. \tag{5.2.12}$$

Since this equation is independent of the sign of the rotation vector $\boldsymbol{\omega}$, the centrifugal force $\mathbf{F}_{\text{centr}}$ is always directed outward, regardless of the direc-

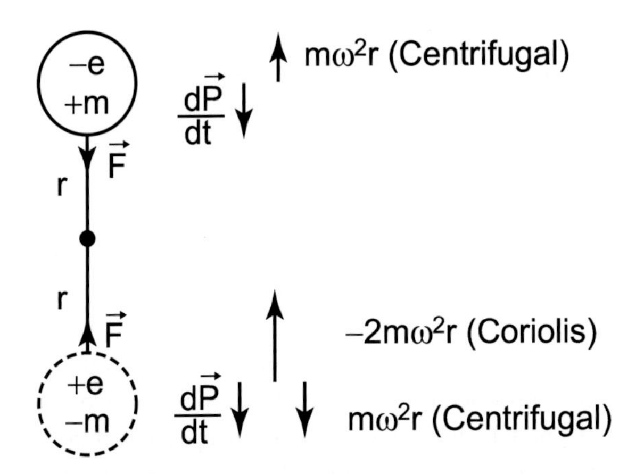

Fig. 5.2.2 The zeron *P–H* pair as observed in the rotating frame of reference. The centrifugal and electrostatic forces cancel out for the particle state *P*. They do not cancel out for the hole state *H*, since the mechanical momentum vector is reversed. But the state *H* has an additional Coriolis force that is caused by mass streaming through *H* in the direction opposite to that of the rotating system. This Coriolis force cancels out the other two momentum components, thus leading to stable rotation.

tion of the streaming mass motion. Thus the H^+ momentum components that correspond to the electrostatic attraction and the centrifugal force are both directed outward. Since the position of H^+ in Fig. 5.2.2 is stationary, it might seem that there is no Coriolis force. But the masses that form the boundary of the hole H^+ are not stationary: mass is streaming in the direction opposite to that of the rotating coordinate system. This gives rise to a Coriolis force [118]

$$\mathbf{F}_{\mathrm{cor}} = -2m\boldsymbol{\omega} \times \mathbf{v} \tag{5.2.13}$$

which is directed inward with a magnitude of $2m\omega^2 r$. This inward Coriolis force counterbalances both the outward centrifugal force and the outward electrostatically induced force. Thus the hole state H^+ follows the same trajectory as the mass state P^-.

Now let us reproduce the zeron as an ensemble of k rotating P–H pairs. Each particle state P_k has a mass m_k and carries an integer negative charge e. Equations (5.2.5)–(5.2.7) become

$$\Delta_k \equiv e^2/2r_k \,, \tag{5.2.14}$$

$$m_k\omega^2 r_k = e^2/4r_k^2 \,, \tag{5.2.15}$$

$$\hbar/2k = m_k r_k^2 \omega \,. \tag{5.2.16}$$

Solving this set of equations gives

$$m_k = m/k^3\,; \quad r_k = kr\,; \quad \hbar\omega = \sum_k \Delta_k\,, \tag{5.2.17}$$

where m and r are the quantities that were defined in the single-pair limit of Eqs. (5.2.5)–(5.2.7). The total particle mass $m_P = \Sigma m_k$ that is contained in this ensemble of k pairs is smaller than the single-pair mass m by a factor of k^2. This is due to the fact that the pair radius r_k is increased over the single-pair radius by a factor of k, which permits more efficient use of the mass in generating centrifugal force and angular momentum. By taking k large enough, we can drive the mass m_P down to any required limit.

If we were to reproduce these results using positive mass-negative mass pairs instead of particle–hole pairs, the negative-mass quanta would keep running off in the wrong direction (Eq. (5.2.2)), and no stable motion would be possible. Also, if we were to use P–H particle–hole pairs in stationary (non-rotating) configurations, they would quickly recombine. Thus *rotating* P–H pairs uniquely lead to a stable excitation state.

The angular momentum of a revolving mass is given by the product $\mathbf{r} \times \mathbf{P}$, where $\mathbf{r}$ is the radius vector and $\mathbf{P}$ is the instantaneous linear momentum vector. We can see from Fig. 5.2.1 that $\mathbf{P}$ for the particle state P is directed along the direction of rotation, but $\mathbf{P}$ for the hole state H is oppositely directed. Thus their angular momenta have opposite signs, and their total angular momentum essentially vanishes: the rotating zeron is a spin-0 entity. The linear momentum of the traveling zeron is also zero, which makes the zeron difficult to detect.

If we picture the zeron as an element of the de Broglie phase wave of an electron, then its total energy must be smaller than the electron energy by a factor of at least 10^{-5}, as shown in Fig. 5.1.2. The angular frequency associated with the de Broglie wave is $\omega_D = mc^2/\hbar$ where mc^2 is the electron energy, so the electrostatic potential energy Δ in Eq. (5.2.8) is equal to the electron energy. Hence this large electrostatic energy must be canceled out by using slightly different masses for P and H in the P–H zeron pair [115]. Since the m masses associated with P and H are much larger than the electron mass, as shown in Eq. (5.2.9), this unbalance causes only a small perturbation in the rotational motion. (This argument scales accordingly when applied to an ensemble of k pairs.) Hence the freedom exists in the zeron model to portray the zeron as an essentially zero angular momentum entity that has only an infinitesimal mass and carries the de Broglie wave frequency. Thus the zeron has the required properties to mathologically represent the electron wave excitation quantum that is displayed in

Fig. 5.1.1 (with momentum nw), which is the scattering diagram that leads to the PSR equations of perturbative special relativity.

If the zeron is used to serve as an element in an electron de Broglie phase wave, it must carry the same rotational frequency ω as that associated with the de Broglie phase wave. This fixes the frequency ω in Eqs. (5.2.6) and (5.2.7), and hence also the mass m and radius r. The de Broglie phase wave frequency is given as $\omega_D = \mathrm{V}/\lambda\!\!\!^{-}$, where $\mathrm{V} = c^2/\mathrm{v}$ is the phase velocity, v is the particle velocity, $\lambda\!\!\!^{-} = \hbar/m_e\mathrm{v}$ is the de Broglie wavelength, and m_e is the electron mass. These equations yield $\hbar\omega_D = m_e c^2$. Thus the electrostatic energy of the zeron — the Planck energy $\Delta = \hbar\omega$ — is equal to the relativistic mass $m_e c^2$ of the electron that produced it, as we mentioned above. Since the total energy of the zeron has to be much smaller than that of the electron in order to reproduce the de Broglie phase velocity (Fig. 5.1.2), the negative hole energy has to be large enough to balance out these positive energy components.

As we have just demonstrated, the zeron (by construction) and the de Broglie phase wave have the same angular velocity, $\omega = \omega_D = mc^2/\hbar$, where mc^2 is the relativistic energy of the electron. What is the rotational frequency of the electron itself? This follows from the limiting angular velocity of the relativistically spinning sphere, $\omega_s = c/r$, and the Compton radius of the sphere, $r = \hbar/m_e c$, which give $\omega_s = m_e c^2/\hbar$. Hence the spinning electron, the electron phase wave, and the zeron particle–hole phase wave element all have the same angular velocity, which is in line with the identification of a cloud of zerons as the constituents of the de Broglie phase wave.

The other use we have for the zeron P–H model is to employ it in constructing a mathological model for the photon. Albert Einstein has issued a warning in this regard:

> *All the fifty years of conscious brooding have brought me no closer to the answer to the question, 'What are light quanta?' Of course today every rascal thinks he knows the answer, but he is deluding himself.* [119]

Keeping this admonition in mind, we nevertheless forge ahead. It has been demonstrated that a beam of circularly polarized photons produces a macroscopically observable rotation in a quarter-wave plate [120], so the spin 1 of the photon is a real physical quantity that can logically be represented in terms of rotating masses. Electromagnetically, Bateman demonstrated in 1923 that the properties of the photon can be represented by a moving electric dipole [121]. These ideas were developed later by Bonner [122], and a discussion of electric charges moving close to the velocity c is given in Jackson's *Classical Electrodynamics* [123].

The photon has balanced particle–antiparticle symmetry, whereas the P^-–H^+ zeron discussed above does not. Also, the photon has spin 1 and the zeron has spin 0. Thus there are essential differences between them. The vacuum state itself appears to have balanced particle–antiparticle symmetry. Hence anti-zerons can logically be produced in the same manner as zerons. The mathological solution to the photon problem is to excite both a zeron P^-–H^+ pair and an anti-zeron P^+–H^- pair, and then merge them

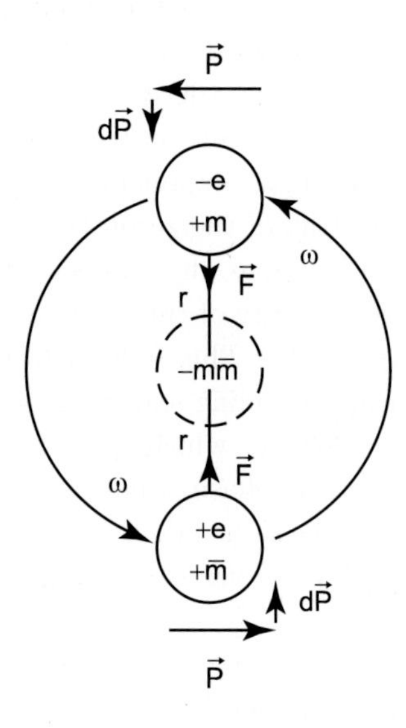

Fig. 5.2.3 The photon, shown reproduced as a matching set of P–H and $\bar{P}$–$\bar{H}$ particle–anti-particle pairs. The two hole states coalesce together at the center, and the P^- and $\bar{P}^+$ particle and antiparticle states rotate around the center in the same manner as the zeron particle state shown in Fig. 5.2.1. The angular momentum of the spinning P^- and $\bar{P}^+$ particles is 1 $\hbar$, and the total energy E_γ of the photon (by construction) is the electrostatic energy of the rotating pair. The required rotational velocity ω is in agreement with the Planck equation $E_\gamma = \hbar\omega$ The coalesced hole states play the role of a (non-interacting) renormalization constant in QED, which is assigned a negative mass value that cancels out the large positive contributions from the rotating masses P and $\bar{P}$ that are required to reproduce the angular momentum 1 $\hbar$ of the photon. This traveling electric dipole reproduces the main features of a circularly-polarized photon [121–123]. As discussed in the text, the photon may actually be reproduced as an ensemble of P–H and $\bar{P}$–$\bar{H}$ pairs, which greatly reduces the required masses of the P and $\bar{P}$ particles.

together so that the H^+ and H^- hole states overlap and coalesce, which gives the photon configuration shown in Fig. 5.2.3. This configuration reproduces the standard spectroscopic features of the photon. The positive-mass particle states P^- and P^+ each obey Eqs. (5.2.5)–(5.2.10) listed above for single-pair configurations, and Eqs. (5.2.14)–(5.2.17) for multi-pair ensembles, and they each contribute $\frac{1}{2}\hbar$ to the angular momentum. The overlapping hole states masses H^+ and H^- cancel out the energy contributions of the rotating P^- and P^+ masses (by construction), so the total energy of the system is the Planck energy $\Delta = \hbar\omega$, which corresponds to the electrostatic potential energy of the configuration. The effective charges of the hole and anti-hole states cancel out. Thus the hole states, apart from serving to help bind the zeron and anti-zeron together, function mainly to cancel out the masses that are necessary in order to produce the spin angular momentum. As we briefly speculate in Sec. 5.3, these hole states may play a role in some

Table 5.2.1 The mathology of the de Broglie electron phase wave.

Lemma 1. Electrons and other particles create de Broglie phase waves as they move through space. A phase wave can be pictured as a manifold of discrete phase wave elements denoted as "zerons" (since they seem to have no detectable properties). If a zeron and an electron are represented as relativistic objects with energies nc^2 and mc^2, respectively, and if their energy (or mass) ratio is $n/m < 10^{-5}$, then the calculated zeron forward velocity in special relativity is $V^f_{\text{zeron}} \cong c^2/v_{\text{electron}}$ for all values of the zeron scattering angle ϕ. Thus the de Broglie phase wave is accurately planar. This special relativity result can be written as $\mathbf{V} \cdot \mathbf{v} \cong c^2$, which is a generalization of the de Broglie equation $Vv = c^2$. The velocity V^f is superluminal, so the zeron (a tachyon) formally has an imaginary rest mass, but its energy nc^2 is real and positive. The scattering (creation) process conserves 4-momentum.

Lemma 2. The zeron can be reproduced as a rotating P–H particle–hole pair created out of the background substrate by the motion of the particle. The zeron rotational motion is essential for its stability, which requires an outward centrifugal force to counterbalance the inward electrostatic force between the P^- particle and the H^+ hole. Stable zerons can be obtained for any value of the angular velocity ω. Setting $\omega = \omega_D$, where $\omega_D = mc^2/\hbar$ is the de Broglie angular velocity, gives a total electrostatic energy for the zeron that is equal to the electron energy mc^2. The rotating mass P in the zeron is larger than the electron mass by a factor of $2/\alpha^2 = 37,558$ in the single P–H pair limit, but grows smaller by a factor of $1/k^2$ for an ensemble of k P–H pairs. The negative hole masses H must be assumed to be of the right value to cancel out most of the positive masses, leaving the zeron with a total effective mass that is positive but much smaller (by at least 10^{-5}) than the electron mass.

Table 5.2.2 The mathology of "hole" states and "particle–hole pairs."

Lemma 3. A hole state H has no properties other than its null occupation number: it represents an available degree of freedom for its neighbors. A mechanical push from the left pushes a left neighbor into the hole, thus creating a momentum impulse to the right, and also the movement of H to the left, in the direction opposite to the impulse. Hence H *moves* mechanically like a "negative mass." It also exhibits a "negative energy" with respect to the continuum mass of the substrate. If the substrate is negatively charged, the vacancy H represents a positive "effective charge." An external electrostatic field that moves positive charges to the right will not operate on H directly, since it has no charge, but will instead pull a negatively charged right neighbor into the hole, thus moving H to the right, in the direction of the electrostatic force. Hence H *moves* electromagnetically like a positive mass, but with the associated momentum directed to the left, like a negative mass.

Lemma 4. A "particle" state P^- is a negatively charged mass quantum that is excited out of the "substrate" of space, leaving behind a corresponding "hole" or vacancy H^+ in the negatively charged substrate, where the hole actually carries no charge, but acts as an "effective" charge. An "anti-particle" state P^+ is an excitation out of the positively charged "anti-substrate," leaving behind an "anti-hole" H^-. The particle P can have any mass value, depending on the excitation, and P–H pairs seem to be permanently bound together by the electric charge in a "gluon-like" manner. These P–H pairs are suggestive of the zero-point fluctuations of the vacuum state. A zeron is reproduced as an ensemble of P–H pairs and/or anti-pairs. A photon is reproduced as an ensemble of "P–H pair plus P–H anti-pair" excitation units, where the H^+ and H^- hole and antihole states in each unit are superimposed and locked together at the center, and the P^- and P^+ masses are rotating around them in a positronium-like orbit. This generic photon model reproduces the main mechanical and electromagnetic spectroscopic features of the photon, and its particle and hole contributions are suggestive of the mathematical structure of renormalization singularities in QED. P–H systems can plausibly travel at luminal and superluminal velocities without violating the bounds imposed by special relativity, since the diverging P and H mass values cancel out. No direct evidence for the existence of spatial substrates has been obtained, but the photon does exist, and its properties seem to be explainable only with this type of P–H model.

of the renormalization singularies that arise in the integrations of quantum electrodynamics.

In this section we have introduced rotating P–H particle–hole pairs as the basis states for the zeron that mathologically constitutes a de Broglie phase wave element, and also for the zeron–anti-zeron "bound state" that mathologically forms the photon. The de Broglie phase wave itself is created as a field of synchronized zerons and/or anti-zerons, and electromagnetic wave packets presumably consist of a somewhat similar synchronized zeron–anti-zeron cloud embedded with synchronized photons. The synchronization process itself may involve yet another unfamiliar concept — the polarization of the underlying vacuum state [124]. With this remark, we end our digression into the area of particle waves.

The results we have obtained in Secs. 5.1 and 5.2 are summarized in the lemmas of Tables 5.2.1 and 5.2.2.

5.3 The Mathology of QED Renormalization: Bare Masses as "Hole" States

Quantum electrodynamics is one of the most successful mathematical formalisms ever used in physics, as we discussed in Sec. 4.2. It describes the interactions between electrons and photons to unprecedented levels of accuracy — one part in 10^{10}. Hence it is undeniably correct. And yet it seems to contain a mathematical flaw. It requires balancing infinities, and using a free normalization constant to cancel them out. This procedure is known as "renormalization theory."

An overview of renormalization theory is given in Section II.D of Gottfried and Weisskopf [125]. Detailed calculations are given in Bjorken and Drell [126], and also in Jauch and Rohrlich [127], who comment as follows [128]:

> *This daring manipulation of infinite quantities raises many questions which are not yet fully understood. From a pragmatic point of view this procedure is perhaps justifiable. But there is no doubt that the necessity for such drastic steps is a clear indication of the need for further revision of the fundamental concepts on which this theory is built.*

An interesting description of the renormalization process is contained in Chapter 4 of Feynman's *QED* [6], where he describes the physical content of the procedure without using equations. On page 128 of this book he comments as follows:

> *The shell game that we play to find n and j (the bare mass and bare charge) is technically called 'renormalization.' But no matter however clever the word, it is what I would call a dippy process!*

In Veltman's book *Facts and Mysteries in Elementary Particle Physics* [8], he mentions that H. A. Kramers introduced the concept of renormalization at the Shelter Island Conference in 1948, and he describes this work as follows [129]:

> *Kramers suggested that one must clearly separate the "bare" mass (the mass of the electron not including the contribution of the field) and the physical mass, that what you see experimentally. ... That the calculation of the field energy produces infinity is regrettable, but by choosing a bare mass of minus that same infinity (plus something extra) the correct experimental result can be reproduced. That is the kind of thing theorists do: sweeping infinities under the rug, smuggling them away. Ugly as it is, the theorists ... produced a calculation of the Lamb shift that agreed with the experimental results of Lamb.*

Later on in his book, Veltman continues this discussion [130]:

> *... if the diagrams give an infinite contribution let us make the basic value also infinite, but with the opposite sign so that the combination comes out to the experimentally observed result. Nonsense minus nonsense gives something ok.*
>
> *This scheme for getting rid of infinities is called renormalization. It is by itself far from satisfactory. No one thinks that the basic quantities are actually infinite.*

The above discussion of the concept of renormalization is well known. The purpose in repeating it here is to show, by comparison, that in evaluating the particle and hole masses for the zeron and photon models defined in Sec. 5.2, we follow a procedure which closely parallels that of Kramers' renormalization procedure. To demonstrate this, let us do the calculation of the zeron in the de Broglie phase wave calculation of Secs. 5.1 and 5.2, using a 1 MeV relativistic electron as the generating particle. As the electron travels through space, it creates and scatters a zeron, as displayed in Fig. 5.1.1. In order to move with a forward velocity equal to the de Broglie phase velocity, the zeron must have an energy which is no larger than 10^{-5} that of the electron (Fig. 5.1.2) (it can of course be much smaller than that value). Assume that the zeron has an energy $nc^2 = 1$ eV (Eqs. (5.1.10) and (5.1.12)), which is 10^{-6} that of the electron. The rotational frequencies of the electron, de Broglie phase wave and zeron (Fig. 5.2.1) are all equal.

The electrostatic energy $\Delta = e^2/2r$ of the zeron is 1 MeV (Eqs. (5.2.5) and (5.2.8)). The energy mc^2 of the rotating particle P^- in the zeron is 37,558 MeV (Eq. (5.2.9)) for the limiting case where there is a single P–H pair. Thus the total zeron energy contained in these three terms is 1 eV + 1 MeV + 37,558 MeV = 37,559.000001 MeV. Since the net zeron energy must be 0.000001 MeV, the hole state H^+ has a postulated energy of $-37{,}559.000000$ MeV. This hole energy plays the role of the "bare mass" energy in the zeron, and is analogous to the "bare mass" of the electron in renormalization theory: in order to achieve the desired answer, we adjust (renormalize) the (unknown) energy of the hole state H^+ so as to give a total energy of 1 eV for the zeron. If we take the more realistic case where the zeron is an ensemble of k P–H pairs, the energy of the rotating particles decreases as $1/k^2$, so an ensemble of 200 pairs would put the rotating energy below the 1 MeV electrostatic energy.

The analogy between particle–hole states and QED renormalization theory becomes closer when we use the photon (Fig. 5.2.3) instead of the zeron (Fig. 5.2.1) for comparison purposes. The physical photon probably does not resemble the diagram in Fig. 5.2.3, but that diagram, in its single-excitation-unit and multi-excitation-unit configurations, where $(P - H) \cdot (\bar{H} - \bar{P})$ is an "excitation unit", is the only mathological representation known to the author which accurately reproduces the spectroscopy of a circularly polarized photon. We might think of it as a "mathological Feynman diagram" for the photon: it represents a prescription for calculating the properties of the photon. Let us take the example of a 1 MeV photon, which uses the dynamics of the positive mass state in the 1 MeV zeron. The electrostatic energy of the photon is $\Delta_{\text{ph}} = 1$ MeV. The magnetic energy is smaller, and can be neglected here. The total energy of the two rotating masses is 75,116 MeV in the single-excitation-unit limit. This decreases as $1/k^2$ in an ensemble of k excitation units. Their combined angular momentum is $\hbar$. These P^- and P^+ masses, and their associated electric and magnetic fields, furnish the radiative interactions of the photon. They thus represent the infinite integrals of the QED calculations. The two merged hole states at the center of the photon have an assigned mass of $-75{,}116$ MeV, and they represent the "bare masses" associated with the QED interaction. The overall mass of the photon is the 1 MeV electrostatic mass, which represents the 1 MeV of energy required to produce the photon in the first place. The forward momentum of the photon arises just from this electrostatic mass, since the forward momenta of the two mass states and two hole states cancel out. Hence the particle–hole

5.4 Vacuum-State Zero-Point Fluctuations as Energy-Conserving Particle–Hole Pairs

When a P^{-}–H^{+} particle–hole pair is created, the pair will quickly annihilate if it is not set into rotation at the proper frequency — where the inward electrostatic attraction balances the outward centrifugal force. Thus *spinning* P–H pairs are required for mathological zeron and photon models. However, there is one area where *non-spinning* P–H excitations may play a role — in zero-point vacuum state fluctuations. The effect of these fluctuations is observed, for example, in the Casimir effect, wherein a pair of closely-spaced parallel plates in a vacuum experience an attractive force. The theoretical formalism for handling zero-point fluctuations is denoted as *stochastic electrodynamics* (SED) [131].

The equation for the Casimir force is

$$F = -\frac{\pi hcA}{480d^4}, \tag{5.4.1}$$

where A is the area of a plate and d is the distance between the plates, in units of cm. Thus the force F varies inversely with the fourth power of the distance. A remarkable aspect of this equation is that the strength of the force is given by Planck's constant h, which appears here in a purely classical equation (SED is a c-number theory that contains no non-commuting operators). The Casimir effect is of importance in that it demonstrates macroscopically that the vacuum state is in fact filled with a background of zero-point radiation [132]. The spectral shape of the zero-point fluctuations is

$$\rho(\omega) = \left(\frac{\omega^2}{\pi^2 c^3}\right) \cdot \frac{1}{2}\hbar\omega \sim \omega^3, \tag{5.4.2}$$

where ω is the radiation frequency. This is the QED expression for the radiation in the case where each oscillator is assigned a zero-point energy of $\frac{1}{2}\hbar\omega$. The cubic dependence on ω is necessary in order to obtain the inverse quartic dependence of the Casimir force on the separation distance d.

The relevance of these results to the present discussion is that since there is no upper bound on ω, these equations predict that the vacuum state has infinite energy, which raises cosmological problems. This infinite zero-point radiation energy is reminiscent of the infinite radiative terms in the QED calculations discussed in Sec. 5.3, and the cure in these two cases may be the same. The conventional way of portraying the zero-point fluctuations is via the creation of virtual electron–positron pairs that violate energy conservation, and hence must be quickly reabsorbed. The expectation value of this process is positive, and it leads to the infinite energy of the vacuum state. But suppose that instead of virtual off-shell electron pairs, the vacuum fluctuations are in fact energy-conserving particle–hole pairs which quickly recombine due to their internal attractive electrostatic force. This on-shell process should resolve the infinite energy problem, in the same way that it can be resolved in QED renormalization theory.

These ideas are, at this point, just conjectures, but they are based on the notion that if we ever want to obtain a mathological description of a photon or a radiative wave element, we must employ the use of matching particle and hole states, and these particle–hole pairs may have a variety of applications. Of course, this then raises the question of what the substance is that the hole is *in*: the vacuum state must not be truly *empty*. We are in a sense just passing the problem of infinities off from one conceptual level to another. Cosmologists may eventually have useful information for us about the properties of the vacuum state, as we discuss briefly in Chapter 7.

Before leaving the field of elementary particle mathology, we consider one final topic — the nature of the fine structure constant α, the elementary particle generator.

Chapter 6 The Mathology of the Fine Structure Constant $\alpha = e^2/\hbar c$

6.1 The Mystery of the Numerical Value $\alpha \cong 1/137$

The close relationship between elementary particle masses and elementary particle lifetimes (mass stability) enabled us to use the observed α-quantization of particle *lifetimes* as the motivation for investigating the possibility of a reciprocal α-quantization of particle *masses.* This investigation revealed that particle masses are in fact α-quantized in the lowest-mass levels — namely, the triad formed by the electron pair, the pion, and the muon pair — and that this *first-order* α-quantization extends on up to the supersymmetric excitation quantum X and the higher-mass ϕ, J/ψ and Υ threshold-state particles. A *second-order* α-quantization boosts the masses of the 316 MeV u and d quarks by a factor of 137 up to 43,182 MeV, where they are denoted as u^α and d^α "α-quarks." These α-quarks combine together to reproduce the isotopic spin states and average mass of the $W^\pm$, Z^0 isotopic spin doublet, and also the mass of the top quark t (Table C1 in Appendix C). The discovery of this mass α-quantization is a worthwhile goal in itself, but it has further theoretical implications with respect to the fine structure constant α, which acts as the coupling constant between the electron pair and pion, and between the electron and muon. The constant α is itself one of the greatest mysteries in modern physics, as Richard Feynman summarized in the quotation [16] given near the end of Sec. 1.3. Continuing on here with this same quotation, we have Feynman expressing his frustration at this situation in the following words:

> *You might say the 'hand of God' wrote that number [the constant α], and 'we don't know how He pushed His pencil.' We know what kind of a dance to do experimentally to measure this number very*

accurately, but we don't know what kind of a dance to do on a computer to make this number come out — without putting it in secretly!

A good theory would say that e [Feynman's notation for α] is 1 over 2 pi times the square root of 3, or something. There have been, from time to time, suggestions as to what e is, but none of them has been useful. [16]

The definition of the dimensionless fine structure constant is $\alpha \equiv e^2/\hbar c \cong 1/137$. The numerical value of this constant is 1/137.03599911 [10]. In spite of its extreme accuracy, this numerical value does not convey much physical insight as to its true significance. The constants e, $\hbar$ and c appear in many areas of physics, and they do not *per se* reveal much information. However, the way they are combined together may furnish some clues. The function of α with respect to the massive elementary particles is to act as an operator on electrons and positrons and generate higher-mass basis states denoted as "α-masses." A first-order "α-leap" produces α-enhanced masses of the forms $m_b = m_e/\alpha$ (Eq. (3.4.1)) and $m_f = 3m_e/2\alpha$ (Eq. (3.4.2)), where m_e is the electron mass, and where m_b and m_f are boson (spin-0) and fermion (spin-1/2) mass quanta, respectively. A second order α-leap produces the α-mass $m_f^\alpha = 3m_e/2\alpha^2$ (Eq. (3.20.5)). These mass-generation processes suggest that a mass value m should logically appear as a factor in α. In order to accomplish this, we first write α in the form

$$e^2/r \cong (1/137)(\hbar c/r)\,. \tag{6.1.1}$$

The left side of Eq. (6.1.1) represents an electrostatic energy. We then use the Compton radius equation

$$r = \hbar/mc \tag{6.1.2}$$

to replace the factor of r on the right-hand side of Eq. (6.1.1). This gives

$$e^2/r \cong (1/137)(mc^2)\,. \tag{6.1.3}$$

The radius r on the left side of Eq. (6.1.3) is the Compton radius that corresponds to the mass value m on the right side of Eq. (6.1.3). Putting dimensions into Eq. (6.1.1), we have

$$e^2/r = (197.327/137.036)(1/r) = 1.440/r \text{ MeV/fermi}\,, \tag{6.1.4}$$

where the radius r is measured in fermi units of 10^{-13} cm. The Compton radius is

$$r = \hbar c/mc^2 = 197.327/mc^2 \text{ fermi}\,, \tag{6.1.5}$$

where the mass-energy term mc^2 is in units of MeV. If we take mc^2 to be the average pion mass, $\bar{\pi}$ = 137.27 Mev, we obtain a pion Compton radius r of 1.438 fermi. This value for r gives an electrostatic energy e^2/r in Eq. (6.1.3) of 1.001 MeV. This value is about 2% below the energy of an unbound electron–positron pair, and it reflects the ~2% hadronic binding energy in the pion (see Sec. 3.15). We have thus tied together the mass of the particle–anti-particle-symmetric pion with the mass of a particle–anti-particle-symmetric electron–positron pair. According to the present-day paradigm of the Standard Model, these two masses should be unrelated.

The calculation of the "intrinsic" mass of the pion that should be inserted into Eq. (6.1.5) is obscured by both binding energy and isotopic spin mass-splitting effects. If we use the $\pi^{\pm}$ mass in Eq. (6.1.5), we obtain a pion Compton radius of 1.414 fermi, which gives an electrostatic energy in Eq. (6.1.3) of 1.019 MeV. This differs from the 1.022 MeV rest-mass energy of an electron–positron pair by just 0.3%. (If we insert the muon mass into Eq. (6.1.5), the resulting radius is 1.868 fermi, which is equal to 2/3 of the classical electron radius $R_0 = e^2/m_ec^2 = 2.818$ fermi, to an accuracy of 0.7%.) The main point we are making in this calculation is not about precise numerical values, but rather about the fact that by combining the fine structure constant $\alpha = e^2/\hbar c$ with the Compton radius $r = \hbar/mc$, we obtain a theoretical expression (Eq. 6.1.3) that ties together the mass of the pion with the mass of an electron pair. This point is of interest in connection with the definition of the fundamental mass excitation $m_b = m_e/\alpha = 70.025$ MeV in Eq. (3.4.1). But it is also of interest with respect to the definition of the constant α. We can rewrite Eq. (6.1.3) in the form

$$(e^2/r)/(mc^2) \cong (1/137)\,. \qquad (6.1.6)$$

We now have the numerical factor ~1/137 expressed as the ratio of an electrostatic energy divided by a mass energy — an electric charge energy over a pion energy, where the electrostatic term consists of two electric charges e separated by a Compton radius r that corresponds to the mass of the pion. This suggests that the factor ~1/137 is in some manner related to the "geometries" of both an electron charge pair and a pion, where this "geometry" can be a function of charge structures or mass structures or both. Thus the theoretical explanation for the numerical value ~1/137 must come from a clear understanding of particle mass structure and charge structure.

In QED the coupling constant α gives the interaction strength for an electron to produce a photon [15]. In the α-quantized mass systematics of the present paper, we have phenomenologically extended this QED result by having the constant α couple to an electron–positron pair and thereby generate the pion, the muon–anti-muon pair, the proton–antiproton pair, and higher excited states. These new results do not constitute a theory, but merely an observation as to what is going on in all of the high energy accelerators. Hopefully the concept of the reciprocal α-quantization of elementary particle lifetimes and masses will serve to open new avenues for exploring these mysteries.

6.2 The Phase Transitions $\alpha_{1,2,3,4} \equiv \alpha_e$, α_μ, α_q, α_γ of the Mass Generator α

One way of investigating the mathology of the fine structure constant α with respect to mass generation is to follow through various scenarios in which this mass generation occurs. Four scenarios we can study are (1) the conversion of electrostatic energy into the 0.5 MeV electron mass, (2) the conversion of electron mass into $m_b = 70$ MeV and $m_f = 105$ MeV particle α-masses, (3) the α-enhancement of the $q \equiv (u, d)$ Standard Model nucleon quarks, and (4) the creation of particle masses m_P in the P–H particle–hole production process for zerons and photons. We can think of these four different mass-creation processes as "phase changes" from one "form" of energy to another.

In *Scenario 1*, electron–positron pair production occurs when a sufficient amount of electrostatic field energy is concentrated in a small enough volume for the energy density and total energy to be raised to the threshold for the conversion of this energy into electron–positron "matter." The electron and positron have to be produced in pairs, but we will for the sake of simplicity concentrate on just the electron. As a model for the electron, we use the Compton-sized relativistically spinning sphere (RSS) model described in Chapter 4, which has an equatorial point charge e. Let us assume that the initial state in this scenario consists of a spread-out field of electrostatic self-energy. We enclose part of the field in a sphere which has a radius r of the right size to enclose a total electric charge e. The self-energy of the field in the sphere is $E = Ae^2/r$, where A is a constant of order unity [133]. For the purposes of the present discussion, where we include particle spin energy, we will assume that $A = 2/3$. If we shrink the sphere down to the radius $R_e = 2/3\ R_0$, where $R_0 = 2.82 \times 10^{-13}$ cm is the "classical electron radius," then the enclosed energy is $E = 0.511$ MeV $= m_e c^2$, which is equal to the energy of a spinning electron. At this critical point an energy "phase transition" occurs in which three things happen: (a) the total electric charge in the compressed field collapses to a massless zero-energy point charge e; (b) the available electrostatic field energy is converted into the spinning "mechanical" mass m_e of the electron, (c) the mass m_e expands to occupy a much larger volume of space. The radius of the compressed electrostatic field at the point of the phase transition is $R_e = (2/3)\ (e^2/m_e c^2)$. The Compton radius of the expanded RSS mechanical mass that represents the final-state electron is $R_C = \hbar c/m_e c^2$. The ratio of these two radii is $R_C/R_e = (3/2)(\hbar c/e^2) = 3/2\alpha \cong 3/2 \times 137$. Hence the α-mediated phase transition has expanded the radius of the "energy volume" by a factor of $3/2\alpha \cong 206$. The process of (a) charge collapse to a massless point, (b) energy conversion from electric charge field energy to mechanical electron mass, and (c) expansion of the system by a linear factor of 206 is what we collectively denote as the "$\alpha_1 \equiv \alpha_e$" phase transition listed in the heading of Sec. 6.2. (The energy $e^2/R_0 = (2/3)m_e c^2$ that is contained inside a sphere of radius R_0 — the classical electron radius — is equal to the non-spinning mechanical rest mass of the electron.)

Scenario 2 involves the production of a pion or a muon–anti-muon pair. We will for simplicity focus on muon–anti-muon production, and we consider just the muon channel. In α-quantization terms, we study the creation of an $m_f = 105$ MeV fermion α-mass by the action of the mass generator α on an electron ground state. There are two alternative scenarios we can advance for this generation process. In *Scenario 2A* we start with a spherical electrostatic field containing a charge e that has been compressed down to the critical radius $R_e = \frac{2}{3}R_0$, where it corresponds to an energy of 0.511 MeV. Instead of converting this compressed electrostatic field into an electron, we further compress its radius by a factor of $(1+3/2\alpha)$, which increases its energy to $0.511 \times (1 + 3/2\alpha) = 105.549$ MeV. At this second critical point it makes a phase transition into a muon. This phase transition follows the same three basic steps as in Scenario 1: (a) the charge collapses to a massless point; (b) the electrostatic field energy is converted into mechanical muon mass; (c) the system expands by a linear factor of $(1 + 3/2\alpha)$. Thus the muon emerges from this conversion process with a Compton radius R_C that is equal to the radius R_e, and with a spinning muon mass that is equal to the electron mass m_e plus the α-mass m_f. (The spinless α-mass m_b is obtained by compressing the electrostatic field down only to the radius R_0, where it has an energy of $2/3 \times 0.511$ MeV, and then applying an additional compression factor of $(1 + 3/2\alpha)$, where the conversion to "spinless muon matter" or "pion matter" takes place. The $J = 0$ α-mass m_b is in essence the mass of a "non-spinning muon." It is the non-spinning form of the $J = 1/2$ α-mass m_f.)

Scenario 2B is a somewhat different route to the same conclusion. Our starting point here is the electron that was produced in Scenario 1, which has mass m_e and Compton radius $R_C = 3.86 \times 10^{-11}$ cm. We assume as an *ansatz* that "electron mechanical matter" is compressible, and that its compressed energy varies inversely with its radius, just as the self-energy of the electrostatic field does. We then compress it down to the radius $R_e = R_C/(1 + 3/2\alpha)$, so that its energy is increased to $m_e(1 + 3/2\alpha)$. At that critical point the "electron mechanical mass" is converted into "muon mechanical mass," which is stable at the higher mass density of the muon. The Compton radius of the muon is the same as the compressed radius of the electron, so the electron-to-muon mass transformation is accomplished with no change in the size of the system.

The Scenario 2 electron-to-muon conversion process, in either its 2A or 2B form, corresponds to the "$\alpha_2 \equiv \alpha_\mu$" phase transition, which produces the muons, $m_\mu \cong m_f + m_e$ and $\bar{m}_\mu \cong \bar{m}_f + \bar{m}_e$, and also the pion, $m_\pi \cong m_b + m_e + \bar{m}_b + \bar{m}_e$. This α-generation process is successively repeated in the creation of the u, d, s, c, b quarks and antiquarks as multiples of the m_f and $\bar{m}_f$ masses (Figs. 0.2.4, 0.2.5 and 0.2.23), the creation of the pseudoscalar mesons as multiples of the m_b and $\bar{m}_b$ masses (Figs. 0.2.11 and 0.2.24), and the creation of hybrid excitations such as the K^* meson and Σ hyperons (Sec. 3.25), which contain both m_f and m_b mass units.

Scenario 3 involves the Standard Model u and d quarks, which have equal masses in our isotopic-spin-averaged constituent-quark mass formalism. We denote these collectively as $q \equiv (u, d)$ "nucleon quarks," since these q quarks reproduce the proton and neutron — the two nucleon states. The q quarks form the "ground state" for an additional mass α-enhancement by a factor of $1/\alpha$, which increases their masses by a factor of 137 and boosts them into the realm of the ultramassive W and Z gauge bosons and top quark t. This *second-order* α-leap (the *first-order* α-leap was from the electron to the muon and then to the u and d quarks) generates the α-quarks $q^\alpha \equiv (u^\alpha, d^\alpha) = q \equiv (u, d)/\alpha$. These q^α quarks reproduce the W–Z average mass and the top quark mass (Secs. 0.10 and 3.20). The phase transition $\alpha_3 = \alpha_q$ takes place during proton–anti-proton collisions at the Fermilab tevatron. A proton and an antiproton collide at energies of 1 TeV each. At these energies, the three q quarks inside the proton and the three

$\bar{q}$ quarks inside the anti-proton are relativistically flattened into disks, and the collision process actually occurs between individual quarks. When a quark and an antiquark meet head-on, the impact compresses them into a much higher energy density, and when this impact energy reaches (roughly speaking) a level that is 137 times the energy of the initial state, a phase transition occurs that results in the formation of u^α and d^α α-quarks. This is the $\alpha_3 = \alpha_q$ phase transition.

Scenarios 1–3 are products of the present studies, since the fine structure constant α has not previously been considered for theories of elementary particle mass generation. Scenario 4, which we now briefly discuss, is on a different footing with respect to elementary particle paradigms, since it involves the well-recognized interaction between an electron and a photon. In quantum electrodynamics (QED), α is the coupling constant that ties together electrons and photons, and the perturbation expansions used in QED are ordered in powers of α. In a figure caption in his book *QED*, Feynman describes the coupling between an electron and a photon in a very vivid manner:

> *The experimentally measured amplitude for an electron to couple with a photon [is] a mysterious number, e [Feynman's notation here for* α*], [which] is a number determined by experiment that includes all the "corrections" for a photon going from point to point in space-time ....* [15]

The constant α represents the summation over all of the perturbation contributions, and it gives the experimental amplitude for the electron–photon interaction. In Scenario 4 we invoke the circularly polarized photon model of Sec. 5.2 that correctly reproduces the spectroscopic features of the photon, and we examine its energy-to-mass conversion process.

Scenario 4 involves the excitation process for creating a photon or zeron, which are pictured as ensembles of P–H particle–hole perturbations of the vacuum state that are caused by the motion of an electron. Two common processes in which this occurs are (1) the acceleration or deceleration of a moving electron, in which *bremsstrahlung* photons (braking radiation) are emitted, and (2) the transition from one orbit to another in an atomic nucleus, with the associated emission of a photon. In both of these processes, momentum components are transferred that serve to generate the spin angular momentum $\hbar$ of the photon. The vacuum state is pictured in this scenario as two superimposed three-dimensional mosaics: one contains negatively charged "particle" mass elements P^-, and the other contains positively charged "anti-particle" mass elements $\bar{P}^+$. As a charged particle moves through these mosaics, it forces mass elements P out of their positions in the mosaics, leaving behind corresponding mosaic vacancies or "holes" H. Randomly produced P–H pairs will quickly reunite, but a stable configuration occurs for P–H pairs that are set into rotation at the angular velocity where their outward centrifugal forces balance the inward Coulomb attraction, as described in Sec. 5.2. A zeron, the basis state or wave element in a particle wave, consists of single particle or antiparticle P–H pairs (Figs. 5.2.1 and 5.2.2), whereas a photon basis state requires matching particle and antiparticle sets of P–H pairs, with the hole states H and $\bar{H}$ superimposed at the axis of the photon (Fig. 5.2.3). Zerons and photons are logically generated as synchronized ensembles of pair basis states. The "phase change" that occurs here is the transformation of kinetic energy from the moving electron into "particle" and "hole" mass components in the zeron or photon. The relevance of the fine structure constant α in this process comes from the QED calculations, which are based on the α-coupling between an electron and a photon. The particle mass elements P that are produced in the photon P–H pairs are not quantized, but can have a continuum of values, in the sense that photons of all energies seem to occur (at least within the resolution of our measurements). The exact nature of these masses P, which relate to the "fabric" of space, is unknown. In fact, the nature of the "mechanical" mass that appears in the electron and other massive particles is essentially unknown, as we discuss in Sec. 7.2. The "photon mass quantum" P represents the "$\alpha_4 \equiv \alpha_f$" phase transition listed in the heading of Sec. 6.2.

The mathological photon model has special significance for the present α-quantized studies of particle mass generation, because the photon mass quantum P represents a second-order α^{-2} leap in mass value from the electromagnetic mass Δ of the photon (Eq. (5.2.9)), which is its observed mass. This large "virtual" positive mass P is of course counterbalanced by the large "virtual" negative mass H of the hole that is left behind when P is created. This second-order mass α^{-2} leap in the photon — a gauge boson — is analogous to the similar α^{-2} leap that occurs in going from an electron $e\bar{e}$ pair to a nucleon quark $q\bar{q}$ pair and then to a nucleon α-quark $q^\alpha\bar{q}^\alpha$ pair that reproduces the W and Z gauge bosons, and also the top quark $T = t\bar{t}$ vector boson. These analogous α^{-2} mass leaps, which are displayed in Table 5.3.1, serve to "justify" the placement of the massless photon in the same gauge boson family as the supermassive W and Z particles.

The non-electromagnetic mass components of elementary particles — the "mechanical" masses — are essentially passive. They carry the energy and inertial properties of a particle (and possibly particle–anti-particle properties), but the interactive forces come from the electric charge elements of the particle. Hence these mechanical masses are difficult to probe experimentally. We may have to wait for astrophysicists to obtain information in the large domain of space, and then try to apply this information to the much smaller domain of the elementary particle.

The essence of the fine structure constant α seems to be its linkage of the "spherical geometry" of electrostatic energy, $E = e^2/r_e = mc^2$ to the "spherical geometry" of mechanical energy, $\hbar c/r_m = mc^2$, which is defined by the Compton radius of a relativistically spinning sphere. The ratio of these two geometries is $r_e/r_m = e^2/\hbar c = \alpha$.

6.3 Three Configurations of the Multiform Electric Charge *e*

In Chapters 2 and 3 we studied the lifetimes and masses of the elementary particles. Then in Chapter 4 we moved on to the spectroscopy of the electron. In Chapter 5 we ventured into the rather shadowy world of electron waves and the electromagnetic photon. A common thread in these studies is the appearance of the fine structure constant α. In Chapter 6 we turned to an analysis of the constant $\alpha \cong 1/137$ itself. To complete this audacious journey, we finally examine perhaps the most basic entity of all — the electric charge e.

Studies of electricity started with the production of static charges by rubbing two suitable materials together, which was known in antiquity. These studies were continued by Benjamin Franklin, for example in his famous "kite in the stormy cloud" experiments, which led to invention of the lightning rod. Franklin came up with the crucial idea that electricity occurs in two forms — positive and negative. Although we have gone on to make very precise measurements of the charge e, and to study it in a wide variety of situations, we still do not have much of an idea as to what the charge e really *is*.

Classically, electric charges have been studied theoretically as distributed clouds of charge density ρ, which are characteristically calculated in the form of a uniform sphere or a uniform surface charge on a sphere. The energy content of the charge cloud is obtained by dividing it into small elements and calculating the Coulomb potential between each pair of elements. The uniform sphere model gives rise to the classical electron radius $R_0 = Ae^2/m_ec^2$, where R_0 is the spherical radius of a charge cloud that contains a total charge e, m_ec^2 is the electron energy, and A is a constant of order unity [133]. If the electric charge does actually exist in the form of a spread out charge distribution — which is the first "multiform shape" state that we can attribute to it — then the interesting experimental fact is that whenever attempts are made to enclose or capture portions of this charge cloud, the charges that are finally measured are always integral multiples of the electric charge e. We can describe this result by asserting that the electric charge e is capable of being spread out in space, but it always remains as an indivisible amount of charge e. Hence when we collect portions of the charge cloud, we are actually collecting a number of discrete "charge balloons," each of which has total charge e. Of course, if we are dealing with macroscopic systems, the space may actually be studded with a mosaic of point charges e whose spacing is too small to detect.

The second multiform state we can attribute to the electric charge e is that of a massless point charge on the equator of a relativistically spinning mechanical mass. This is the electron or muon model described in Sec. 4.4. Experimentally, the electrostatic "size" of the electron has never been determined, but has been pushed down in scattering experiments to an upper limit of no more than about 10^{-17} cm. If we say that this is the size of the electron charge e, and that its self-energy is calculated as described above for the classical electron radius R_0, then its calculated energy is much larger than the total energy of the electron, which is impossible. Hence this is not a valid way of calculating the self-energy of the charge on the electron, and we must arbitrarily select a value. The spectroscopy of the spinning sphere model of the electron, where the charge sits right on the relativistically moving equator, requires the self-energy of the charge to be zero. Hence a zero-self-energy point-like charge is the second "multiform shape" we attribute to the charge e.

The first multiform shape we postulated for the charge e was as a cloud of charge with a calculable self-energy, which we obtained from the classical theory of electrostatic fields. The second multiform shape was as a point charge e with no self-energy, which we deduced from the measured properties of the electron or muon. The third multiform shape we postulate comes from the measured properties of the proton. Unlike the electron and muon, which have no detectable inner structure, the proton appears to contain three internal quarks, $p \equiv uud$, with electric charges $+\frac{2}{3}e$, $+\frac{2}{3}e$,

$-\frac{1}{3}e$, respectively. These charges can also be thought of as the configurations $(+1-\frac{1}{3})e$, $(+1-\frac{1}{3})e$, $-\frac{1}{3}e$. Experimental evidence for three discrete quarks was provided by the Bjorken and Feynman analyses of high energy scattering experiments (Sec. 1.4), which showed that the data could be accounted-for by point-like scattering from three different locations within the proton, and that this scattering was in agreement with the $+\frac{2}{3}e$ and $-\frac{1}{3}e$ charge assignments for the u and d quarks, respectively. Thus the charge on the proton, unlike the charge on the electron, is not localized at one spot, but is spread out into three separate spots, each one of which appears to be point-like. The fragmented electric charge produces the unbreakable gluon forces that characterize hadronic interactions. This fragmented "gluon cloud" is the third multiform shape we assign to the electric charge e.

The proton contains both positive and negative fractional electric charges, so the manner in which these fractional charges correspond to the original integer charge (or charges) on a "proto-proton" is not immediately clear. We know, for example, that a neutral "proto-neutron" seems to contain both positive and negative integer charges. In Sec. 3.15 we denoted the fractional charges that are related to an initial integer charge as a "gluon cluster" (GC). The ~3% hadronic binding energy that appears in low-mass hadronically bound systems (Fig. 3.15.1) is attributed to attractive GC–GC forces. The occurrence of fractional electric charges and unbreakable "rubber-band" gluon forces between these charges was not anticipated. In fact, it is the hope of finding just this kind of unexpected surprise that makes the study of the elementary particle such a worthwhile endeavor.

The three multiform shapes for the electric charge that we can deduce from present-day physics are the extended and self-interacting charge cloud of classical physics, the zero-energy point charge on the leptons, and the point-like fractional charges (with their associated gluon fields) on the hadrons. The essential point about all of these charge configurations is that they all lead to observed charges which are integer multiples of e. Thus the charge e seems to be able to change shapes and do various things, but it never permanently breaks apart. This of course gives us no clue as to what electricity actually *is*, but hopefully its manifestations will provide clues for future research.

Chapter 7 Ramifications

7.1 Cosmological Masses

In the present book we have been studying elementary particles, which make up the observable world. But we now know from recent astrophysical discoveries that this *baryonic matter* seems to be only about a twentieth of the actual matter in the universe. Studies on the rotational motion of large spiral galaxies, which are held together by gravity, reveal that their observable matter is too small by an order of magnitude to supply the needed gravitational attraction. Hence these galaxies must also possess clouds of invisible *dark matter* that contain the requisite gravitational mass. Current estimates are that this dark matter constitutes about a quarter of the total mass of the universe, and is gravitationally lumped together in the galaxies. Its composition is a matter of conjecture. Even more mysterious is the concept of *dark energy*. The universe is expanding outward in all directions, propelled by the force of the initial "big bang" and moving against the pull of its own gravity. This raises the question as to whether this expansion is constant or even increasing, in which case the universe will simply fly apart, or whether the expansion is slowing down, in which case the universe will ultimately stabilize or suffer a gravitational collapse. Recent astrophysical evidence suggests that the expansion rate is actually increasing, and cosmologists account for this fact by introducing dark energy, which has the property that it is gravitationally repulsive instead of attractive, and thus is spread out uniformly throughout the cosmos. This dark energy constitutes the remaining three quarters of the mass in the universe, and its composition and repulsive mechanism are unknown. [134]

A somewhat related cosmological problem is the nature of space itself. When the wave-like nature of light and electromagnetism was first discovered, physicists assumed that space must be full of *ether*, since a wave must be a wave in something. However, Michelson and Morley demonstrated the surprising result that no detectable effects of the motion of this ether relative to the motion of the earth could be observed. This led to the disappearance of the ether. Waves could travel without any propagating medium. Space was simply a void. But developments in twentieth and twenty first century physics have now resurrected a more material concept of space. In quantum electrodynamics, space contains vacuum fluctuations in which virtual electron–positron pairs are continually being created and annihilated. The charge on an electron in space polarizes these pair fluctuations and thereby changes the effective value of the electron charge. This "vacuum polarization" produces measurable effects which are handled by "renormalization" theory [72], as we discussed in Sec. 5.3. The QED calculations that have emerged from renormalization theory, and the matching experimental values, are among the most accurate results ever obtained in physics. These vacuum fluctuations also produce the Casimir effect that was discussed in Sec. 5.4. A more indirect indication of a material space is suggested by the general theory of relativity, which attributes gravity to a curvature of space — the curvature ought to be a curvature of "something." Today the vacuum state is regarded as a dynamic entity, a restless sea of appearing and disappearing electrical charges, the zero-point fluctuations of stochastic electrodynamics — SED.

There is one feature of the present α-quantized mass formalism — the relativistically spinning sphere RSS — that might have relevance to these cosmological masses. In addition to its role in accounting for the 3/2 mass ratio between spin-1/2 and spin-0 states, the RSS has another intriguing property: as it loses energy it *expands in size* in order to conserve angular momentum. That is, it retains its Compton radius $R_C = \hbar/M_s c$, which varies inversely with its spinning mass M_s. The elementary particle excitation sequences that we displayed in Chapter 3 start with the creation of an electron–positron pair out of the vacuum state, which suggests that a Dirac sea of "proto-electron–positron" (PEP) pairs may be spread throughout the cosmos, since this electron-pair creation process can take place anywhere. Suppose that the "fabric of space" is composed of this PEP sea, which seems analogous to the particle and anti-particle three-dimensional "mosaics" invoked for photon production in Scenario 4 of Sec. 6.2. Then if electron pairs are excited out of the PEP sea, used to create various higher-mass particles as allowed by the available energy, and start decaying back down to the ground state, they can lose some of this energy through the creation of stable protons (baryonic matter), and the emission of low-energy photons (the microwave background) and neutrinos or other neutral particles (dark matter). If they then cascade back down into the PEP sea at a somewhat lower energy than when they emerged, they will expand beyond their original size and thereby expand the PEP fabric of space (dark energy).

7.2 The "Mechanical" Mass of the Elementary Particle

The present book is primarily about masses. It is also about the fine structure constant α, since α generates most of these masses; and it is about lifetimes, since the lifetimes displayed in Chapter 2 tell us that the masses are α-generated. In Chapter 3 we searched for basic α-quantized mass building blocks for elementary particles, and we found three of them — the spinless boson mass quantum $m_b = m_e/\alpha$, the spin-1/2 fermion mass quantum $m_f = (3/2)\, m_b$, and the spin-1/2 fermion superheavy mass quantum $m_f^\alpha = (3/2\alpha)\, m_b$, where m_e is the electron mass. This led to to the identification of the electron as the elementary particle ground state from which all higher-mass states are generated. In Chapter 4 we tied the α-quanta m_b and m_f together as spinless and relativistically spinning forms of the same fundamental 70 MeV building block. We also demonstrated that the *electromagnetic* mass of the electron is very small — only about 0.1% of the total — and is tucked away in the magnetic field that creates the anomalous magnetic moment. Abraham Pais summarized this situation as follows [55]

> *...the mass of the electron is certainly not purely electromagnetic in nature. But we still do not know what causes the electron to weigh.*

In fact, no purely electromagnetic structure is stable, so we need a non-electromagnetic framework for the electron just to hold it together. If the mass of the electron is not electromagnetic, and is not (at least in the present author's opinion) gravitational, it must be "mechanical." The purpose of the present section is to point out that this mechanical mass can have properties which are quite different from any of our familiar macroscopic masses.

As we discussed in Sec. 7.1, astrophysicists have recently introduced a *dark matter* that composes most of the mass of spiral galaxies, and which gravitationally holds the galaxies together. This dark matter is *not* believed to consist of the baryonic matter that composes (e.g.) our solar system. Astrophysicists have also introduced a gravitationally repellant *dark energy* that composes three-fourths of the total mass of the universe, and whose composition is completely unknown. Thus our knowledge of the various types of mass that make up the universe is clearly rather rudimentary. To this list of unknown types of mass we can add the *mechanical* mass of the elementary particle, whose properties we now consider [135].

Electrons interact only electromagnetically, so the mechanical mass of the electron must be essentially *non-interacting* — neutrino-like in nature. Because of its similar spectroscopic properties, the mechanical mass of the *muon* must also be neutrino-like. These relativistically spinning particles do not fly apart, so their masses must have a high tensile strength (or the relativistic stresses are not that large, as discussed in Sec. 4.3). The equations for the relativistically spinning sphere in Chapter 4 assume that the rest mass of the sphere is uniformly distributed, and is probably a *continuum*. By way of contrast, the mechanical masses of our macroscopic world (e.g. iron) are in fact mostly empty space — widely separated atoms or molecules held together electromagnetically. Thus a continuous mass distribution is not a familiar state of matter.

The main conceptual problem with the idea of a Compton-sized electron that contains a point electric charge is to understand how it can scatter in a point-like manner. Chasle's theorem [136] may come to our aid here. It states the following:

An external force applied to a rigid body can be separated into two components:

(1) a translational force that acts through the mass center;
(2) a torque that acts around the mass center.

Thus if the mechanical mass of the electron is "rigid," it may scatter in a way that conceals its true (large) size. The concept of a rigid body raises conceptual problems in relativistically rotating systems [137], but the salient facts we know about the electron are that (a) it does scatter in a manner that is accurately described by point-like equations, and (b) all of the spectroscopic information we can bring to bear about the electron indicates that it (like all other elementary particles we can measure) is in fact Compton-sized.

A property of the mechanical mass of the electron which may be less controversial is that it carries the lepton number of the electron, just as the mass of the electron neutrino carries the electron lepton number. Also, masses come in particle and anti-particle configurations. We do not really understand what physical property the lepton number represents. We also do not understand what physical property the strangeness quantum number represents, but it seems to be tied in with the particle and antiparticle aspects of the mass configuration.

The "particle" and "hole" states that we used in Sec. 5.2 to represent the zeron and photon are also mass "states" that need to be stirred into the elementary particle mixture of masses. These are familiar concepts in solid state physics, where we know what kind of medium contains them. They may need to become familiar concepts in particle physics if we can figure out what kind of "medium" the vacuum state is composed of.

The truth about the mechanical mass of the electron is that we really know very little about it — except that it does exist. Thus we have no quantitative results to present here. But if the electron is truly the ground state from which all of the higher masses are generated, then it behooves us to obtain as much understanding in this area as it is possible to achieve. This topic is discussed in more detail elsewhere [135, 138].

7.3 Three Deficiencies in the Standard Model Treatment of Particle Masses

There is nothing more difficult to take in hand, more perilous to conduct, or more uncertain in its success than to take the lead in the introduction of a new order of things, because the innovator has for enemies all those who have done well under the old condition, and lukewarm defenders in those who may do well under the new.

Niccolo Machiavelli, Il Principe (1513)

One of the major accomplishments in 20th century elementary particle physics was the identification of the u, d, s, c, b, t set of fractionally charged fermion quark states. These quarks combine to accurately reproduce the isotopic spin states of a wide variety of particles, and they dominate the lifetimes of the longer-lived states. However, the quark mass values, and hence the mass values of the particles they create, have remained a mystery, as we discussed in Chapter 1. Apart from the difficulties posed by the non-observability of individual quarks (except the t), there are three theoretical reasons for this impasse. (1) By failing to recognize that the weakly interacting leptons and strongly interacting hadrons share a common mass formalism, and by insisting on treating them as separate entities, theorists have lost the hadron ground state. (2) By overlooking the well-documented scaling of long-lived particle lifetimes in powers of the fine structure constant $\alpha = e^2/\hbar c$, theorists have also overlooked an analogous α-scaling of particle masses. (3) By attempting to use the same u, d and s quark masses for both the low-mass pseudoscalar mesons and the higher-mass nucleons and hyperons, theorists have trapped themselves in the pseudoworld of current-mass quarks, as opposed to the empirically-suggested real world of constituent-mass quarks. The Standard Model, as presently constituted, does not adequately deal with particle masses. The resolution of this impasse involves the recognition of the fact that the constant α couples to the electron to generate not only the zero-mass photons of quantum electrodynamics (QED), but also the massive leptons and the hadrons of quantum chromodynamics (QCD). This role as a universal particle coupling constant is a large part of what we describe here as "The Power of α."

The successes of the Standard Model have been so impressive in so many areas that they form the framework upon which we have erected our understanding of the elementary particle. Before the Standard Model arrived on the scene, physicists were confronted with a dizzying array of particles that seemed to lack any kind of organizing principle. The Standard Model has provided this needed organization. In particular, the use of fractionally charged quarks has led to isotopic spin rules for charge states, and also rules for the flavor quantum numbers, that have been confirmed by many subsequent discoveries of predicted particle states. The Standard Model "works," in the same sense that QED "works." Its successes are undeniable. Thus if we phenomenologically investigate the systematics of the elementary particle states, we expect to find agreement with the Standard Model, and in large measure we do. For example, the α-spaced lifetime groupings of the long-lived ($\tau > 10^{-21}$ s) lifetimes in Chapter 2, which represent a result that is outside of the present-day Standard Model systematics, cleanly sort out the various s, c, b quark flavors, and thus reinforce their physical relevance.

The area in which the achievements of the Standard Model have been more modest is that of particle masses, where many recent publications have pointed out deficiencies: masses remain a mystery. In the present book we have used the global systematics of particle lifetimes and particle masses to address this problem, and we have thereby arrived at what seem to be a couple of key results: (a) the long-lived threshold-state particles have α-dependent lifetimes and masses; (b) from the standpoint of masses, leptons and hadrons must be treated together in a common phenomenological formalism. These correspond to points (2) and (1), respectively, in the list at the beginning of this section. The question then arises as to what changes should be made in the Standard Model. One change that is required is in the treatment of the low-mass pseudoscalar mesons — the Standard Model PS octet. We have already addressed this topic in Sec. 2.4 (with respect to lifetimes) and in Sec. 3.21 (with respect to masses). In view of the ramifications of these results, we briefly recapitulate this work, and we also consider the historical background against which this Standard Model PS octet was created.

A crucial difference between the present α-quantized approach and the Standard Model approach is in the Standard Model use of the same spin-1/2 u and d quarks to represent the low-mass pseudoscalar mesons as are used for the higher-mass proton and neutron. This Standard Model procedure gives the correct isotopic spin states for the pseudoscalar mesons, which is an enormous accomplishment that does not yet emerge in an obvious way from the use of α-quantized masses. However the u and d quarks can only reproduce both the pion and the proton if their intrinsic quark masses are small with respect to the masses of these particles, so that the difference between the proton and pion masses can be attributed to their different gluon currents. Historically, the QCD formalism for the PS octet was devised during the 1960's [36], when the quark masses were believed to be so large that nothing much could be said theoretically about their absolute values. The implementation of "asymptotic freedom," where quark masses in their ground states have very small binding energies, changed this situation, but it did not change the formulation of the PS octet. Thus the u and d SM quarks are now regarded as *current* quarks [10] with small intrinsic masses that can be made to match both pion masses and nucleon masses by assuming drastically different gluon-field masses in the two cases. By way of contrast, the α-quantized mass states of Figs. 0.2.6–0.2.10 in the present paper correspond to *constituent* quarks, with their large intrinsic masses creating the observed masses of the particles. The strong "rubber band" binding energy that gives rise to "asymptotic freedom" arises from interactions between the *fractional charges*, with their associated gluon fields, rather than interactions between the *masses*.

The difference between the current-quark representation in the Standard Model and the constituent-quark representation in the present α-quantized model raises the general question as to whether we can in fact employ different mass representations for essentially the same set of quarks. This question was addressed in Sec. 3.27, where it was noted, as pointed out by Gottfried and Weisskopf [2], that the mass systematics of the quark model is not in general determined by the other properties of these quarks, so that different mass representations can in fact be employed.

In connection with the idea of using other quark properties to deduce their mass values, it is interesting to note that if we use experimental baryon and hyperon magnetic moments to deduce appropriate magnetic moments for the u, d, and s quarks, we can in turn insert these values into the magnetic moment equation $\mu = e\hbar/2mc$ and obtain the corresponding effective quark masses. This procedure gives [11]

$$m_u = 338 \text{ MeV}\,, \quad m_d = 322 \text{ MeV}\,, \quad m_s = 510 \text{ MeV}\,. \tag{7.3.1}$$

In Sec. 3.23 we employed the following constituent-quark masses,

$$\begin{gathered} m_u = m_d = 315.6 \text{ MeV}\,, \quad m_s = 525.7 \text{ MeV}\,, \\ m_c = 1576.1 \text{ MeV}\,, \quad m_b = 4727.2 \text{ MeV}. \end{gathered} \tag{3.23.5}$$

to produce the mass fits displayed in Fig. 3.23.1. As can be seen by comparing Eqs. (7.3.1) and (3.23.5), the magnetic-moment-deduced quark mass values are very close to the α-quantized m_u, m_d and m_s quark values, so that the experimental bayron and hyperon magnetic moment data are consistent with the use of constituent-quark masses.

In the lifetime and mass studies of the present paper, the PS pseudoscalar mesons — the "PS octet" of quantum chromodynamics plus the η' "singlet" — exhibit the clearest and most accurate examples of α-quantization. They are the "crown jewels" — the best results we can offer. The lifetime results were discussed in Secs. 2.2–2.4, and the mass results were discussed in Secs. 3.7, 3.10 and 3.21. The accuracy of these results should not be surprising, in the sense that these PS mesons are the lowest-mass spinless states in the elementary particle zoo, and they should have simple and relatively stable structures, which they do. The surprising result about these particles is the difficulties they have historically created for the Standard Model, as we have discussed above.

The conclusion we come to is that whereas the u, d and s quark assignments for the isotopic spins of the PS mesons work very well, the mass values do not. In particular, if we follow the guidance offered by the higher-mass states and assume that the Standard Model quarks are in fact constituent quarks, then the mass values for the PS mesons do not work at all. Hence we need a different set of quarks to represent the PS nonet. An even more far-reaching conclusion is that this different set of quarks is based on spinless mass quanta [139], of a type not envisioned in the quark model. Thus we have traveled well beyond the confines of the Standard Model framework. The reward we obtain for this transgression is that we can accurately calculate the absolute masses of the PS mesons without having to resort to adjustable parameters. Furthermore, the mathology of Chapter 4 ties together the spinless and spinning basis state masses, so that they form a unified whole which embraces both leptons and hadrons in a comprehensive mass formalism. If we really want a solution to the mystery of elementary particle masses, it seems that this is the path we must follow. It is a path that is guided by *The Power of* α.

The most incomprehensible thing about the world is that it is comprehensible.

Albert Einstein

Postscript: The Saga of the $m_b = 70$ MeV and $m_f = 105$ MeV Mass Quanta

The concept of a "ground state" in elementary particle physics is based on the idea that there might exist a low-energy mass unit or set of mass units which serve as "building blocks" for the higher-mass particles. The *stable* massive particles include just the electron and proton (plus the bound neutron) and their antiparticles, but a whole array of unstable particles can be created by accelerators, all of which are much more massive than the electron. A study of this entire array may cast some light on the mass relationship between the electron and proton. The electron mass is so small that it is not useful *per se* as a basic mass unit for the unstable particles — a much larger mass quantum seems required. The proton mass can serve as a ground state for hyperon excitations, but not for mesons (although it was used there in the early Fermi–Yang model [140]). The two lowest-mass *unstable* particle excitations are the 140 MeV pion and the 211 MeV muon–antimuon pair. These give rise to two very different candidates for particle basis states: (1) a 70 MeV mass quantum that was suggested theoretically and phenomenologically more than a half century ago, but which is not manifested as an observed particle state; (2) the 105 MeV muon, which was identified in the cosmic ray mass spectrum during this same time period, but which seemingly has no phenomenological or theoretical basis for even existing. In this Postscript we briefly describe their historical emergence. The relativistically spinning sphere saves us from having to choose between them.

A. The History of the 70 MeV Mass Quantum m_b

The modern-day search for a suitable elementary particle mass unit, or, equivalently, a fundamental length, can be traced back to a 1938 paper by

Werner Heisenberg [141]. In it he singled out the velocity of light c and Planck's constant $\hbar$ (or h) as two "universal constants of the first kind." He excluded possible gravitational constants, which probably are not of importance in the microworld of the elementary particle. The constant c serves to delimit the domain of relativistic phenomena, and the constant $\hbar$ serves to delimit the domain of quantum mechanical phenomena. These two constants by themselves are not sufficient to define a fundamental mass or length. As Heisenberg noted, mass m and length R are tied together by the Compton radius equation $R_C = \hbar c/mc^2$. If we can discover a third fundamental constant — one that gives a different relationship between m and R — then we can, for example, calculate the electron mass in terms of these three constants. As a way of obtaining a third "constant of the first kind," Heisenberg proposed using the equation for the "classical electron radius," $R_0 = e^2/m_e c^2$, where $m_e c^2$ is the mass-energy of an electron and e is the unit of electric charge. In this equation, which seems appropriate from the viewpoint of general dimensional arguments, we can select either the length R_0 or the mass m_e as being the relevant fundamental constant. Heisenberg thought that the length $R_0 = 2.82 \times 10^{-13}$ cm was the more meaningful quantity, as reflected in the phrase "universal length" that appears in the title of his paper [52, 141]. If we set $R_C = R_0$ as a fundamental interaction length, we obtain $m = (\hbar c/e^2)m_e \cong 137\ m_e \cong 70$ MeV as the mass quantum m that corresponds to R_C under this hypothesis. This 70 MeV mass quantum is not singled out specifically in the Heisenberg paper. It was brought into focus in a later paper by Nambu [142].

The classical electron radius $R_0 = e^2/m_e c^2$ is the order-of-magnitude radius that we obtain if we compress an electric charge distribution down to the size where its self-energy is equal to the observed mass of an electron. Thus if the electron *is* this ball of electric charge (which by itself is not a stable configuration), then $R_0 = 2.82 \times 10^{-13}$ cm is the radius of the electron (up to a small constant of order unity that depends on the assumed distribution of the charge [133]). However, we now know that the charge distribution in the electron is no larger than about 10^{-17} cm, so the "classical electron radius" R_0 has no direct connection to the structural size of the electron, although it does appear in the Thomson cross section for the scattering of X-rays by electrons.

The first suggestion in the literature that a 70 MeV mass quantum might play a role in elementary particle structure was made in 1952 by Yoichiro Nambu, in a short letter to *Progress of Theoretical Physics* [142]. In this letter he summarized the known experimental particle data in a

Table P1 The particle data set listed in Nambu's 1952 table [142], shown with the measured experimental mass in MeV and updating comments. The calculated particle mass is 137 × n × m_e, where m_e = 0.511 MeV is the electron mass (particle (1) in the table).

Particle	Mass no. n	Experimental mass (MeV)	Comments
(1) lepton	0	~0	(not relevant here)
(2) photon	0	0	(not relevant here)
(3) μ	$1\frac{1}{2}$	107	muon
(4) π	2	$141(\pi^{\pm})$	pion
(5) V_{02}	6	409	(not verified experimentally)
(6) τ	7	494	kaon (modern designation)
(7) x		511–767	(not verified experimentally)
(8) nucleon	$13\frac{1}{2}$	939, 940	proton, neutron
(9) V_{01}	16, $16\frac{1}{2}$	$Q = 35$, 70	early Λ sightings
(10) V^*	$17\frac{1}{2}$	$Q = 280$	early Λ sighting

table which listed ten particle states, of which two have since disappeared, and two — the electron and photon — are not applicable to the present discussion. The data from Nambu's table are displayed here in Table P1. At the time of Nambu's letter, only four subsequently-verified elementary particles apart from the electron had accurately-measured masses. As listed in Table P1 they are: (3) the 107 MeV muon; (4) the 141 MeV charged pion; (6) the 494 MeV τ meson (now known as the kaon [143]); (8) the 939 and 940 MeV nucleons N (the proton and neutron). Nambu noted that if we select half the pion mass, or 70 MeV, as a basic mass unit (the boson mass quantum m_b of the present studies), then the two bosons in Table P1 (particles 4 and 6) are integer multiples of 70 MeV ($\pi = 2$, $\tau = 7$), and the two fermions (particles 3 and 8) are half-integer multiples ($\mu = 3/2$, nucleon $N = 27/2$). The newly-discovered V particles (now known as Λ hyperons) had Q-values in mass units of both 35 MeV (a half-integer multiple) and 70 MeV. The electron mass itself did not relate to the 70 MeV mass, and was attributed in Nambu's paper to a mass "fine structure" that also embraces the proton-neutron mass splitting and the $\pi^{\pm}$-π^0 mass splitting.

If we make a phenomenological assessment of the mass values displayed in Table P1, there are three choices we can plausibly make for a fundamental mass quantum Q.

(i) The fundamental mass is $Q_{1/2} = 35$ MeV. Then π, τ, μ and N are all integer multiples of $Q_{1/2}$.
(ii) The fundamental mass is $Q_1 = 70$ MeV. Then π and τ are integer multiples of Q_1, and μ and N are half-integer multiples of Q_1.
(iii) There are two fundamental masses, $Q_1 = 70$ MeV and $Q_{3/2} = 105$ MeV. Then π and τ are integer multiples of Q_1, and μ and N are integer multiples of $Q_{3/2}$.

Nambu pointed out [142] that a 70 MeV mass "incidentally agrees with Heisenberg's natural unit," which is $m_e/\alpha = 70.03$ MeV, where m_e is the electron mass and α is the fine structure constant. He thus suggested that 70 MeV was his favored choice as a fundamental mass. But he then had to add a 35 MeV mass unit to fit the fermion masses and the Q-values of the V particles. He made the comment [142] that "we have adopted the view that the heavy V_0 particles have two kinds of Q-values, namely $\sim$35 MeV (1/2 mass unit) and $\sim$70 MeV (1 m.u.)."

A "fundamental mass" is logically defined as the smallest mass unit in the system, especially if the higher masses emerge as multiples of this mass unit. Under this definition, the fundamental mass is indivisible (since smaller mass units would be more fundamental). Thus Nambu's selection of a 70 MeV fundamental mass unit (choice (ii) above) means that a 35 MeV mass quantum must in some sense serve as an independent fundamental mass unit. Hence choice (ii) actually leads to two fundamental mass quanta, 70 MeV for bosons and 35 MeV for fermions. Choice (iii) is similar to (ii), but with 105 MeV for fermions.

There was no discussion in Nambu's paper of the spin value of the 70 MeV mass quantum. Since the electron spin is 1/2, and the scaling factor $\alpha^1 \cong 137$ is a pure number, the Heisenberg $m_e/\alpha = 70$ MeV "natural unit" should plausibly be regarded as a *spin*-1/2 *fermion* mass quantum. However, since Nambu's "τ" meson (the kaon [143]), which is an integral-spin boson, contains an *odd* number of Nambu mass quanta, the Nambu 70 MeV mass unit should be an *integral-spin boson* (the quantum m_b in the present paper). But if it is a boson, it cannot reproduce the masses of the muon and nucleon, which are half-integral-spin fermions. Without the systematics of the relativistically spinning sphere (Sec. 4.3), which was unknown at that time, Nambu had no way to tie the boson and fermion mass quanta together. (The relativistically spinning sphere model actually selects 70 MeV and 105 MeV — not 70 MeV and 35 MeV — as the two fundamental masses, as specified in choice (iii) above.) Also, Nambu had no lifetime information on these particles that would help in determining their dependence on α. Thus the Nambu conjecture about the 70 MeV mass quantum was based mainly on its fits to two particles — the pion and the kaon. In summing up his paper [142], Nambu commented that "this rule is purely of an empirical nature, and might turn out to be entirely illusory or accidental ...". This line of research does not appear to have been followed up later by Nambu. He instead turned to work on what we now call the Standard Model, where he made several important contributions [144]. The Standard Model, it should be noted, does *not* use the 70 MeV mass quantum as a basic unit, does *not* employ the fine structure constant α in hadron physics, and does *not* treat hadrons and leptons in the same mass formalism [10].

A decade later David Jackson, in his 1962 book *Classical Electrodynamics*, inserted the following prescient footnote [145]:

> *The fact that* $(\hbar/m_\pi c) \simeq 1/2(e^2/m_e c)$, *corresponding to the pi-meson mass being 2 × 137 times the electron mass, is another of those numerological coincidences which may ultimately have some deep significance.*

Jackson did not mention Nambu's paper.

In a 1970 paper [146], Robert Pease applied the methods of diophantine quantization to the *quadratic* Gell-Mann–Okubo mass relation $\pi^2 + 3\eta^2 = 4K^2$. A simple diophantine solution was obtained in which $\pi = 2m_b$, $\eta = 8m_b$, and $K = 7m_b$, where m_b is the 70 MeV mass quantum. This same solution of course leaps out from visual inspection in a linear representation of the masses (Figs. 0.2.12 and 0.2.13). One difficulty with the *quadratic* mass approach, which Pease states was originally suggested by Feynman, is that it cannot adequately accommodate the η' meson, which is handled very naturally in the linear mass representation of Fig. 0.2.12. Mass formulas are also discussed by David Akers in a 1994 paper [70]. Tables C1 and C2 in Appendix C show examples of particle mass values that are obtained using the mass quantum m_b.

My personal involvement in the search for a fundamental elementary particle mass quantum commenced the third week of June, 1969, as described in the Preface to this book. The initial stages of this work were centered on the 105 MeV muon mass, and the 70 MeV mass quantum emerged from these studies in the Fall of 1969 with the development of the equations for the relativistically spinning sphere. I discovered Nambu's paper in the Spring of 1971, and my first published mention of his work is in Ref. [16].

The starting point for my work on elementary particle masses in 1969 was the opinion I formed, after a week of perusing the experimental data, that the muon mass $m_\mu \cong 105$ MeV seems to be a fundamental elementary particle mass quantum, and that the internal hadron binding energies appear to be weak — on the order of a few percent. This basic muon-like mass quantum is denoted in the present work as the *half-integral-spin fermion* excitation quantum m_f, which adds to the electron mass to reproduce the muon mass. The u and d constituent quarks each have masses of about $3m_f$, and three-quark proton has a mass of $9m_f$ (Table C1). But this idea raised a problem: the muon appears to be a point-like object, whereas the proton is known to be Compton-sized. How can nine point-like muons merge to form an extended proton? This led to the following question: "Is there a model that one can construct for the muon?" The obvious first choice for such a model is a spinning sphere. But if we treat the sphere classically, its calculated moment-of-inertia does not give the correct value for the gyromagnetic ratio between the magnetic moment and the spin of the muon. However, as I discovered, if (1) the rotation is calculated relativistically, (2) the sphere is assigned the Compton radius of the muon and spun at the relativistic limit (where the equator is moving at, or infinitesimally below, the speed of light), and (3) a zero-mass point electric charge is placed on the equator, then the muon spin and magnetic moment emerge calculationally at their correct values. This is the result I was looking for. But this calculation also gave an unexpected bonus. It turned out that the relativistic mass increase of the spinning sphere over its non-spinning value is a factor of 3/2, independently of its radius. Thus if we stop a 105 MeV fermion mass quantum m_f from spinning, it assumes its rest mass value of $2/3 \times 105 = 70$ MeV, and it becomes a spin-0 *boson* mass quantum m_b. This ties in with the fact that the 140 MeV pion, with its balanced particle–antiparticle symmetry, contains a matching $m_b\bar{m}_b$ pair of mass units. Hence a 70 MeV mass quantum m_b accurately reproduces not only the muon (in its $\mu = m_f$ spin-1/2 form), but also the pion (in its $\pi = m_b\bar{m}_b$ spin-0 form), if we assume modest $m_b\bar{m}_b$ binding energies. This is how the 70 MeV mass quantum m_b emerged in my work. The later discovery of the observations by Nambu, Jackson and Pease served to reinforce results that were already quite firmly in place.

In the years following these discoveries, I have looked in every book on special relativity that I could find in the libraries, and I have never seen any reference to the relativistically spinning sphere model (Chapter 4). I have also had the opportunity to meet personally with two very prominent authors of textbooks on relativity, and neither of them expressed any familiarity with these ideas.

One conceptual problem with the 70 MeV mass quantum m_b is that it, unlike its 105 MeV counterpart m_f, does not correspond to a physically

Fig. P1. The author's license plate as photographed in 1978.

Fig. P2. The same license plate photographed in 1990, after its removal from the car, which was no longer operational.

Fig. P3. The author's license plate in 2005.

observed particle. However, there have been indirect manifestations of this mass unit (where we now depart from the realm of physics). Figure P1 is a photo of my automobile license plate, which carried a 1979 expiration sticker. Figure P2 is a later photo of the same license plate, now bearing a 1990 sticker, and with the license plate removed from the automobile, which had unfortunately met its demise. Figure P3 shows a second-generation, and next century, sighting in the year 2005. Further information can be obtained at the website *70mev.org*.

Just as this book was about to be finalized, a possible direct manifestation of the 70 MeV mass quantum appeared on the arXiv website [147]. (We have now returned to the realm of physics.) This was the discovery of the charm baryon Ω_c^* excited state, which decays radiatively to the Ω_c^0 ground state. A precision measurement of the $\Omega_c^* - \Omega_c^0$ mass difference yielded the value 70.8 ± 1.0 (stat.) $\pm$ 1.1 (syst.) MeV. Both of these charm baryons have the quark flavor content *ssc*, so the radiative decay $\Omega_c^* \to \Omega_c^0 + \gamma$ may reflect an intrinsic *s*-quark substructure (see Sec. 3.26). An analogous baryon radiative decay is $\Sigma^0 \to \Lambda^0 + \gamma$, which yields a mass difference of 76.959 MeV [10]. Both the Σ^0 and Λ^0 have the quark flavor content *uds*. These two baryon radiative decays are each suggestive of a 70 Mev m_b constituent-quark mass structure. As an interesting sidelight, and a commentary on the sophistication that goes into the hardware and software of these very detailed elementary particle experiments, we note that this arXiv preprint [147] contains the names of 609 physicists from 81 different institutions (the BABAR Collaboration), who carried out the Ω_c^*

experiment. It is the experimentalists who will finally determine the details of the elementary particle mass structure.

B. The History of the 105 MeV Mass Quantum m_f

The mu meson, or *muon*, has posed problems ever since its identification more than a half century ago. At first it was thought to be the particle predicted by Yukawa that is required in his theory to mediate the strong nuclear force. However, when the muon was detected and measured in cosmic ray experiments, it turned out to have roughly the mass predicted by Yukawa, but it interacted only electromagnetically (weakly), whereas the Yukawa particle should have a much stronger and shorter-ranged force. This paradox was resolved by the subsequent discovery of the pi meson, or *pion*, which has a mass close to that of the muon, and which *does* interact strongly. After its discovery, the pion was hailed as the Yukawa particle, and was incorporated into the family of hadronic elementary particles. The muon was now viewed as a particle without a purpose, and it was excluded from the hadronic elementary particle zoo and treated as a leptonic outcast. This situation was summarized very succinctly at that time by I. I. Rabi in his oft-quoted rhetorical question about the muon:

Who ordered that?

A half century later, in the May 2006 issue of Physics Today, Frank Wilczek updated this situation as follows:

> *Quark and lepton masses ... have eluded calculation despite decades of intense effort. ... When the muon was discovered, I. I. Rabi asked, 'Who ordered that?' His question continues to resonate, undamped, with ever-increasing amplitude.* [148]

This present-day assessment by Wilczek contains two interesting and possibly related elements: (1) the existing theories of elementary particle masses do not encompass either the quarks or the leptons, and thus are (at least in the opinion of the present author) manifestly incomplete — some crucial element is missing; (2) the collection of observed elementary particle states consists almost entirely of unstable particles, and yet the longest-lived (next to the neutron) and lowest-mass unstable particle — the muon — which ought to be one of the most important particle states, does not seem to fit in at all, and logic demands "with ever-increasing amplitude" that it *ought* to fit in. Hence we have on the one hand a particle mass puzzle with a missing piece, and on the other hand a prominent particle with no theory to encompass it. This situation calls to mind a basic investigative principle in detective stories, which goes roughly as follows: if all alternative explanations have been excluded, then whatever remains, no matter how improbable, must be (tentatively, at least) regarded as correct.

Ergo, maybe the muon is the missing part of the mass puzzle!

The challenge here, which was first tendered by Rabi, and which is now forcefully restated again by Wilczek, is to provide a rationalé for the existence of the muon. In the present book *The Power of* α, this challenge has, in my opinion, been accepted and met. The response to the Rabi and Wilchek challenge is summarized in the Introduction to the present book, where it is contained in the "muon-quantized" quark and particle masses displayed in the "muon mass tree" of Fig. 0.2.23, and in the $m_f \cong 105$ MeV excitation systematics of Table 0.10.1. These muon results are expanded in more detail in the mass discussions of Chapter 3, and in Tables C1 and C2 of Appendix C.

The mass excitations that are exhibited in the Fig. 0.2.23 muon mass tree include four Standard Model quark masses,

$$q = (u, d)\,,\ s\,,\ c\,,\ b\,,$$

and seven key elementary particle masses,

$$\mu\,,\ p\,,\ \tau\,,\ \phi\,,\ J/\psi\,,\ B_c\,,\ \Upsilon\,,$$

which are accurately reproduced as multiples of the *first-order* α-quantized muonic mass quantum $m_f = (3/2)\ (m_e/\alpha) \cong 105$ MeV, with one or two electron masses m_e added in. Also included in the Fig. 0.2.23 muon mass tree are the two higher-mass particle states, the isotopic-spin-averaged gauge boson doublet,

$$\overline{WZ}\,,$$

and the Standard Model top quark,

$$t\,,$$

which are accurately reproduced as multiples of the *second-order* α-quantized muonic mass quantum $m_f^\alpha \equiv m_f/\alpha \cong 14{,}394$ MeV. The only adjustable parameter in these mass tree calculations is the number of m_f or m_f^α mass quanta to use in each case, and these numbers form clear-cut patterns that do not represent random choices. The average calculated mass accuracy of the nine particle states (including the t quark) displayed in the muon mass tree is 0.83%, with no binding energy corrections applied. In detail, the three low-mass unpaired-quark particle states μ, p and τ, which require no binding energy corrections, are reproduced to an average mass accuracy of 0.48%, and the four high-mass particle and quark states B_c, Υ, $\overline{WZ}$ and t where the binding energy seems to essentially vanish (Figs. 3.15.1 and 3.15.2), are reproduced to an average mass accuracy of 0.28%. Since all nine of these states are reproduced *entirely* from mass quantum m_f or m_f^α (plus one or two electron masses m_e), they serve to delineate the indispensable role played by the muon in elementary particle mass generation. This is especially true when we consider the fact that some of the mass ratios given here have not been obtained by any other means, as was eloquently enunciated by Wilczek in the quotation at the beginning of this section [148].

The massive W and Z gauge bosons and top quark t, which occupy the upper branches of the muon mass tree, have three features that are of special phenomenological interest.

(i) They extend the domain of the muon mass quantum $m_f \cong 105$ MeV upward in energy, but not in the form of increased multiples of this mass unit. Instead, they occur as multiples of the "α-enhanced" mass quantum $m_f^\alpha \cong 105 \times 137$ MeV, which is a much larger but still "muon-based" excitation unit.

(ii) They extend the elementary particle mass range upward so that it now ranges over two powers of $\alpha^{-1} \cong 137$ with respect to the electron ground state, and thus to an extent mirrors the reciprocal lifetime scaling that extends over many powers of α. Further upward mass scaling may be hard to obtain due to physical and fiscal limitations, but downward mass scaling into the realm of neutrino masses is a possibility.

(iii) The W, Z and t particle and quark states represent "New Physics," which arises due to their unanticipated and highly accurate mass equation $m_w + m_z = m_t$. It has always been the hope (and the rationale) of particle physicists that in building larger and more expensive accelerators, they will not only verify the already existing theories, but will also obtain some New Physics results in the form of unexpected "surprises," which will serve to guide them to more accurate formulations of their theories. The W, Z and t mass relationship constitues such a *surprise*: it establishes a mass link between the W, Z gauge boson doublet and the t quark, which were thought to be unrelated entities. It also calls into question the nature of the $W^{\pm}$, Z^0 doublet itself. Should it be regarded as a $W^{\pm}$, W^0 isotopic spin doublet, whose cental (average) mass value has real physical significance (as in Fig. 0.2.23)?

As mentioned in the Preface, and also in Sec. A of the Postscript, I started my phenomenological studies of elementary particle masses in June 1969 by spending a week poring over the Review of Particle Physics. I emerged from this immersion in the data with two basic ideas: (1) the muon mass is a fundamental mass unit for quarks and particles; (2) quark binding energies are small (a few percent). I could not have imagined then that in the year 2006, thirty seven years later, I would be still be working on these same two assumptions (plus some help from the coupling constant α). This work has led to a *muon mass tree* that now stretches all the way from the 105 MeV muon mass to the 172,500 MeV top quark mass. The upper branches of this muon tree, which encompass the energy region above 12 GeV, were discovered and set in place only after I attended a number of colloquia devoted to top quark physics in the Spring of 2006 [149]. The muon mass unit, in its α-quantized form $m_f = (3/2)(m_e/\alpha)$, is central to these muon excitation-tree results, which provide the phenomenological framework for the lepton masses, and for the Standard Model fermion quark masses and associated hadron masses. This should be sufficient to allay the concerns voiced by Prof. Rabi so many years ago.

In this Postscript we have been following the histories of the m_b, m_f and m_f^{α} mass units as basic elementary particle excitation quanta. In Tables C1 and C2 of Appendix C we display the mass equations for these quanta, and for the particles they form, which are expressed there as functions of just the electron mass m_e and its operator, the fine structure constant α, which plays the role of an elementary particle mass generator. Hence, in the final analysis, it is the ubiquitous *electron* and its antimatter partner the *positron* who emerge as the ultimate elementary particle ground states for all baryonic matter. [150]

I have been pursuing this approach to elementary particle structure for more than a third of a century. The present book, *The Power of* α, summarizes the results that have thus far emerged from these studies. Appendix D gives an annotated bibliography of my books and papers on elementary particles. During the years 1970–1972 I delivered invited colloquia on this work at 22 universities around the US and in Canada, traveling roughly sixty thousand miles (at their expense) in the process. These colloquia, plus a few later ones, are listed in Table P2. I also presented papers at conferences, laboratories, and American Physical Society meetings, as listed in Table P3. During this period of time the elementary particle experiments and the associated theorical framework have continued to expand in size, complexity and accuracy until they now stand collectively as one of the monumental achievements in the field of science. It has been a real privilege to be able to witness these developments, and to a modest extent participate in them.

Table P2 Invited colloquia on elementary particles by Malcolm Mac Gregor.

1. Univ. of California at Los Angeles, Jan. 22 (1970)
2. Univ. of California, Berkeley, March 9 (1970)
3. Univ. of Florida, Oct. 1 (1970)
4. Florida State Univ., Oct. 2 (1970)
5. Univ. of Minnesota, Oct. 7 (1970)
6. Univ. of Wisconsin, Oct. 8 (1970)
7. Purdue Univ., Oct. 9 (1970)
8. Univ. of Arizona, Oct. 19 (1970)
9. Rice Institute, Nov. 4 (1970)
10. Univ. of Michigan, Jan. 19 (1971)
11. Ohio State Univ., Jan. 25 (1971)
12. Ohio Univ., Jan. 26 (1971)
13. Wayne State Univ. (Detroit), Jan. 27 (1971)
14. Notre Dame Univ., Jan. 28 (1971)
15. Arizona State Univ., Feb. 18 (1971)
16. New Mexico Univ., Feb. 19 (1971)
17. Univ. of Western Ontario, April 13 (1971)
18. Virginia Tech, Oct. 28 (1971)
19. Univ. of Virginia, Oct. 29 (1971)
20. Univ. of Oregon, Feb. 24 (1972)
21. Oregon State Univ., Feb. 25 (1972)
22. Univ. of California at Davis, Oct. 25 (1972)
23. Univ. of California, Berkeley, Jan. 24 (1974)
24. Wayne State Univ. (Detroit), Sept. (1992)
25. Univ. of Cincinnati, Sept. 23 (1994)

Table P3 Talks on elementary particles by Malcolm Mac Gregor.

1. American Physical Society, Boulder, Colorado, Oct. 30 (1969)
2. American Physical Society, Los Angeles, Dec. 30 (1969)
3. Theoretical Division, Lawrence Livermore Lab, Feb. 21 (1970)
4. American Physical Society, Washington, D.C., April 27–30 (1970)
5. American Physical Society, Austin, Texas, Nov. 5 (1970)
6. Coral Gables High Energy Conference, Jan. 20–22 (1971)
7. Theoretical Division, Lawrence Livermore Lab, April 20 (1971)
8. American Physical Society, Washington, D.C., April 27–30 (1971)
9. Theoretical Division, Lawrence Livermore Lab, Feb. 22 (1972)
10. American Physical Society, Berkeley, California, Dec. 28 (1973)
11. Syracuse Symposium, May 2–4 (1975)
12. Radiochemistry Division, Lawrence Livermore Lab, Sept. 16 (1975)
13. Vanderbilt High Energy Conference, March 2 (1976)
14. Theoretical Division, Lawrence Livermore Lab, June 5 (1979)
15. American Physical Society, Washington, D.C., April 30 (1986)
16. American Physical Society, San Francisco, Jan. 28 (1987)
17. American Physical Society, Arlington, Virginia, April 20, (1987)
18. American Physical Society, Washington, D.C., April 16 (1990)
19. American Physical Society, Washington, D.C., April 20 (1992)
20. American Physical Society, Washington, D.C., April 20 (1994)

Appendices

Appendix A. The 157-Particle Lifetime Data Compilation

This compilation is based on RPP2006 [10]. The *mean life* or *lifetime* τ is shown for each particle. The experimentally measured full widths $\Gamma = \hbar/\tau$, which are used to calculate the mean lives τ for the short-lived particles, are also displayed. Only the particles with well-determined lifetimes (not just upper limits) are included. The 36 long-lived threshold-state particles, which are the ones with lifetimes $\tau > 1$ zs (10^{-21} s), are listed first, followed by 5 other particles of special interest. These are arranged in the order of decreasing lifetimes. The remaining 116 short-lived particles are arranged in the order of particle type and increasing mass. A few particles that have multiple charge states with similar masses and lifetimes are displayed with isotopic-spin-averaged lifetime values. Appendix B gives the experimental mass values for the particles of Appendix A.

	Mean life τ **(sec)**	**Full width** Γ **(MeV)**	**Particle**
	Thirty six particles with lifetimes $\tau > 1$ **zs**		
1	8.8570E+02		neutron
2	2.1970E−06		muon
3	5.1140E−08		K_L^0
4	2.6033E−08		$\pi^\pm$
5	1.2385E−08		$K^\pm$
6	2.9000E−10		Ξ^0
7	2.6310E−10		Λ^0
8	1.6390E−10		Ξ^-

	Mean life τ (sec)	Full width Γ (MeV)	Particle
Thirty six particles with lifetimes $\tau > 1$ zs (*Cont.*)			
9	1.4790E−10		Σ^-
10	8.9530E−11		K_S^0
11	8.2100E−11		Ω^-
12	8.0180E−11		Σ^+
13	1.6380E−12		$B^\pm$
14	1.5300E−12		B^0
15	1.4660E−12		B_s^0
16	1.3900E−12		Ξ_b
17	1.2300E−12		Λ_b^0
18	1.0400E−12		$D^\pm$
19	5.0000E−13		$D_s^\pm$
20	4.6000E−13		$B_c^\pm$
21	4.4200E−13		Ξ_c^+
22	4.1010E−13		D^0
23	2.9060E−13		τ
24	2.0000E−13		Λ_c^+
25	1.1200E−13		Ξ_c^0
26	6.9000E−14		Ω_c^0
27	8.4000E−17		π^0
28	5.0632E−19	1.300E−03	η
29	7.4000E−20		Σ^0
30	3.2392E−20	2.0320E−02	$\Upsilon(3S)$
31	2.0582E−20	3.1980E−02	$\Upsilon(2S)$
32	1.2185E−20	5.4020E−02	$\Upsilon(1S)$
33	7.0472E−21	9.3400E−02	$J/\psi(1S)$
34	6.8564E−21	9.6000E−02	$D^*(2010)^\pm$
35	3.2424E−21	2.0300E−01	η'
36	1.9532E−21	3.3700E−01	$\psi(2S)$
Five miscellaneous particles with lifetimes $\tau < 1$ zs			
37	1.5451E−22	4.2600E+00	$\phi(1020)$
38	7.7528E−23	8.4900E+00	$\omega(782)$
39	2.5812E−23	2.5500E+01	$\eta_c(1S)$
40	3.0743E−25	2.1410E+03	$W^\pm$
41	2.6379E−25	2.4952E+03	Z^0

	Mean life τ **(sec)**	**Full width** Γ **(MeV)**	**Particle**
	One hundred sixteen particles with lifetimes $\tau < 1$ **zs**		
	Thirty five nonstrange mesons		
42	4.4057E−24	1.4940E+02	$\rho(770)$
43	9.4030E−24	7.0000E+01	$f_0(980)$
44	8.7762E−24	7.5000E+01	$a_0(980)$
45	1.8284E−24	3.6000E+02	$h_1(1170)$
46	4.6353E−24	1.4200E+02	$b_1(1235)$
47	3.5541E−24	1.8520E+02	$f_2(1270)$
48	2.7199E−23	2.4200E+01	$f_1(1285)$
49	1.1967E−23	5.5000E+01	$\eta(1295)$
50	6.1515E−24	1.0700E+02	$a_2(1320)$
51	2.1940E−24	3.0000E+02	$\pi_1(1400)$
52	1.2881E−24	5.1100E+01	$\eta(1405)$
53	1.1989E−23	5.4900E+01	$f_1(1420)$
54	3.0615E−24	2.1500E+02	$\omega(1420)$
55	2.4838E−24	2.6500E+02	$a_0(1450)$
56	4.4776E−24	1.4700E+02	$\rho(1450)$
57	7.5657E−24	8.7000E+01	$\eta(1475)$
58	6.0386E−24	1.0900E+02	$f_0(1500)$
59	9.0166E−24	7.3000E+01	$f_2'(1525)$
60	2.9254E−24	2.2500E+02	$\pi_1(1600)$
61	3.6365E−24	1.8100E+02	$\eta(1645)$
62	2.0896E−24	3.1500E+02	$\omega(1650)$
63	3.9179E−24	1.6800E+02	$\omega_3(1670)$
64	2.5414E−24	2.5900E+02	$\pi_2(1670)$
65	4.3881E−24	1.5000E+02	$\phi(1680)$
66	4.0883E−24	1.6100E+02	$\rho_3(1690)$
67	2.6328E−24	2.5000E+02	$\rho(1700)$
68	4.8045E−24	1.3700E+02	$f_0(1710)$
69	3.1798E−24	2.0700E+02	$\pi(1800)$
70	7.5657E−24	8.7000E+01	$\phi_3(1850)$
71	1.3945E−24	4.7200E+02	$f_2(1950)$
72	3.2585E−24	2.0200E+02	$f_2(2010)$
73	2.1029E−24	3.1300E+02	$a_4(2040)$
74	2.9254E−24	2.2500E+02	$f_4(2050)$
75	4.4175E−24	1.4900E+02	$f_2(2300)$
76	2.0634E−24	3.1900E+02	$f_2(2340)$

	Mean life τ (sec)	Full width Γ (MeV)	Particle
	Eleven strange kaons K		
77	1.3021E−23	5.0550E+01	$K^*(892)$
78	7.3135E−24	9.0000E+01	$K_1(1270)$
79	3.7828E−24	1.7400E+02	$K_1(1400)$
80	2.8371E−24	2.3200E+02	$K^*(1410)$
81	2.2697E−24	2.9000E+02	$K_0^*(1430)$
82	6.3442E−24	1.0375E+02	$K_2^*(1430)$
83	2.0441E−24	3.2200E+02	$K^*(1680)$
84	3.5388E−24	1.8600E+02	$K_2(1770)$
85	4.1397E−24	1.5900E+02	$K_2^*(1780)$
86	2.3848E−24	2.7600E+02	$K_2(1820)$
87	3.3243E−24	1.9800E+02	$K_4^*(2045)$
	Thirteen charm mesons D and ψ		
88	3.2265E−23	2.0400E+01	$D_1(2420)^0$
89	2.6328E−23	2.5000E+01	$D_1(2420)^\pm$
90	1.8284E−23	3.6000E+01	$D_2^*(2460)$
91	4.3881E−23	1.5000E+01	$D_{s2}(2573)^\pm$
92	6.3290E−23	1.0400E+01	$\chi_{c0}(1P)$
93	7.3956E−23	8.9000E−01	$\chi_{c1}(1P)$
94	3.1952E−23	2.0600E+00	$\chi_{c2}(1P)$
95	4.7015E−23	1.4000E+01	$\eta_c(2S)$
96	2.8618E−23	2.3000E+01	$\psi(3770)$
97	2.2697E−23	2.9000E+01	$\chi_{c2}(2P)$
98	8.2276E−24	8.0000E+01	$\psi(4040)$
99	6.3904E−24	1.0300E+02	$\psi(4160)$
100	1.0616E−23	6.2000E+01	$\psi(4415)$
	Three bottom mesons Υ		
101	3.2108E−23	2.0500E+01	$\Upsilon(4S)$
102	5.9837E−24	1.1000E+02	$\Upsilon(10860)$
103	8.3318E−24	7.9000E+01	$\Upsilon(11020)$
	Thirteen nonstrange baryons		
104	2.1940E−23	3.0000E+02	$N(1440)P_{11}$
105	5.7236E−24	1.1500E+02	$N(1520)D_{13}$
106	4.3881E−24	1.5000E+02	$N(1535)S_{11}$

	Mean life τ (sec)	Full width Γ (MeV)	Particle
	Thirteen nonstrange baryons (*Cont.*)		
107	3.9892E−24	1.6500E+02	$N(1650)S_{11}$
108	4.3881E−24	1.5000E+02	$N(1675)D_{15}$
109	5.0632E−24	1.3000E+02	$N(1680)F_{15}$
110	6.5821E−24	1.0000E+02	$N(1700)D_{13}$
111	6.5821E−24	1.0000E+02	$N(1710)P_{11}$
112	3.2911E−24	2.0000E+02	$N(1720)P_{13}$
113	1.3164E−24	5.0000E+02	$N(2190)G_{17}$
114	1.6455E−24	4.0000E+02	$N(2220)H_{19}$
115	1.3164E−24	5.0000E+02	$N(2250)G_{17}$
116	1.0126E−24	6.5000E+02	$N(2600)I_{1,11}$
	Ten nonstrange baryons Δ		
117	5.5781E−24	1.1800E+02	$\Delta(1232)P_{33}$
118	1.8806E−24	3.5000E+02	$\Delta(1600)P_{33}$
119	4.5394E−24	1.4500E+02	$\Delta(1620)S_{31}$
120	2.1940E−24	3.0000E+02	$\Delta(1700)D_{33}$
121	1.9946E−24	3.3000E+02	$\Delta(1905)F_{35}$
122	2.6328E−24	2.5000E+02	$\Delta(1910)P_{31}$
123	3.2911E−24	2.0000E+02	$\Delta(1920)P_{33}$
124	1.8284E−24	3.6000E+02	$\Delta(1930)D_{35}$
125	2.3095E−24	2.8500E+02	$\Delta(1950)F_{37}$
126	1.6455E−24	4.0000E+02	$\Delta(2420)H_{3,11}$
	Thirteen strange hyperons Λ		
127	1.3164E−23	5.0000E+01	$\Lambda(1405)S_{01}$
128	4.2193E−23	1.5600E+01	$\Lambda(1520)D_{03}$
129	4.3881E−24	1.5000E+02	$\Lambda(1600)P_{01}$
130	1.8806E−23	3.5000E+01	$\Lambda(1670)S_{01}$
131	1.0970E−23	6.0000E+01	$\Lambda(1690)D_{03}$
132	2.1940E−24	3.0000E+02	$\Lambda(1800)S_{01}$
133	4.3881E−24	1.5000E+02	$\Lambda(1810)P_{01}$
134	8.2276E−24	8.0000E+01	$\Lambda(1820)F_{05}$
135	6.9285E−24	9.5000E+01	$\Lambda(1830)D_{05}$
136	6.5821E−24	1.0000E+02	$\Lambda(1890)P_{03}$
137	3.2911E−24	2.0000E+02	$\Lambda(2100)G_{07}$
138	3.2911E−24	2.0000E+02	$\Lambda(2110)F_{05}$
139	4.3881E−24	1.5000E+02	$\Lambda(2350)H_{09}$

	Mean life τ (sec)	Full width Γ (MeV)	Particle
	Nine strange hyperons Σ		
140	1.7742E−23	3.7100E+01	$\Sigma(1385)P_{13}$
141	6.5821E−24	1.0000E+02	$\Sigma(1660)P_{11}$
142	1.0970E−23	6.0000E+01	$\Sigma(1670)D_{13}$
143	7.3135E−24	9.0000E+01	$\Sigma(1750)S_{11}$
144	5.4851E−24	1.2000E+02	$\Sigma(1775)D_{15}$
145	5.4851E−24	1.2000E+02	$\Sigma(1915)F_{15}$
146	2.9919E−24	2.2000E+02	$\Sigma(1940)D_{13}$
147	3.6567E−24	1.8000E+02	$\Sigma(2030)F_{17}$
148	6.5821E−24	1.0000E+02	$\Sigma(2250)$
	Four strange hyperons Ξ		
149	6.2985E−23	9.5000E+00	$\Xi(1530)^0 P_{13}$
150	2.7425E−23	2.4000E+01	$\Xi(1820)D_{13}$
151	1.0970E−23	6.0000E+01	$\Xi(1950)$
152	3.2911E−23	2.0000E+01	$\Xi(2030)$
	One strange hyperon Ω		
153	1.1967E−23	5.5000E+01	$\Omega(2250)$
	One charm baryon Λ_c		
154	1.8284E−22	3.6000E+00	$\Lambda_c(2593)^+$
	Three charm baryons Σ_c		
155	2.9716E−22	2.2150E+00	$\Sigma_c(2455)$
156	4.2465E−23	1.5500E+01	$\Sigma_c(2520)$
157	9.9729E−24	6.6000E+01	$\Sigma_c(2800)$

Appendix B. The Elementary Particle Mass Compilation

Appendix B is the mass counterpart of Appendix A, and is also based on RPP2006 [10]. It includes mainly the particles listed in Appendix A plus the stable electron and proton. The mass calculations described in the present book do not include the mass splittings that arise due to the different charge states (isotopic spins) of a particle, so only isotopic-spin-averaged charge-independent (CI) mass values are used. These averaged mass values are also displayed here.

Mass (MeV)	Particle	CI mass	CI state
	Long-lived threshold-state particles		
Leptons			
0.510998918	$e^{\pm}$		
105.658369	$\mu^{\pm}$		
1776.99	$\tau^{\pm}$		
Pseudoscalar mesons			
134.9766	π^0	137.27	π
139.57018	$\pi^{\pm}$		
493.677	$K^{\pm}$	495.66	K
497.648	K^0		
547.51	η		
957.78	η'		
Baryons and hyperons			
938.272029	p	938.92	n
939.565360	n		
1115.683	Λ		
1189.37	Σ^+		
1192.642	Σ^0	1193.15	Σ
1197.449	Σ^-		

Mass (MeV)	Particle	CI mass	CI state
Long-lived threshold-state particles			
Baryons and hyperons (*Cont.*)			
1314.83	Ξ^0	1318.07	Ξ
1321.31	Ξ^-		
1672.45	Ω^-		
Charm mesons			
1864.5	D^0	1866.9	D
1869.3	$D^\pm$		
1968.2	$D_s^\pm$		
2006.7	D^{*0}	2008.35	D^*
2010.0	$D^{*\pm}$		
3069.619	$J/\psi(1S)$		
3686.093	$\psi(2S)$		
Charm baryons			
2286.46	Λ_c		
2467.9	Ξ_c^+	2469.45	Ξ_c
2471.0	Ξ_c^0		
2697.5	Ω_c		
3518.9	Ξ_{cc}		
Bottom mesons			
5279.0	$B^\pm$	5279.2	B
5279.4	B^0		
5367.5	B_s		
6286.0	B_c		
9460.30	$\Upsilon(1S)$		
10023.26	$\Upsilon(2S)$		
10355.2	$\Upsilon(3S)$		

Mass (MeV)	Particle	CI mass	CI state
	Long-lived threshold-state particles		
Bottom baryons			
5624.0	Λ_b		
	Miscellaneous particles		
782.65	ω		
1019.460	ϕ		
2980.4	$\eta_c(1S)$		
80403.0	$W^{\pm}$	85795.3	$\overline{WZ}$
91187.6	Z^0		
>114400.0	Higgs		
	Short-lived excited-state particles		
Nonstrange mesons			
775.5	$\rho(770)$		
980.0	$f_0(980)$		
984.7	$a_0(980)$		
1170.0	$h_1(1170)$		
1229.5	$b_1(1235)$		
1275.4	$f_2(1270)$		
1281.8	$f_1(1285)$		
1294.0	$\eta(1295)$		
1318.3	$a_2(1320)$		
1376.0	$\pi_2(1400)$		
1409.8	$\eta(1405)$		
1426.3	$f_1(1420)$		
1425.0	$\omega(1420)$		
1474.0	$a_0(1450)$		
1459.0	$\rho(1450)$		
1476.0	$\eta(1475)$		
1507.0	$f_0(1500)$		
1525.0	$f_2'(1525)$		
1653.0	$\pi_1(1600)$		
1617.0	$\eta_2(1645)$		

Mass (MeV)	Particle	CI mass	CI state
	Short-lived excited-state particles		
Nonstrange mesons (*Cont.*)			
1670.0	$\omega(1650)$		
1667.0	$\omega_3(1670)$		
1672.4	$\pi_2(1670)$		
1680.0	$\phi(1680)$		
1688.8	$\rho_3(1690)$		
1720.0	$\rho(1700)$		
1718.0	$f_0(1710)$		
1812.0	$\pi(1800)$		
1854.0	$\phi_3(1850)$		
1944.0	$f_2(1950)$		
2011.0	$f_2(2010)$		
2001.0	$a_4(2040)$		
2025.0	$f_4(2050)$		
2297.0	$f_2(2300)$		
2339.0	$f_2(2340)$		
Strange mesons *K*			
893.83	$K^*(892)$ or $K^*(894)$		
1272.0	$K_1(1270)$		
1402.0	$K_1(1400)$		
1414.0	$K^*(1410)$		
1414.0	$K_0^*(1430)$		
1425.6	$K_2^*(1430)^{\pm}$		
1432.4	$K_2^*(1430)^0$		
1717.0	$K^*(1680)$		
1773.0	$K_2(1770)$		
1776.0	$K_2^*(1780)$		
1816.0	$K_2(1820)$		
2045.0	$K_4^*(2045)$		
Charm mesons *D*			
2422.3	$D_1(2420)^0$	2422.85	$D_1(2420)$
2423.4	$D_1(2420)^{\pm}$		
2461.1	$D_2^*(2460)^0$	2460.05	$D_2^*(2460)$
2459.0	$D_2^*(2460)^{\pm}$		

Mass (MeV)	Particle	CI mass	CI state

Short-lived excited-state particles

Charm mesons D (*Cont.*)

Mass (MeV)	Particle	CI mass	CI state
2112.0	$D_S^*(2112)^\pm$		
2535.35	$D_{S1}(2536)^\pm$		
2573.5	$D_{S2}(2573)^\pm$		

Charm excitations J/ψ

Mass (MeV)	Particle	CI mass	CI state
3414.76	$\chi_{c0}(1P)$		
3510.66	$\chi_{c1}(1P)$		
3556.2	$\chi_{c2}(1P)$		
3638.0	$\eta_c(2S)$		
3771.1	$\psi(3770)$		
3929.0	$\chi_{c2}(2P)$		
4039.0	$\psi(4040)$		
4153.0	$\psi(4160)$		
4421.0	$\psi(4415)$		

Bottom excitations Υ

Mass (MeV)	Particle	CI mass	CI state
10579.4	$\Upsilon(4S)$		
10865.0	$\Upsilon(10860)$		
11019.0	$\Upsilon(11020)$		

Baryon excitations N

Mass (MeV)	Particle	CI mass	CI state
1440.0	$N(1440)P_{11}$		
1520.0	$N(1520)D_{13}$		
1535.0	$N(1535)S_{11}$		
1655.0	$N(1650)P_{11}$		
1675.0	$N(1675)D_{15}$		
1685.0	$N(1680)F_{15}$		
1700.0	$N(1700)D_{13}$		
1710.0	$N(1710)P_{11}$		
1720.0	$N(1720)P_{13}$		
2190.0	$N(2190)G_{17}$		
2250.0	$N(2220)H_{19}$		
2275.0	$N(2250)G_{17}$		
2600.0	$N(2600)I_{1,11}$		

Mass (MeV)	Particle	CI mass	CI state
Short-lived excited-state particles			
Baryon excitations Δ			
1232.0	$\Delta(1232)P_{33}$		
1600.0	$\Delta(1600)P_{33}$		
1630.0	$\Delta(1620)S_{31}$		
1700.0	$\Delta(1700)D_{33}$		
1890.0	$\Delta(1905)F_{35}$		
1910.0	$\Delta(1910)P_{31}$		
1920.0	$\Delta(1920)P_{33}$		
1960.0	$\Delta(1930)D_{35}$		
1930.0	$\Delta(1950)F_{37}$		
2420.0	$\Delta(2420)H_{3,11}$		
Hyperon excitations Λ			
1406.0*	$\Lambda(1405)S_{01}$		
1519.5	$\Lambda(1520)D_{03}$		
1600.0	$\Lambda(1600)P_{01}$		
1670.0	$\Lambda(1670)S_{01}$		
1690.0	$\Lambda(1690)D_{03}$		
1800.0	$\Lambda(1800)S_{01}$		
1810.0	$\Lambda(1810)P_{01}$		
1820.0	$\Lambda(1820)F_{05}$		
1830.0	$\Lambda(1830)D_{05}$		
1890.0	$\Lambda(1890)P_{03}$		
2100.0	$\Lambda(2100)G_{07}$		
2110.0	$\Lambda(2110)F_{05}$		
2350.0	$\Lambda(2350)H_{09}$		
Hyperon excitations Σ			
1382.8	$\Sigma(1385)^{+}P_{13}$	1384.6	$\Sigma(1385)$
1383.7	$\Sigma(1385)^{0}P_{13}$		
1387.2	$\Sigma(1385)^{-}P_{13}$		
1660.0	$\Sigma(1660)P_{11}$		
1670.0	$\Sigma(1670)D_{13}$		
1750.0	$\Sigma(1750)S_{11}$		

Mass (MeV)	Particle	CI mass	CI state

Short-lived excited-state particles

Hyperon excitations Σ (*Cont.*)

Mass (MeV)	Particle	CI mass	CI state
1775.0	$\Sigma(1775)D_{15}$		
1915.0	$\Sigma(1915)F_{15}$		
1940.0	$\Sigma(1940)D_{13}$		
2030.0	$\Sigma(2030)F_{17}$		
2250.0	$\Sigma(2250)$		

Hyperon excitations Ξ

Mass (MeV)	Particle	CI mass	CI state
1531.80	$\Xi(1530)^0 P_{13}$	1533.4	$\Xi(1530)$
1535.0	$\Xi(1530)^- P_{13}$		
1823.0	$\Xi(1820)D_{13}$		
1950.0	$\Xi(1900)$		
2025.0	$\Xi(2030)$		

Hyperon excitation Ω

Mass (MeV)	Particle	CI mass	CI state
2252.0	$\Omega(2250)^-$		

Charm baryon excitation Λ_c

Mass (MeV)	Particle	CI mass	CI state
2595.4	$\Lambda_c(2593)^+$		

Charm baryon excitations Σ_c

Mass (MeV)	Particle	CI mass	CI state
2454.02	$\Sigma_c(2455)^{++}$	2453.89	$\Sigma_c(2455)$
2453.76	$\Sigma_c(2455)^0$		
2518.4	$\Sigma_c(2520)^{++}$	2518.2	$\Sigma_c(2520)$
2518.0	$\Sigma_c(2520)^0$		
2801.0	$\Sigma_c(2800)^{++}$	2798.3	$\Sigma_c(2800)$
2792.0	$\Sigma_c(2800)^+$		
2802.0	$\Sigma_c(2800)^0$		

Top quark t

Mass (MeV)	Particle	CI mass	CI state
172500.0	t		

Appendix C. Calculated and Experimental α-Quantized Particle "Ground State" Masses

Table C1 contains calculated and experimental mass values for 33 fundamental elementary particle states, which include the leptons, the Standard Model constituent quarks, the "ground-state" excitations in which each quark combination first appears, the pseudoscalar mesons, and the supermassive quarks and gauge boson doublet. The experimental masses for isotopic spin multiplets are averaged. The calculated mass values are all expressed in terms of the electron mass m_e and the powers of α, together with numerical factors that are applied to this mass in order to generate the higher-mass states. The particle states below 12 GeV are generated by single-α-leap excitations, and they feature the boson mass quanta $m_b = m_e/\alpha$ and $m_f = 3m_e/2\alpha$. The particle states above 12 GeV are generated by double-α-leap excitations, and they feature the fermion mass quantum $m_f^\alpha = 3m_e/2\alpha^2$. No binding energy corrections are applied. It can be seen from the quoted errors (which are expressed as percentages with respect to the experimental mass values) that the paired quark–anti-quark meson states below 4 GeV require a 2–3% hadronic binding energy corrections, whereas the paired quark–anti-quark meson states above 4 GeV have essentially zero binding energies. The Λ and Ξ hyperons have errors which suggest they might require ~3.6% hadronic binding energies, whereas the proton p appears right on the mass shell. The highest-mass state — the top quark t — is generated from the lowest-mass state — the electron — by a double α-leap excitation with a mass accuracy of 0.13%, which indicates that this *constituent-quark* mass systematics, at least for these basic particle excitations, is valid over the entire range of elementary particle masses. The mass equations displayed in Table C1 demonstrate that the electron functions as the fundamental ground state for all of the massive elementary particles, with the coupling constant α playing the role of the mass generator.

Table C2 contains the m_f excitations from Table C1 that have calculated masses which are smaller than the experimental masses, and it displays the number of excitation quanta m_b that must be added to these excitations in order to make the calculated masses approximately equal to or slightly larger than the experimental masses. These "hybrid" particle excitations, which contain both m_b and m_f quanta in the same particle, are then used in Table 3.15.2 of Sec. 3.15 to obtain the calculated hadronic binding energy (HBE) values that are displayed as x's in Fig. 3.15.2.

Table C1

	Ground state $m_e = 0.510998918$	Particle mass generator $\alpha = 1/137.03599911$		Calculated mass	Exper. mass	Error (%)	Error (MeV)
	α-generated mass quanta	(All masses in MeV/c^2)					
	m_b	m_e/α		70.025			
	m_f	$(3/2)(m_e/\alpha)$		105.038			
	m_f^α	$(3/2)(m_e/\alpha^2)$		14,393.968			
	Platform state	single-α-leap equations	$\#m_f$				
1	$\mu = m_e + m_f$	$m_e[1 + (3/2\alpha)]$	1	105.549	105.658	−0.10	−0.109
2	$q \equiv (u, d)$	$m_e[1 + (3/2\alpha)(1 + 2)]$	3	315.625			
3	s	$m_e[1 + (3/2\alpha)(1 + 4)]$	5	525.700			
4	$p = qqq$	$m_e[1 + (3/2\alpha)(1 + 8)]$	9	945.85	938.27	+0.81	+7.58
5	τ	$m_e[1 + (3/2\alpha)(1 + 16)]$	17	1786.16	1776.99	+0.52	+9.17
6	c	$m_e[1 + (3/2\alpha)(5 \times 3)]$	15	1576.079			
7	b	$m_e[1 + (3/2\alpha)(5 \times 3 \times 3)]$	45	4727.215			
8	$\phi = s\bar{s}$	$2m_e[1 + (3/2\alpha)(5)]$	10	1051.40	1019.46	+3.13	+31.94
9	$J/\psi = c\bar{c}$	$2m_e[1 + (3/2\alpha)(15)]$	30	3152.16	3096.92	+1.78	+55.24
10	$\Upsilon = b\bar{b}$	$2m_e[1 + (3/2\alpha)(45)]$	90	9454.43	9460.30	−0.062	−5.87
11	$\omega = q\bar{q}$	$2m_e[1 + (3/2\alpha)(3)]$	6	631.25	782.65	−19.35	−151.40
12	$K^* = q\bar{s}$	$m_e[1 + (3/2\alpha)(3 + 5)]$	8	840.81	893.83	−5.93	−53.02
13	$D = q\bar{c}$	$m_e[2 + (3/2\alpha)(3 + 15)]$	18	1891.70	1866.9	+1.33	+24.80
14	$B = q\bar{b}$	$m_e[2 + (3/2\alpha)(3 + 45)]$	48	5042.84	5279.2	−4.48	−236.36
15	$D_s = s\bar{c}$	$m_e[2 + (3/2\alpha)(5 + 15)]$	20	2101.78	1968.2	+6.79	+133.58

Table C1 (*Continued*)

	Ground state $m_e = 0.510998918$	Particle mass generator $\alpha = 1/137.03599911$		Calculated mass	Exper. mass	Error (%)	Error (MeV)
			$\#m_f$				
16	$B_s = s\bar{b}$	$m_e[2 + (3/2\alpha)(5 + 45)]$	50	5252.92	5367.5	−2.14	−114.50
17	$B_c = c\bar{b}$	$m_e[2 + (3/2\alpha)(15 + 45)]$	60	6303.29	6286	+0.28	+17.29
18	$\Lambda = qqs$	$m_e[1 + (3/2\alpha)(3 + 3 + 5)]$	11	1155.93	1115.68	+3.61	+40.25
19	$\Sigma = qqs$	$m_e[1 + (3/2\alpha)(3 + 3 + 5)]$	11	1155.93	1193.15	−3.12	−37.22
20	$\Xi = qss$	$m_e[1 + (3/2\alpha)(3 + 5 + 5)]$	13	1366.00	1318.07	+3.64	+47.93
21	$\Omega = sss$	$m_e[1 + (3/2\alpha)(5 + 5 + 5)]$	15	1576.08	1672.45	−5.76	−96.37
22	$\Lambda_c = qqc$	$m_e[1 + (3/2\alpha)(3 + 3 + 15)]$	21	2206.31	2286.46	−3.51	−80.15
23	$\Xi_c = qsc$	$m_e[1 + (3/2\alpha)(3 + 5 + 15)]$	23	2416.38	2469.45	−2.14	−53.07
24	$\Xi_{cc} = qcc$	$m_e[1 + (3/2\alpha)(3 + 15 + 15)]$	33	3466.76	3518.9	−1.48	−52.14
25	$\Omega_c = ssc$	$m_e[1 + (3/2\alpha)(5 + 5 + 15)]$	25	2626.46	2697.5	−2.63	−71.04
26	$\Lambda_b = qqb$	$m_e[1 + (3/2\alpha)(3 + 3 + 45)]$	51	5357.44	5624.0	−4.74	−266.56
			$\#m_b$				
27	$\pi = m_b\bar{m}_b$	$2m_e[1 + (1/\alpha)]$	2	141.07	137.27	+2.77	+3.80
28	$\eta = 4m_b\bar{m}_b$	$2m_e[1 + (1/\alpha)(1 + 3)]$	8	561.23	547.51	+2.51	+13.72
29	$\eta' = 7m_b\bar{m}_b$	$2m_e[1 + (1/\alpha)(1 + 6)]$	14	981.38	957.78	+2.46	+23.60
30	$K = 7m_b$	$m_e[1 + (1/\alpha)(1 + 6)]$	7	490.69	495.66	−1.0	−4.97
		double-α-leap equations	$\#m_f^\alpha$				
31	$q^\alpha \equiv (u^\alpha, d^\alpha)$	$m_e[1 + (3/2\alpha^2)(3)]$	3	43,182.4			
32	$\overline{WZ} = q^\alpha\bar{q}^\alpha$	$2m_e[1 + (3/2\alpha^2)(3)]$	6	86,364.8	85,795.3	+0.66	+569.5
33	$t = 4q^\alpha$	$m_e[1 + (3/2\alpha^2)(3)(4)]$	12	172,728.1	172,500	+0.13	+228.1
	$T = t\bar{t}$		24	345,456.3			

Table C2

	Ground state $m_e = 0.510998918$	Particle mass generator $\alpha = 1/137.03599911$	Table C1 $\#m_f$	Added $\#m_b$	Calculated mass	Exper. mass	Error (%)
	α-generated mass quanta	(All masses in MeV/c^2)					
	m_b	m_e/α			70.025		
	m_f	$(3/2)(m_e/\alpha)$			105.038		
	m_f^α	$(3/2)(m_e/\alpha^2)$			14,393.968		
	Platform state	single-α-leap equations	$\#m_f$	$\#m_b$			
11	$\omega = q\bar{q} + m_b\bar{m}_b$	$2m_e[1 + (3/2\alpha)(3) + 1/\alpha]$	6	2	771.30	782.65	−1.45
12	$K^* = q\bar{s} + m_b$	$m_e[1 + (3/2\alpha)(3+5) + 1/\alpha]$	8	1	910.84	893.83	+1.90
14	$B = q\bar{b} + 2m_b\bar{m}_b$	$m_e[2 + (3/2\alpha)(3+45) + 4/\alpha]$	48	4	5322.94	5279.2	+0.83
16	$B_s = s\bar{b} + m_b\bar{m}_b$	$m_e[2 + (3/2\alpha)(5+45) + 2/\alpha]$	50	2	5392.97	5367.5	+0.47
19	$\Sigma = qqs + m_b$	$m_e[1 + (3/2\alpha)(3+3+5) + 1/\alpha]$	11	1	1225.95	1193.15	+2.75
21	$\Omega = sss + 2m_b$	$m_e[1 + (3/2\alpha)(5+5+5) + 2/\alpha]$	15	2	1716.13	1672.45	+2.61
22	$\Lambda_c = qqc + 2m_b$	$m_e[1 + (3/2\alpha)(3+3+15) + 2/\alpha]$	21	2	2346.36	2286.46	+2.62
23	$\Xi_c = qsc + 2m_b$	$m_e[1 + (3/2\alpha)(3+5+15) + 2/\alpha]$	23	2	2556.43	2469.45	+3.52
24	$\Xi_{cc} = qcc + 2m_b$	$m_e[1 + (3/2\alpha)(3+15+15) + 2/\alpha]$	33	2	3606.81	3518.9	+2.50
25	$\Omega_c = ssc + 2m_b$	$m_e[1 + (3/2\alpha)(5+5+15) + 2/\alpha]$	25	2	2766.51	2697.5	+2.56
26	$\Lambda_b = qqb + 4m_b$	$m_e[1 + (3/2\alpha)(3+3+45) + 4/\alpha]$	51	4	5637.54	5624.0	+0.24

Appendix D. Annotated Bibliography of the Author's Elementary Particle Publications

Books on elementary particles

1. *The Nature of the Elementary Particle* (Springer-Verlag, Berlin, 1978).
2. *The Enigmatic Electron* (Kluwer Academic, Dordrecht, 1992).
3. *The Power of α* (World Scientific, Singapore, 2007).

Annotated elementary particle bibliography

1. UCRL-71842 (Rev.) Part I, Aug. 8 (1969) (unpublished), "Elementary Particle States."

This paper introduced the idea that quark binding energies are very small (~1%), and it identified the muon mass and pion mass as fundamental light-quark building blocks. The necessity for having a neutral muon mass quantum $\mu = 105$ MeV was pointed out. (With the subsequent development, a few months later, of the relativistic equations for a spinning sphere, the neutral muon mass was later identified as the spin-1/2 form of the neutral spin-0 mass quantum $M = 70$ MeV.)

2. UCRL-71842 (Rev.) Part II, Aug. 8 (1969) (unpublished), "Meson and Baryon Resonances."

This paper introduced the idea that baryons and mesons occur in nuclear-physics-type rotational bands, with an $\ell(\ell+1)$ splitting of the levels. The meson rotational bands that were identified in this 1969 paper differ substantially from later (1978) assignments, but the baryon rotational bands are very similar. (After the discovery of the high-mass b-quark states, further work, as published in Ref. [42] in 1990, led to the abandonment of the general concept of particle rotational bands.)

The ideas discussed in Refs. [1] and [2] are documented in the following two A. P. S. abstracts:

3. Bull. Am. Phys. Soc. **14**, 1178 (1969) (Boulder, CO meeting), "A Nuclear Physics Approach to Elementary Particle Structure."

4. Bull. Am. Phys. Soc. **14**, 1206 (1969) (Los Angeles, CA meeting), "Elementary Particle Structure from the Viewpoint of Nuclear Physics."

5. UCRL-72128, Nov. 5 and 14 (1969) (unpublished), "Classical Models for Elementary Particles."

This report contains the equations for a relativistically spinning sphere, which are applied to the calculation of the spin angular momentum and magnetic moment of the muon and electron. In particular, the spinless 70 MeV mass quantum M is related to the spinning 105 MeV muon. This report was submitted for publication in December 1969, and was published in Lett. Nuovo Cimento **4**, 211 (1970). The report UCRL-72187, Dec. 10 (1969), is a later version of this work, and UCRL-72287, Jan. 27 (1970) and UCRL-72288, Jan. 29 (1970) contain further extensions of these ideas.

6. Nuovo Cimento Letters **4**, 211–214 (1970), "Models for Particles."

This paper contains the material described above in UCRL-72128.

7. UCRL-72645 (Rev. I), Aug. 18 (1970) (unpublished), "Hyperon Spectroscopy."

This report contains a summary of meson and baryon resonances as reproduced using a fundamental mass quantum M in both its spin-0 (71 MeV) and spin-1/2 ($\sim$106 MeV) configurations. The assumed binding energies are small ($<5\%$).

8. UCRL-72646, Aug. 14 (1970) (unpublished), "A Model for Elementary Particle States."

This report is a comprehensive treatment of the meson and baryon resonances that were summarized in UCRL-72645.

9. Nuovo Cimento Letters **4**, 1043–1049 (1970); Erratum, Nuovo Cimento Letters **1**, 307 (1971), "Nucleon and Hyperon Structure."

This paper contains light-quark systematics for the baryon resonances, including S-state energy level diagrams and baryon rotational bands.

10. Nuovo Cimento Letters 4, 1249–1258 (1970), "Meson and Kaon Structure."

This paper contains light-quark systematics for the meson resonances, including S-state energy level diagrams, and it has an updated method for sorting meson resonances into rotational bands.

11. Nuovo Cimento Letters **4**, 1309–1315 (1970), "Extension of Kemmer-Symmetric Coupling to Meson Decays."

particle lifetimes. When this paper was originally issued in preprint form, the CERN preprint librarian notified the author that he had received 65 requests for copies of the paper.

23. Phys. Rev. D **9**, 1259–1329 (1974), "Light-quark hadron spectroscopy: experimental systematics and angular momentum systematics," 71 pages, denoted as Paper I.

24. Phys. Rev. D **10**, 850–883 (1974), "Light-quark hadron spectroscopy: a geometric quark model for S-states," 34 pages, denoted as Paper II.

These two papers are the published version of the report on the light-quark model that was initially submitted to The Physical Review in July 1971. Because of its great length, the original document was split into two parts. Even so, the first of these two papers is possibly the longest paper ever accepted for publication in The Physical Review, especially by a single author. These two papers contain a complete account of all the work done by the author on the light-quark model between the years 1969 and 1974. The work since that time has centered mainly on extensions in the area of nuclear physics, and on the incorporation of the subsequently-discovered New Particles into the systematics of the model.

25. Phys. Rev. D **12**, 1492–1494 (1975), "The mass of the $\psi(3095)$ as an $N\bar{N}$ resonance."

This paper identifies the J/ψ and ψ' New Particles as $N\bar{N}$ excitations, and it shows that they are the $n = 5$ and $n = 6$ terms in the previously-published $(333)^n$ meson excitation series. It also mentions the fact that the lifetime of the J/ψ fits into the scaling in powers of α which was previously established on the basis of the lifetimes of the long-lived Old Particles.

26. Phys. Rev. D **13**, 574–590 (1976), "Lifetimes of SU(3) groups and ψ particles as a scaling in powers of α."

This paper continues the study of elementary particle lifetimes by incorporating the ψ particles, and also by extending the span of lifetimes to include the short-lived "resonances" as well as the long-lived "particles." A detailed discussion is given as to the validity or nonvalidity of applying phase space corrections to these elementary particle lifetimes.

27. Phys. Rev. D **14**, 1323–1334 (1976), "Two related questions: Centrifugal barriers and the spin of the Ω^-."

This paper suggests that the centrifugal barrier concept used by elementary particle physicists to make decay rate connections may be invalid. As a semi-related topic, the paper also discusses the question of the spin of the Ω^- hyperon.

28. Phys. Rev. D **14**, 1463–1466 (1976), "Mass spectra of the new particles and the ψ' radiative decays."

This paper extends the systematics of the new ψ particles within the framework of the light-quark model.

29. Phys. Rev. Lett. **42**, 1724–1728 (1979), "Interpretation of *p-p* Dibaryon Resonances at 2140, 2260 and 2430 MeV."

This paper identifies a possible dibaryon rotational band.

30. Phys. Rev. **20**, 1616–1632 (1979), "*p-p* resonances: A link between nuclear and hadronic excitations."

This paper discusses rotational systematics in nuclear and hadronic excitations.

31. Nuovo Cimento **58A**, 159–192 (1980), "Can 35 Pionic Mass Intervals among Related Resonances be Accidental?"

This paper points out in detail the large number of $\pi \cong 140$ MeV mass intervals that are displayed within groups of related narrow-width particle states.

32. Lett. Nuovo Cimento **30**, 417–420 (1981), "Do Dewan–Beran Relativistic Stresses Actually Exist?"

This is a discussion of the problem of relativistic stresses in rotating systems.

33. Lett. Nuovo Cimento **31**, 341–346 (1981), "The Electromagnetic Scaling of Particle Lifetimes and Masses."

This paper reviews the scaling of elementary particle lifetimes in powers of α, and then extends the discussion to include the mass systematics of these same particles.

34. Nuovo Cimento **69A**, 241–294 (1982), "Do Spinless Constituent Quarks Exist?"

This paper gives a comprehensive review of the mass systematics of the Standard Model quarks from a constituent-quark viewpoint, and it shows that they cannot by themselves reproduce the full spectrum of elementary particle states. However, when they are supplemented by a set of spinless 70 MeV mass quanta, then the particle masses can be accurately accounted for. Other properties such as magnetic moments are also addressed.

35. Lett. Nuovo Cimento **43**, 49–54 (1985), "Generalization of the Postulates of Special Relativity."

This paper discusses the relativistic properties of particle–wave systems and gives Møller's derivation of the de Broglie phase wave. It also describes computer calculations for the relativistic transformation properties of a relativistically spinning sphere.

36. Lett. Nuovo Cimento **44**, 697–704 (1985), "A Dynamical Basis for the de Broglie Phase Wave."

This paper gives the relativistic equations for the acceleration of a spatial mass quantum by a moving electron, and it shows that in the perturbative limit where the excitation energy of the quantum is much less than that of the incident electron, the forward velocity of the excitation quantum is equal to the de Broglie phase velocity. This result holds for all excitation scattering angles.

37. Bull. Am. Phys. Soc. **31**, 844 (1986) (Washington, D.C. meeting), "Kinematic Production of the de Broglie Phase Wave."

38. Bull. Am. Phys. Soc. **32**, 31 (1987) (San Francisco meeting), "Point-like Scattering by a Large Electron."

39. Bull. Am. Phys. Soc. **32**, 1022 (1987) (Arlington, VA meeting), "The anomalous magnetic moment of the electron as a magnetic self-energy effect."

40. Found. Phys. Lett. **1**, 25–45 (1988), "A Particle-Wave Steering Mechanism."

This paper gives a derivation of the special relativistic equations in the perturbative limit for the acceleration of a spatial excitation by a moving electron, and it discusses a possible steering mechanism associated with the de Broglie phase wave.

41. Found. Phys. Lett. **2**, 577–589 (1989), "On the Interpretation of the Electron Anomalous Magnetic Moment."

This paper discusses the anomalous magnetic moment of the electron and shows that it is logically associated with the self-energy of the electron's magnetic field. Calculations by Fermi and Rasetti, and by Born and Schrödinger, when extended to a Compton-sized current loop, demonstrate that the external magnetic field energy is of the same magnitude as the magnetic moment anomaly. This result is also indicated by the calculation of the self-inductance of a thin current loop.

42. Nuovo Cimento **103A**, 983–1052 (1990), "An Elementary Particle Constituent-Quark Model."

This comprehensive paper contains an updating of particle lifetimes, and of the constituent-quark approach to particle masses, with new information from the b-quark states included. The experimental fact that the very-narrow-width particle states extend up to the highest masses indicates that quark masses also extend up to these energies, and it rules against the interpretation of high-mass states as nuclear-physics-type rotational levels. The role of the spinless 70 MeV excitation quanta that are required in this constituent-quark approach is clarified with the identification of a characteristic "supersymmetric" excitation quantum $X = 420$ MeV that dominates many excitation processes. The role of the Standard Model u, d, s, c, b quarks is also readily apparent, especially in the lifetime systematics of the long-lived particles. The observed α-quantization of both lifetimes and masses is a key factor in tying all of these results together.

43. Bull. Am. Phys. Soc. **35**, 949 (1990) (Washington, D.C. meeting), "Experimental Systematics of Particle Masses and Excitation Energies."

44. Bull. Am. Phys. Soc. **37**, 905 (1992) (Washington, D.C. meeting), "Experimental Anomalies in keV Mott Scattering on Al and Cu."

45. Found. Phys. Lett. **5**, 15–23 (1992), "KeV Channeling Effects in the Mott Scattering of Electrons and Positrons."

This paper discusses size effects in models of the electron. If the electron is truly Compton-sized, as seems mandated by its spectroscopic features, then there should be some experimental indications of this large size. One possible method for revealing this is to scattering electrons off a target that is smaller than the electron. There are some Mott scattering experiments on atomic nuclei that might reveal an extended charge structure in the electron if the scattering is carried out at appropriate energies. This paper described extensive computer calculations that were carried out to investigate these ideas.

46. Bull. Am. Phys. Soc. **39**, 1153 (1994) (Washington, D.C. meeting), "The Equations of Motion of a 'Hole' State."

47. Found. Phys. Lett. **8**, 135–160 (1995), "Model Basis States for Photons and 'Empty Waves'."

This paper discusses a key element that is required in order to quantitatively reproduce the spectroscopic properties of a photon: namely, the P–H "particle plus hole" pair, which is pictured as an excitation of the "vacuum state." These P–H pairs are stable only when they are rotating at particular (de Broglie) frequencies. Single P–H pairs, denoted as "zerons," formally have zero linear and angular momentum (to first order). Matching and superimposed particle and anti-particle P–H pairs carry the angular momentum $\hbar$ of the photon, and they reproduce its main electromagnetic characteristics.

48. *The Present Status of the Quantum Theory of Light*, S. Jeffers, S. Roy, J.-P. Vigier and G. Hunter (eds.) (Kluwer Academic, Dordrecht, 1997), pp. 17–35, "The Missing 'P Field' in Electromagnetism and Quantum Mechanics."

This paper introduces the concept of polarized vacuum-state excitations, which serve to synchronize an ensemble of traveling P–H pairs in a representation of a photon or an electromagnetic field.

49. *Causality and Locality in Modern Physics*, G. Hunter, S. Jeffers and J.-P. Vigier (eds.) (Kluwer Academic, Dordrecht, 1998), pp. 359–364, "The Relativistic Kinematics of the de Broglie Phase Wave."

This paper contains the same basic results as those published in Ref. [36].

50. Int. J. Mod. Phys. A **20**, 719–798 (2005) and Addendum, *ibid.* **20**, 2893–2894 (2005), "Electron Generation of Leptons and Hadrons with Conjugate α-Quantized Lifetimes and Masses."

This is a review article that summarizes the experimental evidence for an α-dependence in the lifetimes and masses of the long-lived threshold-state elementary particles. It is the fore-runner for the book *The Power of* α.

51. *What is the Electron?* V. Simulik (ed.) (Apeiron Press, Montreal, 2005), pp. 129–153. "What causes the electron to weigh?"

This chapter in the book discusses the concept of the mass of the electron.

52. arXiv.org/hep-ph/0603201, 24 March 2006, "The top quark to electron mass ratio $m_t = 18m_e/\alpha^2$, where $\alpha = e^2/\hbar c$."

This paper gives the first calculation of the top quark mass as a second-order α-leap from the mass of the electron. The mass equation shown in the title gives a calculated top quark mass of 172.73 GeV, as compared to the current consensus Fermilab CDF/D0 mass value of 172.5 ± 2.3 GeV.

53. arXiv.org/hep-ph/0607233, 20 July 2006, "A 'Muon Mass Tree' with α-quantized lepton, quark and hadron masses."

This paper describes the muon and pion "mass trees" that are displayed in Figs. 0.2.23 and 0.2.24 of the present book.

References

[1] M. H. Mac Gregor, *The Enigmatic Electron* (Kluwer, Dordrecht, 1992).
[2] K. Gottfried and V. F. Weisskopf, *Concepts of Particle Physics, Volume I* (Oxford University Press, New York, 1984).
[3] Ref. [2], p. 100.
[4] L. Lederman with R. Teresi, *The God Particle* (Dell Publishing, New York, 1993).
[5] Ref. [4], pp. 2–3.
[6] R. P. Feynman, *QED, The Strange Theory of Light and Matter* (Princeton University Press, Princeton, 1985).
[7] Ref. [6], p. 152.
[8] M. Veltman, *Facts and Mysteries in Elementary Particle Physics* (World Scientific, Singapore, 2003).
[9] Ref. [8], pp. 281–283.
[10] W.-M. Yao *et al.* (Particle Data Group), "Review of Particle Physics" (RPP), J. Phys. G: Nucl. Part. Phys. **33**, 1–1232 (2006).
[11] Ref. [10], p. 1023.
[12] Ref. [8], p. 67.
[13] H. Fritzsch, *Elementary Particles: Building Blocks of Matter* (World Scientific, New Jersey, 2005).
[14] Ref. [13], p. 57.
[15] Ref. [6], Fig. 78 on p. 126.
[16] Ref. [6], pp. 129–130.
[17] Ref. [4], p. 28.
[18] Ref. [4], p. 29.
[19] M. H. Mac Gregor, "The Fine-Structure Constant $\alpha = e^2/\hbar c$ as a Universal Scaling Factor," Lett. Nuovo Cimento **1**, 759 (1971).
[20] Ref. [1], pp. 49–50.
[21] Ref. [1], Ch. 5.
[22] Ref. [8], Secs. 11.3 and 11.4.
[23] Ref. [1], Ch. 7.
[24] Ref. [1], Ch. 8.

[25] Ref. [1], Ch. 15.
[26] M. H. Mac Gregor, Lett. Nuovo Cimento **4**, 1309–1315 (1970), Table V.
[27] M. Roos *et al.* (Particle Data Group), Phys. Lett. B **33**, 1 (1970).
[28] B. B. Mandelbrot, *The Fractal Geometry of Nature, Updated and Augmented* (Freeman, New York, 1983).
[29] Ref. [28], p. 22.
[30] Ref. [28], p. 154.
[31] Ref. [1], pp. 26–29.
[32] Ref. [13], p. 103.
[33] D. Akers, "Dirac Monopoles and Mac Gregor's Formula for Particle Lifetimes", Nuovo Cimento **105A**, 935–939 (1992), Eq. (6).
[34] R. A. Arndt and M. H. Mac Gregor, "Nucleon-Nucleon Phase Shift Analyses by Chi-Squared Minimization," in *Methods in Computational Physics, Vol. 6: Nuclear Physics*, eds. B. Alder, S. Fernbach and M. Rotenberg (Academic Press, New York, 1966), pp. 253–296.
[35] B. A. Schumm, *Deep Down Things: The Breathtaking Beauty of Particle Physics* (Johns Hopkins Univ. Press, Baltimore, 2004), pp. 135–137.
[36] B. T. Feld, *Models of Elementary Particles* (Blaisdell, Waltham, 1969), pp. 290–291.
[37] Ref. [13], p. 106.
[38] M. H. Mac Gregor, "Electron Generation of Leptons and Hadrons with Conjugate α-Quantized Lifetimes and Masses," Int. J. Mod. Phys. A **20**, Issue No. 4, pp. 719–798 (2005) and "Addendum," *ibid.* **20**, Issue No. 13, pp. 2893–2894 (2005). The present book follows the broad outlines of this review paper, and many of the figures in the book have been copied or abstracted from that paper.
[39] See Ref. [38], Sec. 2.7.
[40] M. H. Mac Gregor, "Hadron Spectroscopy", in *Fundamental Interactions at High Energy*, proceedings of the 1971 Coral Gables Conference, eds. M. Dal Cin, G. J. Iverson and A. Perlmutter (Gordon and Breach, New York, 1971), Vol. 3, *Invited Papers*, ed. M. Hamermesh, pp. 75–154, Table XI on p. 119.
[41] Ref. [26], Table VI; also Ref. [40], Table XII.
[42] M. H. Mac Gregor, "Experimental Systematics of Particle Lifetimes and Widths," Nuovo Cimento **20A**, 471–507 (1974), Fig. 5b on p. 484.
[43] M. H. Mac Gregor, "Light-quark hadron spectroscopy: Experimental systematics and angular momentum systematics," Phys. Rev. D **9**, 1259–1329 (1974), Sec. XII, Fig. 21.
[44] Figs. 2.11.3, 2.11.4 and 2.11.5 are Figs. 3, 4 and 5a, respectively, in Ref. [42].
[45] Ref. [8], p. 63.
[46] M. H. Mac Gregor, "Lifetimes of SU(3) groups and ψ particles as a scaling in powers of α," Phys. Rev. D **13**, 574–590 (1976), Fig. 3.
[47] Ref. [46], Fig. 4.
[48] J. D. Barrow and F. J. Tipler, *The Anthropomorphic Cosmological Principle* (Oxford University Press, New York, 1986), p. 230.

[49] M. H. Mac Gregor, "An Elementary Particle Constituent-Quark Model", Nuovo Cimento **103A**, 983–1052 (1990), Fig. 5.
[50] Ref. [49], Fig. 6.
[51] Fig. 2.11.10 in the present book is taken from Fig. 11 in Ref. [38].
[52] W. Heisenberg, "Über Die in Der Theorie Der Elementarteilchen, Auftretende Universelle Länge", Annalen der Physik **32** (5th Ser.), Issue 1–2, pp. 20–33 (1938).
[53] L. Gray, P. Hagerty and T. Kalogeropoulos, Phys. Rev. Lett. **26**, 1491 (1971).
[54] J. D. Jackson, *The Physics of Elementary Particles* (Princeton Univ. Press, Princeton, 1958), pp. 57–58.
[55] A. Pais, "*Subtle is the Lord ...*" (Oxford University Press, New York, 1982), p. 159.
[56] Ref. [6], p. 145.
[57] B. Hoeneisen, arXiv.org:hep-th/0609080, 12 September 2006, p. 6.
[58] Ref. [35], pp. 317–324.
[59] D. Glenzinski *et al.* (CDF and D0 Collaborations), arXiv.org: hep-ex/0603039, 20 March 2006.
[60] Ref. [35], Ch. 9.
[61] Ref. [36], p. 233.
[62] Ref. [36], p. 292.
[63] Ref. [42]; the CERN librarian said that 65 physicists requested copies of this paper when it appeared there in 1973 in preprint form.
[64] Ref. [43], Secs. VI–IX.
[65] M. H. Mac Gregor, *The Nature of the Elementary Particle* (Springer-Verlag, Berlin, 1978).
[66] Ref. [43], Fig. 5.e.
[67] Ref. [49], Fig. 1.
[68] Particle Data Group, Review of Particle Properties, Phys. Lett. B, 204 (1988).
[69] M. H. Mac Gregor, "Can 35 Pionic Mass Intervals among Related Resonances be Accidental?" Nuovo Cimento **58A**, 159 (1980).
[70] D. Akers, International Journal of Theoretical Physics **33**, 1817–1829 (1994).
[71] P. Palazzi, "Patterns in the Meson Mass Spectrum," Particle Physics Preprint Archive p3a-2004-001, July 28, 2004; "Meson Shells," p3a-2005-001, 1 March, 2005. These papers can be downloaded from http://particlez.org/p3a/.
[72] Ref. [6], p. 128.
[73] P. F. Schewe, B. Stein and D. Castelvecchi, "Indications of a change in the proton-to-electron mass ratio," The American Institute of Physics Bulletin of Physics News, Number 774, April 19, 2006. This news item was communicated to the author by David Akers, who for well over a decade has kept the author informed of interesting developments and reports in this area of physics.
[74] Ref. [4], pp. 114–116.

[75] Ref. [6], p. 7.
[76] J. M. Jauch and F. Rohrlich, *The Theory of Photons and Electrons* (Springer-Verlag, Berlin, 1976).
[77] R. P. Feynman, in *The Quantum Theory of Fields, Proceedings of the Twelfth Conference on Physics at the University of Brussels, October, 1961*, ed. R. Stoops (Interscience, New York), pp. 75–76.
[78] H. Arzelies, *Relativistic Kinematics* (Pergamon Press, Oxford, 1966), Ch. IX.
[79] Ref. [78], bottom of p. 235.
[80] See Ref. [43], Refs. [167–177] therein.
[81] Ref. [43], App. B.
[82] Ref. [65], Ch. 6.
[83] C. Møller, *The Theory of Relativity* (Clarendon Press, Oxford, 1952), p. 318, Eq. (42).
[84] P. Ehrenfest, Phys. Z. **10**, 918 (1909).
[85] A. Einstein, in *The Principle of Relativity* (Dover, New York, 1923), p. 116.
[86] M. H. Mac Gregor, "Do Dewan-Beran Relativistic Stresses Actually Exist?", Lett. Nuovo Cimento **30**, 417 (1981).
[87] L. D. Landau and E. M. Lifshitz, *The Classical Theory of Fields (Revised Second Edition)* (Addison-Wesley, Reading, 1962), pp. 47–48.
[88] E. Dewan and M. Beran, Am. J. Phys. **27**, 517 (1959).
[89] M. H. Mac Gregor, Lett. Nuovo Cimento **4**, 211 (1970).
[90] Ref. [13], p. 106.
[91] In *The Feynman Lectures in Physics*, Vol. II (Addison-Wesley, Reading, 1964), the statement appears on p. 34–3 that "there is no classical explanation" for the electron g-value of 2.
[92] Ref. [1], Ch. 7.
[93] W. R. Smythe, *Static and Dynamic Electricity* (McGraw-Hill, New York, 1939), p. 138.
[94] Ref. [1], pp. 46–47.
[95] A. O. Barut, in D. Hestenes and A. Weingartshofer (eds), *The Electron: New Theory and Experiment* (Kluwer Academic, Dordrecht, 1991), p. 109.
[96] M. H. Mac Gregor, "KeV channeling effects in the Mott scattering of electrons and positrons," Found. Phys. Lett. **5**, 15 (1992).
[97] Ref. [1], Part. IV.
[98] B. E. Lautrup, A. Peterman and E. de Rafael, Physics Reports **3C**, 193 (1972).
[99] F. Rasetti and E. Fermi, Nuovo Cimento **3**, 226 (1926).
[100] J. D. Jackson, *Classical Electrodynamics* (Wiley, New York, 1962), p. 143.
[101] M. Born and E. Schrödinger, Nature (London) **135**, 342 (1935).
[102] Ref. [1], p. 64.
[103] M. H. Mac Gregor, "On the interpretation of the electron anomalous magnetic moment," Found. Phys. Lett. **2**, 577–589 (1989).
[104] Ref. [1], Ch. 8.
[105] H. Margenau, in *Quantum Theory, I. Elements*, ed. D. R. Bates (Academic Press, New York, 1961), p. 6.

[106] Ref. [1], Ch. 5.
[107] M. H. Mac Gregor, "Generalization of the postulates of special relativity," Lett. Nuovo Cimento **43**, 49 (1985) (see Eq. (14) and Table I).
[108] Ref. [1], Ch. 12.
[109] D. Hestenes, in Ref. [95], p. 31.
[110] M. H. Mac Gregor, "A Dynamical Basis for the de Broglie Phase Wave," Lett. Nuovo Cimento **44**, 697 (1985).
[111] M. H. Mac Gregor, "A Particle-Wave Steering Mechanism," Found. Phys. Lett. **1**, 25 (1988).
[112] M. H. Mac Gregor, "The Relativistic Kinematics of the de Broglie Phase Wave," in *Causality and Locality in Modern Physics*, eds. G. Hunter, S. Jeffers and J.-P. Vigier (Kluwer Academic, Dordrecht, 1998), p. 359.
[113] W. Rindler, *Essential Relativity*, 2nd edn. (Springer-Verlag, Heidelberg, 1977), pp. 72 and 91.
[114] R. P. Feynman, *The Character of Physical Law* (MIT Press, Cambridge, 1967), p. 129.
[115] M. H. Mac Gregor, "Model basis states for photons and 'empty waves'," Found. Phys. Lett. **8**, 135–160 (1995).
[116] M. H. Mac Gregor, "Stationary vacuum-polarization 'P-fields': the missing element in electromagnetism and quantum mechanics," in *The Present Status of the Quantum Theory of Light*, eds. S. Jeffers, S. Roy, J.-P. Vigier and G. Hunter (Kluwer Academic, Dordrecht, 1997), pp. 17–35.
[117] Y. Terlitskii, *Paradoxes in the Theory of Relativity* (Plenum Press, New York, 1968), Ch. VI.
[118] H. Goldstein, *Classical Mechanics* (Addison-Wesley, Cambridge, 1950), p. 135; D. W. Sciama, *The Physical Foundations of General Relativity* (Doubleday, Garden City, 1969), pp. 6–11.
[119] R. H. Stuewer, *The Compton Effect: Turning Point in Physics* (Science History Publications, New York, 1975), p. 332.
[120] R. A. Beth, Phys. Rev. **50**, 115 (1936); see, for example, B. Cagnac and J. C. Pebay-Peyroula, *Modern Atomic Physics: Fundamental Principles* (Wiley, New York, 1975), Ch. 11.
[121] H. Bateman, Phil. Mag. **46**, 977 (1923).
[122] W. B. Bonner, Int. J. Theor. Phys. **2**, 373 (1969) and **3**, 57 (1970).
[123] J. D. Jackson, Ref. [100], pp. 380–384.
[124] Ref. [116]; also, UCRL-JC-122763 (1-23-1996).
[125] Ref. [2], pp. 252–273.
[126] J. D. Bjorken and S. D. Drell, *Relativistic Quantum Fields* (McGraw-Hill, New York, 1965), Ch. 19.
[127] Ref. [76], Ch. 9 and 10.
[128] Ref. [76], p. 175.
[129] Ref. [8], p. 253.
[130] Ref. [8], p. 265.
[131] Ref. [1], pp. 26–29.
[132] T. H. Boyer, Amer. J. Phys. **53**(2), 167 (1985); Scientific American **251**, 8 (1985).

[133] Ref. [1], p. 16 and p. 24, Note 1.
[134] For a recent summary, see G. Kane, "The Mysteries of Mass," Scientific American, July (2005), pp. 40–48.
[135] Ref. [1], Ch. 15.
[136] Ref. [118], pp. 124 and 143; also J. L. Synge and B. A. Griffith, *Principles of Mechanics* (McGraw-Hill, New York, 1942), pp. 271–273 and 349–350.
[137] L. D. Landau and E. M. Lifshitz, *The Classical Theory of Fields, Revised Second Edition* (Addison-Wesley, Reading, 1962), pp. 47–48.
[138] M. H. Mac Gregor, "What Causes the Electron to Weigh?" in *What is the Electron?*, ed. V. Simulik, (Apeiron, Montreal, 2005), pp. 129–153.
[139] M. H. Mac Gregor, "Do spinless constituent quarks exist?," Nuovo Cimento **69A**, 241 (1982).
[140] E. Fermi and C. N. Yang, Phys. Rev. **76**, 1739 (1949).
[141] W. Heisenberg, "The universal length appearing in the theory of elementary particles" (English translation), Ref. [52]. The author would like to thank Paolo Palazzi and the CERN Library for supplying an English translation of this paper.
[142] Y. Nambu, Prog. Theor. Phys. **7**, 131 (1952).
[143] See Ref. [54], Secs. 5.2 and 5.3.
[144] See Ref. [8], p. 228.
[145] J. D. Jackson, *Classical Electrodynamics* (Wiley, New York, 1962), p. 590 (footnote).
[146] R. J. Pease, Phys. Rev. D **2**, 1069 (1970).
[147] B. Aubert *et al.* (Babar Collaboration), "Observation of an excited charm baryon Ω_c^* decaying to $\Omega_c^* \gamma$," arXiv.org: hep-ex/0608055, 24 August 2006. The author would like to thank P. Palazzi for calling his attention to this preprint.
[148] F. Wilczek, "Reference Frame, On Absolute Units, III: Absolutely Not?," Physics Today, May (2006), pp. 10, 11.
[149] M. H. Mac Gregor, "The top quark to electron mass ratio $m_t = 18m_e/\alpha^2$," arXiv.org: hep-ph/0603201, 24 March 2006.
[150] Another physical system involving electrons in which ground state problems have arisen is the quantum Hall bar. For discussions of this topic, see Refs. [151–153].
[151] M. H. Mac Gregor, "A unified quantum Hall close-packed composite boson (CPCB) model," Found. Phys. Lett. **13**, 443–460 (2000).
[152] M. H. Mac Gregor, "Quantum Hall Enigmas," in *Gravitation and Cosmology: From the Hubble Radius to the Planck Scale*, eds. R. Amoroso, G. Hunter, M. Kafatos and J.-P. Vigier (Kluwer Academic, Dordrecht, 2002), pp. 337–348.
[153] M. H. Mac Gregor, "Quantum Hall Quantized Cyclotron Entrance Channels," Found. Phys. Lett. **17**, 381–391 (2004).

Index